AF693364

LIBRARY COPY
WITHDRAWN
JUL 8 1976
LANGLEY RESEARCH CENTER
LIBRARY, NASA
HAMPTON, VIRGINIA

FIBER OPTICS

QC 448 . L53

76H-16465

FIBER OPTICS

by M.P. LISITSA,

L.I. BEREZHINSKII,

M.Ya. VALAKH

ISRAEL PROGRAM FOR SCIENTIFIC TRANSLATIONS

NEW YORK • JERUSALEM • LONDON • 1972

© 1972 by
ISRAEL PROGRAM FOR SCIENTIFIC TRANSLATIONS LTD.

Distributed by

KETER INC., NEW YORK
KETER PUBLISHING HOUSE, JERUSALEM
KETER PUBLISHERS LTD., LONDON

ISBN 0 7065 11026
Library of Congress Catalog Card Number 78–184436
IPST Cat No. 2309

This book is a translation from Russian of
VOLOKONNAYA OPTIKA
Izdatel'stvo "Tekhnika"
Kiev 1968

L46183A-002
$20.00

Preface

For a number of years now glass fibers have been used in the manufacture of various elements and components of technical and industrial units. They are very popular in aviation, refrigeration, electrical engineering, and in design of constructions subjected to high thermal loads and electrical tensions. However, for a long time glass fibers remained without use in optico-mechanical and rf electronics industries. Precision optics traditionally used monolithic slabs of optical glass, which provided excellent raw material to be processed and formed into a great variety of optical instruments and optical components—from miniature microobjectives to the enormous lenses of astronomical refractors.

Special composition glasses, in particular those based on semiconductor compounds, are now beginning to be used in the infrared spectrum, significantly broadening the limits of application of glass optics.

Glass fibers were generally judged inadequate for critical scientific and technical applications, e.g., in the design of optical and electron-optical systems. Nevertheless, the last years have witnessed a phenomenally swift penetration of glass fibers into various fields of technology. This material, which is still often treated with amusement and suspicion, has proved to be most useful and promising in various technical applications, including some entirely new fields, such as laser optics. The highly promising technical properties of this material stimulated the development of the new field of fiber optics.

Fiber optics is steadily conquering new positions in measurement techniques, in medicine, biology, astronomy, black-and-white and color television, computers, ultra-high-speed photography, etc. Despite these growing uses of fiber optics, no book has been published so far in the world literature attempting to present a consistent theory of the effects and the processes which take place in individual fibers and in fiber bundles, and highlighting the theoretical treatment by experimental findings. No general review of the applications of fiber elements in optical and electron-optical systems is available, either.

The profusion of papers and articles published all over the world on the problems of fiber optics has long since made the publication of such monographic work absolutely essentially.

The aim of the present authors was to summarize the advances in fiber optics. To reach this goal, we made use of all the available original publications and the most significant promotion material of various manufacturers. The whole literature, unfortunately, was too wide a sea to cover in this one volume.

In addition to chapters dealing directly with the theoretical and experimental aspects of optical fibers and fiber bundles, the book contains an introductory chapter on the basic elements of geometrical and wave optics which are essential for better understanding of the specialized topics.

We did not aim at writing an engineering handbook to facilitate the design and calculation of various fiber systems. We were mainly concerned with the physical processes taking place in individual fibers and in fiber bundles. For this reason, much space and effort have been devoted to the theoretical treatment of the effects and the main attention is focused on those relations which are particularly useful in practical applications. Nevertheless, the book will be most helpful to personnel specifically engaged on the design and applications of fiber systems.

It is our pleasant duty to acknowledge the great help of Prof. M.U. Belyi who offered some valuable comments while reviewing the manuscript, and of V.I. Ponomarenko and A.M. Yaremko who took active part in the discussion of some theoretical sections.

Being the first monograph in the world literature on the subject, our *Fiber Optics* is obviously far from perfect. Critical comments and remarks will be most welcome, and the readers are invited to write to Kiev, 4, Pushkinskaya, 28, *Tekhnika* Publishing House, to the attention of the author.

Contents

1 Elements of Geometrical and Wave Optics

1. Elements of photometry

The *luminous flux* is a fundamental characteristic of radiation which measures the radiant power in terms of the light sensation that it produces in the detector. Irrespective of the particular detector employed, the total luminous flux Φ radiated by a point source O will be defined as the luminous power emitted by the source onto a closed surface around the source. In the case of a transparent medium, the geometry of the enclosing surface is irrelevant, since the total quantity of luminous energy is conserved.

Alongside with the total luminous flux, we will also consider the flux of luminous power through any surface element dS (not closed). This quantity is referred to as the luminous flux $d\Phi$ through a surface element dS. If dS subtends a solid angle $d\Omega$ at the point O (Figure 1), we have

$$d\Phi = I\,d\Omega, \tag{1}$$

where I is the light intensity, i.e., the luminous flux per unit solid angle.

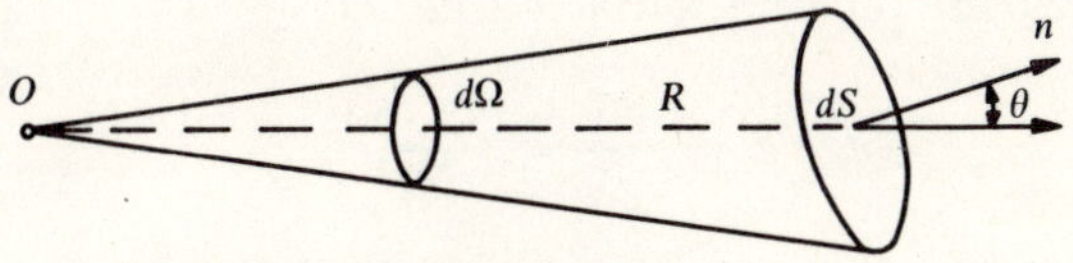

Figure 1
A pencil of rays.

If the normal to the surface element dS makes an angle θ to the axis of the cone and the distance between the light source and the surface element is R, we may rewrite (1) in the form

$$d\Phi = I\,\frac{dS\cos\theta}{R^2}.$$

In the general case, the light intensity is a function of direction. Eq. (1) is then best written in spherical coordinates. Writing $I_{\theta,\phi}$ for the light intensity in a given direction and seeing that $d\Omega = \sin\theta \, d\theta \, d\phi$ (Figure 2), we get

$$d\Phi = I_{\theta,\phi} \sin\theta \, d\theta \, d\phi. \tag{2}$$

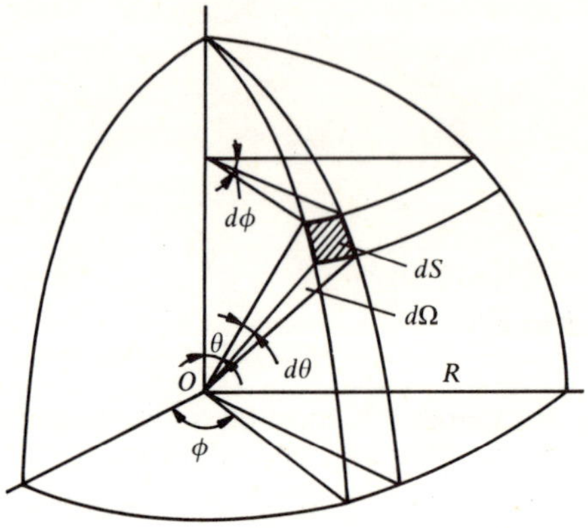

Figure 2
A solid angle in spherical coordinates.

The total flux is obtained by integration,

$$\Phi = \int_0^{2\pi} \int_0^{\pi} I_{\theta,\phi} \sin\theta \, d\theta \, d\phi.$$

In the particular case of $I_{\theta,\phi} = \text{const}$, we have

$$\Phi = 4\pi I.$$

Another important characteristic is the *illuminance* (or the *illumination*) produced by a light flux incident on unit surface area, i.e.,

$$E = \frac{d\Phi}{dS} = \frac{I \cos\theta}{R^2}. \tag{3}$$

The illuminance is thus inversely proportional to the square of the distance from a point source and directly proportional to the cosine of the angle between the light ray and the normal to the illuminated surface.

The *luminous flux density* $\mathcal{R}$ is defined as the total luminous flux emitted by a unit surface area of the luminous surface. The corresponding expression coincides with Eq. (3), but θ in this case is the angle between the direction chosen and the normal to the luminous surface. The concept of luminance is applied not only to self-luminous surfaces, but also to objects illuminated by external sources.

To find the illuminance produced by a real, extended source (as opposed to a point source), its surface is divided into infinitesimal elements and an integral is taken over the entire surface, taking into consideration the angular dependence of light intensity. If the distance R is substantially greater than the source size, the point source approximation (3) is fully applicable [24].

One of the most important characteristics in photometry is the *brightness B* which characterizes the emission of a luminous surface in a given direction. It is defined as the ratio of the light intensity dI in a given direction to the projection of the surface element dS onto the plane perpendicular to that direction:

$$B_\theta = \frac{dI}{dS \cos\theta}.$$

To derive some fundamental relations of photometry, we have to introduce the concept of a light tube. Consider two surface elements dS_1 and dS_2 (Figure 3) in the space around the source. In all likelihood, the luminous flux passing successively through these two surface elements hardly emerges outside the conical surface described using the contour curves of dS_1 and dS_2 as the bases. The part of the space enclosed within this conical surface is called a *light tube.* A light tube is said to have been defined if the contour curves of dS_1 and dS_2, the angles θ_1 and θ_2 between the normals to the tube bases and the tube axis O_1O_2, and the solid angles subtended by the tube bases at O_1 and O_2 are known. If the distance O_1O_2 is R, we have

$$d\Omega_1 = \frac{dS_1 \cos\theta_1}{R^2}; \qquad d\Omega_2 = \frac{dS_2 \cos\theta_2}{R^2}. \tag{4}$$

The luminous fluxes through the tube bases dS_1 and dS_2 can be written in the form

$$d\Phi_1 = B_1 dS_1 \cos\theta_1 d\Omega_2; \qquad d\Phi_2 = B_2 dS_2 \cos\theta_2 d\Omega_1. \tag{5}$$

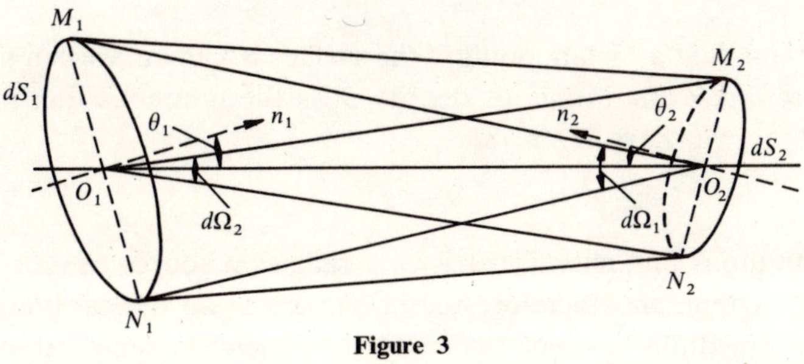

Figure 3
A light tube.

From the definition of a light tube we have the equality $d\Phi_1 = d\Phi_2$, and from (4) and (5) we thus have

$$B_1 \cos\theta_1 \cos\theta_2 \frac{dS_1 dS_2}{R^2} = B_2 \cos\theta_1 \cos\theta_2 \frac{dS_1 dS_2}{R^2}, \tag{6}$$

i.e.,

$$B_1 = B_2 = B.$$

In other words, the brightness of a light beam is constant in any cross section of the light tube. The light tube can be continued to the source and the surface element dS can be selected on the surface of the source. Then the luminous flux $d\Phi$ emitted by this surface element in the direction of the angle θ toward another surface element subtending a solid angle $d\Omega$ will be expressed by Eqs. (5).

Comparison of Eqs. (1) and (5) gives

$$dI = BdS_1 \cos\theta_1 = BdS_2 \cos\theta_2. \tag{7}$$

This is an expression of *Lambert's law*: the light intensity dI radiated in a certain direction by a surface element dS is proportional to the cosine of the angle θ between the beam direction and the normal to the surface. Surfaces which obey Lambert's law are said to be *ideally scattering.* The surface of a uniformly luminous sphere is ideally scattering. The light intensity of the inclined surface elements decreases according to Lambert's law, but this decrease is compensated by the increase in their area subtending the same solid angle, so that all in all the sphere is seen as a uniformly luminous disk.

Eqs. (5) define the elementary unit fluxes in small solid angles. Integration of Eq. (5) expresses the total luminous flux in terms of the surface brightness of a Lambert-law radiator of area dS:

$$\Phi = \int BdS \cos\theta \, d\Omega = BdS \int_0^{2\pi} \int_0^{\pi/2} \cos\theta \sin\theta d\theta d\phi = \pi BdS. \tag{8}$$

If the brightness B at a certain point of the surface is known, we can find the luminous density $\mathscr{R}$. Since the luminous density $\mathscr{R}$ is the luminous flux radiated by a unit surface area, we have from (8)

$$\mathscr{R} = \pi B. \tag{9}$$

The total luminous flux characteristic of a radiation source cannot be increased by any optical system, and therefore the brightness of an optical image formed in the same optical medium in which the source is immersed cannot exceed the source brightness. If, however, the image is formed in a medium with a different refractive index n_2 (n_1 being the refractive index of the object space), we have, neglecting absorbtion and reflection losses [37],

$$B_2 = B_1 \frac{n_2^2}{n_1^2}.$$

The *flux intensity* F is defined as the luminous flux transmitted in a given direction through an apparent cross section of unit area inside a unit solid angle. The flux intensity is the light field analog of surface brightness. It is therefore expressed by formulae which are analogous to those expressing brightness, and is measured in the same units.

2. Formation of an optical image

Consider a *homocentric light beam*, i.e., a light beam whose components all cross at one common center. In general, homocentricity of a beam is broken by an optical system, and it is only in exceptional cases that the emergent rays will also meet at one point. If, however, the rays do meet at one point, the result is a *stigmatic image* of a luminous point. We distinguish between real and virtual images according as the image is formed by the intersection of the rays themselves or by the intersection of their continuations. A luminous point and its stigmatic image are conjugate points of an optical system which converts a divergent homocentric beam into a convergent homocentric beam.

It can be shown [1, 37] that the only optical system which always conserves the homocentricity of a light beam is a system of plane mirrors. All other real optical systems produce images of varying sharpness. A minimum departure from homocentricity is observed if the beam is infinitely narrow and propagates along the optical axis of the system. Beams of this kind are known as *paraxial beams*. They are used in describing so-called ideal optical systems, which assign a single point in the image space to every point in the object space. Although the theory of the ideal optical systems is far-fetched idealization, it is of considerable practical importance, since the main problem of applied optics is now to devise real systems which would be as close to the ideal as possible.

Refraction and reflection by plane and spherical surfaces. Fundamental relations for ideal optical systems

The conditions of reflection and refraction at the interface of two media is described by the two laws of Descartes and Snell:

$$\frac{\sin\theta}{\sin\theta'} = \frac{n_2}{n_1} = n_{21}; \qquad \theta = \theta'', \tag{10}$$

where θ, θ' and θ'' are, respectively, the angle of incidence, the angle of refraction, and the angle of reflection; n_1 and n_2 are the refractive indices of the first and the second medium, respectively.

Since the refractive index of the second medium relative to the first medium n_{21} is equal to $1/n_{12}$, the above relations lead to the reciprocity of the incident and the refracted rays, i.e., reversing the direction of each ray, we do not alter their relative configuration.

We will now show that refraction of a homocentric beam at a plane surface cannot form an image of a point source. Let the refracting plane KL be the interface between two media with refractive indices n_1 and n_2 (Figure 4). A ray from a point S distant $SK = y$ from the surface falls at an angle θ on the interface and is refracted at an angle θ'. Let the continuation of the refracted beam meet the perpendicular dropped from the point S onto the interface at a point S', such that $S'K = y'$. The angles θ and θ' can be found from the triangles SKL and $S'KL$, respectively:

$$\sin\theta = \frac{x}{\sqrt{x^2 + y^2}};$$
$$\sin\theta' = \frac{x}{\sqrt{x^2 + y'^2}}. \tag{11}$$

Using Eqs. (11) and (10), we can establish a relation between the segments y and y':

$$y' = \frac{n^2}{n_1}\sqrt{y^2 + \left(1 - \frac{n_1^2}{n_2^2}\right)x^2}. \tag{12}$$

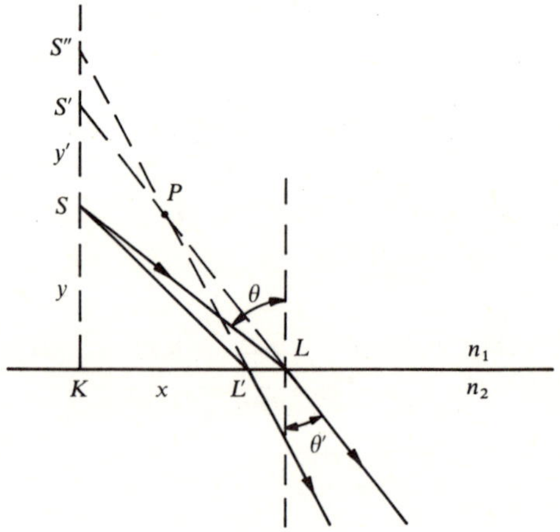

Figure 4

Breakdown of beam homocentricity in refraction at a plane interface.

The expression shows that, depending on the ratio of the refractive indices n_1 and n_2, the distance y' may be larger or smaller than y. The significant point is that y' depends on the value of x, as well as on the refractive indices of the two media and the distance of the luminous point from the refracting plane. The refracted beam is thus no longer homocentric, and the breakdown of homocentricity is more pronounced the greater the solid angle of the incident pencil of rays. For an infinitely narrow pencil, x is approximately constant for all the pencil rays, and the beam remains homocentric.

We see from Figure 4 that different rays from the point source S are refracted through different angles at the interface KL. The continuation of the ray SL' meets the perpendicular erected from K at a point S'', which is different from S'. All the rays propagating in the plane of the drawing between SL and SL' are refracted to produce virtual images of the point S on the segment $S'S''$. This segment is the *first focal line* of the refracted plane beam. The *second focal line* is obtained for the rays which lie outside the plane of the drawing. It passes through the point P at right angles to the plane of the drawing at that point. Refraction which takes place at a plane interface between two media does not produce a stigmatic image of a luminous point.

This example is a typical case of *astigmatism* of a refracting surface. Astigmatism is a kind of aberration whereby a point source is imaged in the form of two mutually perpendicular lines lying in different planes.

Let us now consider the refraction of a light beam by an arbitrary spherical surface (Figure 5). Let C be the center of a spherical surface separating between two media with refractive indices n_1 and n_2. The angles will be reckoned from the radius OC; the clockwise direction is taken as positive. Another highly important factor is to assign consistently correct signs to the various line segments, since it is only in this way that a single-valued solution of the problem can be obtained. The segments are generally measured from the refracting surface. The positive direction is taken to coincide with the direction of light propagation.

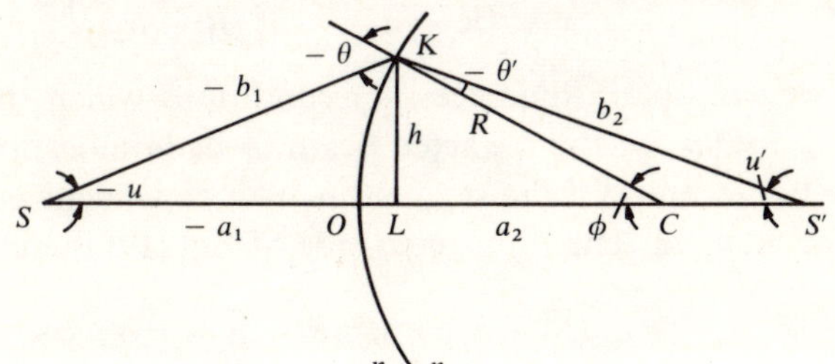

Figure 5

Refraction of light by a spherical surface.

Using the theorem of sines, we have from the triangles SKC and CKS'

$$\frac{-b_1}{\sin\phi} = \frac{R}{\sin(-u)} = \frac{R - a_1}{\sin(-\theta)};$$

$$\frac{b_2}{\sin\phi} = \frac{R}{\sin u'} = \frac{a_2 - R}{\sin(-\theta')}.$$

The law of refraction in our notation can be written in the form

$$n_1 \sin(-\theta) = n_2 \sin(-\theta').$$

From the above relations, we obtain [37]

$$\frac{R}{\sin u'} = \frac{R - a_2}{\sin\theta'}; \tag{13}$$

$$\frac{R}{\sin u} = \frac{R - a_1}{\sin\theta}; \tag{14}$$

$$\frac{b_1}{n_1(a_1 - R)} = \frac{b_2}{n_2(a_2 - R)}. \tag{15}$$

Simple manipulations give the additional relations

$$b_1^2 = (R - a_1)^2 + R^2 - 2R(R - a_1)\cos\phi; \tag{16}$$

$$b_2^2 = (R - a_2)^2 + R^2 - 2R(R - a_2)\cos\phi; \tag{17}$$

$$\phi = u - \theta = u' - \theta'. \tag{18}$$

Squaring (15) and inserting the above expressions for b_1^2 and b_2^2, we find

$$\frac{R^2 + (R - a_1)^2}{n_1^2(R - a_1)^2} - \frac{R^2 + (R - a_2)^2}{n_2^2(R - a_2)^2} =$$

$$= 2R\cos\phi\left(\frac{1}{n_1^2(R - a_1)} - \frac{1}{n_2^2(R - a_2)}\right). \tag{19}$$

From Eq. (19) we can obtain the necessary conditions which should be satisfied by the segments a_1 and a_2 for the refracted beam to be homocentric. The refracted beam will remain homocentric if the segment a_2 is independent of the angle ϕ (or, equivalently, of the angle u). The right-hand side of Eq. (19) therefore must vanish, and we have

$$n_1^2(R - a_1) = n_2^2(R - a_2) \tag{20}$$

and also

$$\frac{R^2 + (R - a_1)^2}{n_1^2(R - a_1)^2} = \frac{R^2 + (R - a_2)^2}{n_2^2(R - a_2)^2}. \tag{21}$$

The solution of these simultaneous equations is greatly simplified if we use the additional condition (15), substituting a_1 and a_2 for b_1 and b_2. This substitution is fully justified for paraxial beams ($u \to 0$). Then

$$\frac{n_1 (a_1 - R)}{a_1} = \frac{n_2 (a_2 - R)}{a_2}. \tag{22}$$

Comparison of (20) and (22) gives $n_1 a_1 = n_2 a_2$. Using this equality, we obtain from (22)

$$a_1 = R + \frac{n_2}{n_1} R;$$

$$a_2 = R + \frac{n_1}{n_2} R.$$

Points on the optical axis defined by these conditions are called *aplanatic points* of the spherical surface. They constitute aberration-free images of each other.

In general, a spherical interface between two media with different refractive indices does not ensure a homocentric refracted beam, even if the incident beam is strictly homocentric. As a result, a point in the object space (the medium with n_1) is imaged by a certain circle of confusion in the image space (the medium with n_2). The confusion is smaller for smaller angles u of the incident pencil. In the paraxial limit ($u \to 0$), a homocentric refracted beam is obtained which produces a point image. In this case, condition (22) is exactly satisfied. This condition is often written in the form

$$n_1 \left(\frac{1}{R} - \frac{1}{a_1}\right) = n_2 \left(\frac{1}{R} - \frac{1}{a_2}\right) = Q \tag{22a}$$

or

$$\frac{n_1}{a_1} - \frac{n_2}{a_2} = \frac{n_1 - n_2}{R}. \tag{22b}$$

The parameter Q, which does not change on refraction, is known as *Abbe's zero invariant*. It can be linked up directly with the law of refraction.

Equations of the form (22) which describe the behavior of paraxial homocentric beams apply to ideal optics, i.e., Gaussian optics. The focal distance of a spherical surface can be found from (22b).

Setting $a_1 = \infty$, we find

$$a_2 = f_2 = \frac{n_2 R}{n_2 - n_1}; \tag{23}$$

and for $a_2 = \infty$, we have

$$a_1 = f_1 = \frac{-n_1 R}{n_2 - n_1}. \tag{24}$$

Therefore, in general, $f_1 \neq f_2$ and $f_2/f_1 = -n_2/n_1$.

Let us now consider the imaging of a three-dimensional object by refraction of narrow beams on a spherical surface. To isolate the paraxial beams (relative to a certain axis), a small-aperture stop D is provided at the center of the sphere (Figure 6). The axes of the homocentric rays from the points S, S_1, and S_2 of a luminous spherical object MN are normal to the surface of the sphere. Since a paraxial geometry has been ensured, each of the refracted beams remains homocentric, forming a corresponding image. If $SO = S_1O_1 = S_2O_2 = a_1$, we have from Eq. (22) $S'O = S'_1O_1 = S'_2O_2 = a_2$, i.e., the image points also form a sphere. Therefore, a luminous plane, whose points all lie at different distances a_1 from the refracting sphere, cannot be imaged without distortion. In other words, only a narrow region of space near the axis of symmetry, for which the condition $a_1 = \text{const}$ holds true and gives a satisfactory image by narrow homocentric beams. Moreover, the axis of every elementary beam must be normal to the refracting surface. Otherwise, the refracted beam is no longer homocentric and an astigmatic image is formed.

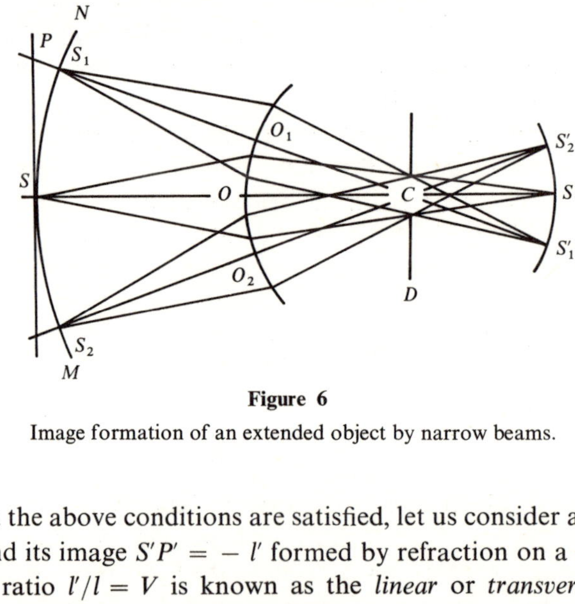

Figure 6

Image formation of an extended object by narrow beams.

Assuming that the above conditions are satisfied, let us consider a linear luminous object $SP = l$ and its image $S'P' = -l'$ formed by refraction on a spherical surface (Figure 7). The ratio $l'/l = V$ is known as the *linear* or *transverse magnification.* It is taken to be positive if the image is direct, and negative for a reversed image.

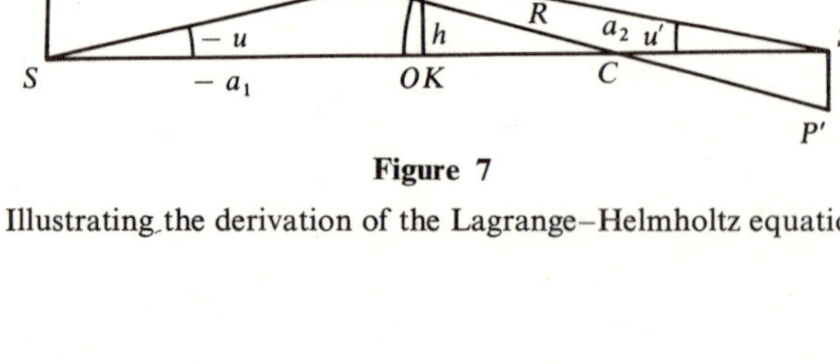

Figure 7

Illustrating the derivation of the Lagrange–Helmholtz equation.

From the similarity of the triangles PSC and $P'S'C$, we have

$$\frac{l'}{l} = \frac{a_2 - R}{a_1 - R}$$

or, using Eqs. (22),

$$\frac{l'}{l} = \frac{n_1 a_2}{n_2 a_1}. \tag{25}$$

Expressing the segment h from the triangles SMK and $S'MK$

$$h = -a_1 \tan(-u) = a_2 \tan u'$$

and seeing that the rays are paraxial, we find

$$\frac{a_2}{a_1} = \frac{u}{u'}. \tag{26}$$

Combining (25) and (26), we obtain the standard Lagrange–Helmholtz equation

$$l n_1 u = l' n_2 u'. \tag{27}$$

For wide paraxial beams, Eq. (27) is replaced by the more general relation

$$l n_1 \sin u = l' n_2 \sin u', \tag{28}$$

which is an expression of Abbe's sine condition.

Alongside with the linear magnification, we often have to consider the *angular magnification*

$$W = \frac{\tan u'}{\tan u} = \frac{a_1}{a_2}. \tag{29}$$

Comparison of Eqs. (25) and (29) establishes an obvious relation between the linear and the angular magnification:

$$WV = \frac{n_1}{n_2}.$$

In the case of several centered spherical surfaces, the image space of any given surface is the object space of the next surface, so that the above relations can be applied without difficulty. In the general case, the Lagrange–Helmholtz condition takes the form of a string of equalities

$$l n_1 u = l' n_2 u' = l'' n_3 u'' = \dots .$$

We will now consider some fundamental relations for spherical lenses, which constitute the simplest centered system comprising two spherical refracting surface.

For a thin lens, we have

$$n_1\left(\frac{1}{a_2}-\frac{1}{a_1}\right)=(n_2-n_1)\left(\frac{1}{R_1}-\frac{1}{R_2}\right),$$

where n_1 is the refractive index of the environment, and n_2 is the refractive index of the lens material.

Introducing the relative refractive index $n = n_2/n_1$, we may write

$$\frac{1}{a_2}-\frac{1}{a_1}=(n-1)\left(\frac{1}{R_1}-\frac{1}{R_2}\right). \tag{31}$$

Taking successively $a_1 = \infty$ and $a_2 = \infty$ in this relation, we obtain the focal distances of the lens

$$f_1=-f_2=\frac{1}{(n-1)\left(\dfrac{1}{R_1}-\dfrac{1}{R_2}\right)}. \tag{32}$$

Eq. (31) thus can be rewritten in the form

$$\frac{1}{a_2}-\frac{1}{a_1}=\pm\frac{1}{f}. \tag{33}$$

The fundamental equations for thick lenses are more complicated [37].

Aberrations of real optical systems

The above relations for ideal optical systems constitute the first approximation only, and they apply with varying degrees of accuracy to real systems. The ray geometry and the ray paths in real optical systems are different from those in ideal systems. Even close points, as long as they are separated by a finite distance, are never reproduced by a real optical system without distortions (or aberrations, as we say in optics). The main reasons for these aberrations are the imperfect paraxiality, the finite beam width, and the dependence of the refractive indices of the optical elements on wavelengths (dispersion).

The simplest error of optical systems is the *spherical aberration*, associated with the finite width of the beam. Even if the luminous object is a point source lying on the optical axis, the finite beam width is enough to produce an astigmatic image (Figure 8). As a result, the point source S is imaged by a certain segment $\bar{S}'S'$ on the optical axis; on the screen we thus see an illuminated circle, whose radius increases as the incident pencil fans out, i.e., as the beam becomes progressively less paraxial. If S' is the ideal image of the point source S, the segment $\bar{S}'S'$ is called the longitudinal spherical aberration. The maximum radius of the light spot on a plane perpendicular to the optical axis at the point S' characterizes the transverse spherical aberration.

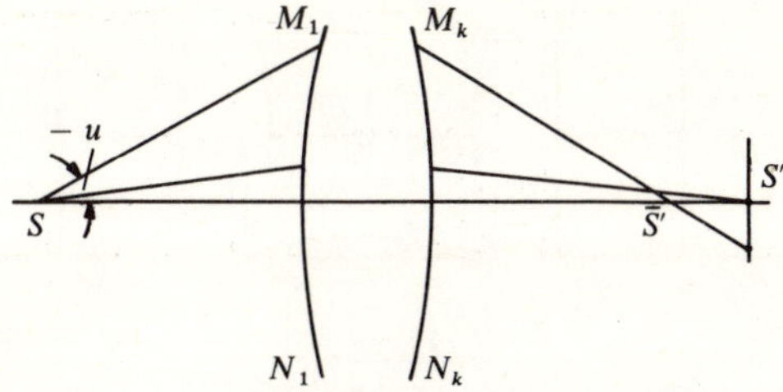

Figure 8
Spherical aberration:
M_1N_1 and M_kN_k are the first and the last refracting interfaces of an optical system.

The spherical aberration for a given beam width depends on the curvatures of the optical surfaces and the refractive indices of the constituent optical elements.

Coma is another type of aberration associated with the finite beam width. It is observed when the luminous object lies off the optical axis. The beam is not conical in the image space, since its rays do not meet at one point. A cross section of this beam cut by a plane normal to the axis at the point S' is generally of a highly complex shape, not unlike the tail of a comet.

Another aberration is known as *distortion.* If the beam axes make a large angle with the optical axis of the system, the image is not entirely like the luminous object. Distortion is associated with the dependence of the linear magnification V of the system on the angle that the beam makes with the optical axis. Indeed, Eq. (25) is valid only for ideal optical systems, i.e., systems where the angles u and ϕ are small enough for the approximation $a_1 = b_1$, $a_2 = b_2$ to apply. In the general case, the magnification V is a function of the angle u.

Distortion is generally defined quantitatively by the equality

$$\beta = \frac{V_u - V_0}{V_0} = \frac{V_u}{V_0} - 1,$$

where V_0 is the magnification for infinitesimal angles u; V_u is the magnification for some finite angle u.

Figure 9 shows the distortion of the image of a square grid in two different cases. If the magnification V increases away from the center of the image field ($\beta > 0$), the straight lines of the grid are curved so that they are convex toward the center. The result is known as *pincushion distortion.* Otherwise ($\beta < 0$), *barrel distortion* is obtained.

Although distortion generally does not affect the sharpness of the image, it may prove to be very harmful in some cases (e.g., in geodetic surveys), requiring appropriate corrections.

Field curvature is another common type of aberration in real optical systems. This aberration, producing a curved image of a plane, is observed for point sources

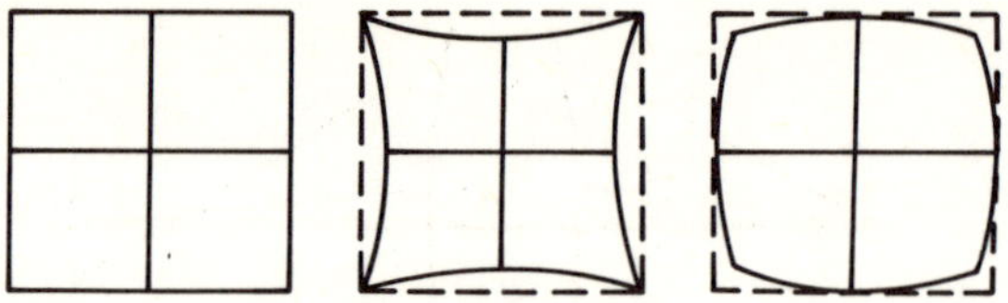

Figure 9

Image distortion:

a the original object—a square grid, *b* pincushion distortion of the image, *c* barrel distortion of the image.

which lie at a finite distance from the optical axis (assuming that these points emit narrow beams). If an extended luminous object is used, e.g., a plane perpendicular to the optical axis of the system, the corresponding beam in the image plane is astigmatic. Only the rays lying in the same section of the beam in the object space meet at one point in the image space. For example, the rays lying in a plane through the optical axis (meridional rays) and all the rays in the perpendicular plane (sagittal rays) converge at two distinct points in the image space. Moreover, the image points are different for rays which make different angles with the central ray of the beam. As a result, the locus of the convergence points of the meridional and the sagittal rays comprises two surfaces which are tangent to the Gaussian image plane (the plane perpendicular to the optical axis at the ideal image point).

Field curvature is particularly harmful in wide-angle systems, e.g., in landscape photography. Slight displacements of the image plane will improve the sharpness of lines extending in one direction, while ruining the sharpness of lines in other directions. A sharp image of a small portion of the plane can be obtained only near the optical axis. Correction of the field curvature requires fairly complex optical systems with lenses of certain radii of curvature, which have to be manufactured from glasses of various grades.

3. Numerical aperture of an optical system

In ideal optical systems, Abbe's sine condition (28) is satisfied. The linear magnification (25) in this case is written in the form

$$\frac{l'}{l} = \frac{n_1 \sin u}{n_2 \sin u'}. \tag{34}$$

The product $n_1 \sin u$ is generally called the numerical aperture of the system, designated A.

The value of the numerical aperture is of importance in calculations of the resolving power of optical systems. As we know, resolution is limited by diffraction effects.

Thus, the minimum distance between two luminous points M and N (Figure 10) which is resolved by a microscope is limited by diffraction effects from the objective rim (or the aperture stop KK'). According to Rayleigh's criterion, the points M' and N' are resolved if the separation l' between these points is less than the radius of the first dark diffraction ring [24], which is expressed in relative units by the equality

$$\phi = \frac{0.61\lambda}{KP}.$$

Seeing that ϕ is small, we may write

$$l' = \phi PM' = 0.61\lambda \frac{PM'}{KP}.$$

Since $KP/PM' = u'$ (in a microscope, the angle u' is generally very small), we finally get

$$l'u' = 0.61\lambda.$$

Hence, using (34), we obtain for $n_2 = 1$

$$l = \frac{l'u'}{n_1 \sin u} = \frac{0.61\lambda}{A}. \tag{35}$$

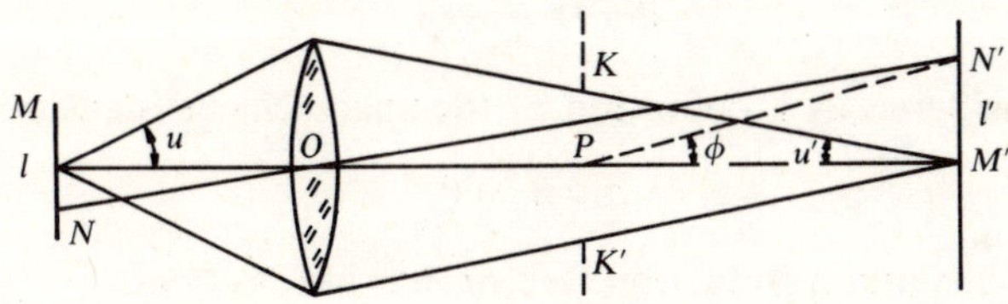

Figure 10

Definition of the resolving power of an objective.

Eq. (35) is valid for self-luminous objects and for those illuminated by an external source. The resolving power of an objective increases as the wavelength decreases and the numerical aperture increases.

4. The electromagnetic theory of light. Maxwell's equations

Numerous optical phenomena cannot be explained on the basis of geometrical optics. Maxwell's electromagnetic theory of light must be used for this purpose. Maxwell managed to combine a great variety of experimental findings into a consistent system of equations, which predicted the existence of electromagnetic waves.

Maxwell's first equation describes the relationship between the current and the magnetic field set up by that current. In this sense, it constitutes a generalization of the well-known law of Biot-Savart and Laplace. In differential form, this equation is written as

$$\text{curl}\, \boldsymbol{H} = \frac{4\pi}{c} \boldsymbol{j}_n + \frac{1}{c} \frac{\partial \boldsymbol{D}}{\partial t}, \tag{36}$$

where $\boldsymbol{H}$ is the magnetic field vector, $\boldsymbol{D}$ is the electrical displacement vector, $\boldsymbol{j}_n$ is the conductivity current density.

The second term on the right is the displacement current density.

In integral form, this equation is written as

$$\oint_L \boldsymbol{H} dl = \frac{4\pi}{c} \int_S \left(\boldsymbol{j}_n + \frac{1}{4\pi} \frac{\partial \boldsymbol{D}}{\partial t} \right) dS, \tag{37}$$

where L is the length of a closed contour; S is the surface subtended by this contour.

Maxwell's second equation

$$\text{curl}\, \boldsymbol{E} = -\frac{1}{c} \frac{\partial \boldsymbol{B}}{\partial t} \tag{38}$$

indicates that a time-variable magnetic field of induction $\boldsymbol{B}$ sets up an electric field of strength $\boldsymbol{E}$.

Eqs. (36) and (38) are generally supplemented by two further equations. One is

$$\text{div}\, \boldsymbol{D} = 4\pi\rho, \tag{39}$$

and it can be considered as a definition of the space charge density ρ. The second equation

$$\text{div}\, \boldsymbol{B} = 0 \tag{40}$$

signifies that the magnetic induction vector has no sources.

Eqs. (36) and (38) link five vectors $\boldsymbol{E}$, $\boldsymbol{H}$, $\boldsymbol{D}$, $\boldsymbol{B}$, and $\boldsymbol{j}_n$. To derive a single-valued solution of any electromagnetic problem with given initial and boundary conditions, three further vector equations are required. For homogeneous isotropic media, these supplementary equations are

$$\boldsymbol{D} = \varepsilon \boldsymbol{E}; \quad \boldsymbol{B} = \mu \boldsymbol{H}; \quad \boldsymbol{j}_n = \frac{\sigma}{4\pi} \boldsymbol{E}, \tag{41}$$

where ε and μ are, respectively, the dielectric constant and the magnetic permeability of the medium and σ is the conductivity of the medium.

Using Eqs. (41), we write Eqs. (36) and (38) for a space without charges and currents in the form

$$\text{curl}\, \boldsymbol{H} = \frac{\varepsilon}{c} \frac{\partial \boldsymbol{E}}{\partial t}; \tag{42}$$

$$\text{curl}\, \boldsymbol{E} = -\frac{\mu}{c} \frac{\partial \boldsymbol{H}}{\partial t}. \tag{43}$$

Applying the curl operator to both sides of Eq. (42) and seeing that $\operatorname{div} \boldsymbol{H} = 0$, we find

$$\operatorname{curl}\operatorname{curl} \boldsymbol{H} = \Delta \boldsymbol{H} = \frac{\varepsilon}{c}\frac{\partial}{\partial t}\operatorname{curl} \boldsymbol{E}.$$

Inserting the expression for curl $\boldsymbol{E}$ from Eq. (43), we obtain the wave equation

$$\Delta \boldsymbol{H} = \frac{\varepsilon\mu}{c^2}\frac{\partial^2 \boldsymbol{H}}{\partial t^2}. \tag{44}$$

We can similarly derive the wave equation for the electric field vector:

$$\Delta \boldsymbol{E} = \frac{\varepsilon\mu}{c^2}\frac{\partial^2 \boldsymbol{E}}{\partial t^2}. \tag{45}$$

These equations suggest that the electromagnetic field generated at any point in space does not remain localized: it propagates with a velocity $v = c/\sqrt{\varepsilon\mu}$. Indeed, in the simplest case, when the vectors $\boldsymbol{E}$ and $\boldsymbol{H}$ are functions of a single coordinate, x say, the wave equations in scalar form take the form

$$\begin{gathered}\frac{\partial_2 E_x}{\partial x^2} = \frac{\varepsilon\mu}{c^2}\frac{\partial^2 E_x}{\partial t^2}; \quad \frac{\partial^2 E_y}{\partial x^2} = \frac{\varepsilon\mu}{c^2}\frac{\partial^2 E_y}{\partial t^2}; \quad \frac{\partial^2 E_z}{\partial x^2} = \frac{\varepsilon\mu}{c^2}\frac{\partial^2 E_z}{\partial t^2};\\ \frac{\partial^2 H_x}{\partial x^2} = \frac{\varepsilon\mu}{c^2}\frac{\partial^2 H_x}{\partial t^2}; \quad \frac{\partial^2 H_y}{\partial x^2} = \frac{\varepsilon\mu}{c^2}\frac{\partial^2 H_y}{\partial t^2}; \quad \frac{\partial^2 H_z}{\partial x^2} = \frac{\varepsilon\mu}{c^2}\frac{\partial^2 H_z}{\partial t^2}.\end{gathered} \tag{46}$$

Direct differentiation will show that any function of the form $f(t \pm x/v)$ with $v^2 = c^2/\varepsilon\mu$ satisfies Eqs. (46). The physical meaning of such a function is well known: this is a plane wave propagating with a velocity v in the positive or the negative direction along the x axis. The magnitude of the vectors $\boldsymbol{E}$ and $\boldsymbol{H}$ is constant over the entire plane perpendicular to the direction of propagation.

In general, Eqs. (44) and (45) are independent. However, there is a definite relation between the electric and the magnetic vectors $\boldsymbol{E}$ and $\boldsymbol{H}$. Inserting the functional solution $f(t \pm x/v)$ in Maxwell's first equation and seeing that $v = c/\sqrt{\varepsilon\mu}$, we find

$$\boldsymbol{i} \times \boldsymbol{H} = \pm\sqrt{\frac{\varepsilon}{\mu}}\,\boldsymbol{E}, \tag{47}$$

where $\boldsymbol{i}$ is the unit vector along the x axis.

Similarly, Maxwell's second equation gives

$$\boldsymbol{i} \times \boldsymbol{E} = \pm\sqrt{\frac{\mu}{\varepsilon}}\,\boldsymbol{H}. \tag{48}$$

We see from these relations that the electric and the magnetic field vectors are perpendicular to each other and to the direction of wave propagation. It should be stressed that this conclusion is valid for plane waves only. *Plane electromagnetic waves are thus transverse waves.*

A plane electromagnetic wave propagating with a velocity v along the x axis is described by a function of the form $f(t \pm x/v)$. In this case, $E_x = 0$ and $H_x = 0$. Since the light wave characteristics are harmonic functions of time, the solution of Maxwell's equations can be written, e.g., in the form

$$\boldsymbol{E} = \boldsymbol{E}_0 \sin \frac{2\pi}{T}\left(t - \frac{x}{v}\right); \quad \boldsymbol{H} = \boldsymbol{H}_0 \sin \frac{2\pi}{T}\left(t - \frac{x}{v}\right),$$

where $\boldsymbol{E}_0$, $\boldsymbol{H}_0$ are amplitudes and T is the oscillation period of $\boldsymbol{E}$ and $\boldsymbol{H}$.

The above solutions correspond to a zero initial phase.

In practice, plane wave solutions are generally sought in complex notation

$$\begin{aligned} \boldsymbol{E} &= \boldsymbol{E}_0 e^{i(\omega t - kr)}; \\ \boldsymbol{H} &= \boldsymbol{H}_0 e^{i(\omega t - kr)}, \end{aligned} \tag{49}$$

where ω is the circular (radian) frequency, $\boldsymbol{k} = (2\pi/\lambda)\, \boldsymbol{n}$ is the wave vector which points in the direction of wave propagation.

The phase velocity of monochromatic waves is $v_{ph} = \omega/k = \lambda/T = v$.

Using Eqs. (49), we can rewrite the wave equations (44) and (45) in the form

$$\Delta \boldsymbol{E} + k^2 \boldsymbol{E} = 0; \tag{50}$$

$$\Delta \boldsymbol{H} + k^2 \boldsymbol{H} = 0, \tag{51}$$

where $k_2 = \omega^2 \varepsilon\mu/c^2$.

It can be readily shown that in the general case of a conducting medium ($\sigma \neq 0$), the propagation constant is expressed in the form

$$k^2 = \frac{\omega^2 \varepsilon\mu - i\omega\mu\sigma}{c_2}. \tag{52}$$

The wave equations (50) and (51) are Helmholtz differential equations. For real k^2, their solutions are expressed in the form of traveling waves.

Let us write Maxwell's equation in scalar form. This form is convenient for the treatment of light wave propagation in narrow-gauge fibers:

$$\frac{\partial E_\gamma}{\partial x_\beta} - \frac{\partial E_\beta}{\partial x_\gamma} = -\frac{1}{c}\frac{\partial B_\alpha}{\partial t}; \tag{53}$$

$$\frac{\partial H_\gamma}{\partial x_\beta} - \frac{\partial H_\beta}{\partial x_\gamma} = \frac{4\pi}{c} j_{n\alpha} + \frac{1}{c}\frac{\partial D_\alpha}{\partial t}; \tag{54}$$

$$D_\alpha = \varepsilon E_\alpha; \quad B_\alpha = \mu H_\alpha; \quad j_{n\alpha} = \frac{}{4\pi} E_\alpha. \tag{55}$$

Eqs. (53) and (54) are applicable for α, β, and γ equal to 1, 2, 3, respectively, and also to 2, 3, 1 and 3, 1, 2, i.e., cyclic permutations of the first triad. This is a more compact form for writing the wave equations, since three scalar equations are actually expressed by one. The coordinate axes are designated x_1, x_2, and x_3.

For a wave propagating along a thin fiber, the plane wave approximation is not always sufficient, and we should therefore seek a solution of Maxwell's equations in a more general form, assuming nothing more than periodicity with time [21],

$$\boldsymbol{E} = \mathscr{E}\, e^{i\omega t} \quad \text{and} \quad \boldsymbol{H} = \mathscr{H}\, e^{i\omega t}, \tag{56}$$

where $\mathscr{E}$ and $\mathscr{H}$ are some vectors with complex components. They are often referred to as the complex electric and magnetic vectors.

Inserting from (49) in Eqs. (53) and (54) and using Eqs. (56), we find

$$\frac{\partial \mathscr{E}_\gamma}{\partial x_\beta} - \frac{\partial \mathscr{E}_\beta}{\partial x_\gamma} = \frac{i\omega\mu}{c}\mathscr{H}_\alpha; \quad \frac{\partial \mathscr{H}_\gamma}{\partial x_\beta} - \frac{\partial \mathscr{H}_\beta}{\partial x_\gamma} = \frac{\sigma - i\omega\varepsilon}{c}\mathscr{E}_\alpha. \tag{57}$$

In what follows, we will use the conventional notation for $\boldsymbol{E}$ and $\boldsymbol{H}$ vectors. This need not cause any confusion, since the solutions will always be multiplied explicitly by the time factor $e^{i\omega t}$ and the complex electric and magnetic vectors $\mathscr{E}$ and $\mathscr{H}$, after all, satisfy the same wave equations (50) as the vectors $\boldsymbol{E}$ and $\boldsymbol{H}$.

5. Energy density and energy flux

Electromagnetic waves propagating in free space or in a material medium transport a certain amount of energy. The total energy W of an electromagnetic field is a sum of the energy densities of the electric and magnetic fields, $(\varepsilon/8\pi)\boldsymbol{E}^2$ and $(\mu/8\pi)\boldsymbol{H}^2$, respectively, i.e.,

$$W = \frac{\varepsilon \boldsymbol{E}^2 + \mu \boldsymbol{H}^2}{8\pi}. \tag{58}$$

Let us derive the law of energy conservation in the case of no currents. Multiplying Eqs. (36) and (38) by $\boldsymbol{E}$ and $\boldsymbol{H}$, respectively, we subtract the two equations term by term:

$$\boldsymbol{E} \operatorname{curl} \boldsymbol{H} - \boldsymbol{H} \operatorname{curl} \boldsymbol{E} = \frac{1}{c}\boldsymbol{E}\frac{\partial \boldsymbol{D}}{\partial t} + \frac{1}{c}\boldsymbol{H}\frac{\partial \boldsymbol{B}}{\partial t}. \tag{59}$$

The left-hand side of Eq. (59) is transformed using the standard vector identity

$$\operatorname{div}(\boldsymbol{a} \times \boldsymbol{b}) = \boldsymbol{b} \operatorname{curl} \boldsymbol{a} - \boldsymbol{a} \operatorname{curl} \boldsymbol{b},$$

so that

$$-\operatorname{div}(\boldsymbol{E} \times \boldsymbol{H}) = \frac{1}{2c}\frac{\partial}{\partial t}(\varepsilon \boldsymbol{E}^2 + \mu \boldsymbol{H}^2)$$

or

$$\frac{\partial W}{\partial t} = - \operatorname{div} \boldsymbol{S}, \tag{60}$$

where

$$\boldsymbol{S} = \frac{c}{4\pi}(\boldsymbol{E} \times \boldsymbol{H}). \tag{61}$$

Finally, integrating Eq. (60) over the volume V, enclosed within a closed surface S', we find

$$\frac{\partial}{\partial t}\int W\,dV = \oint S\,dS'. \tag{62}$$

The left-hand side of Eq. (62) describes the change in total field energy in unit time. Hence, the integral on the right should be considered as the energy flux through a surface S'. The vector $\boldsymbol{S}$, generally called the Poynting vector, therefore represents the energy flux density of the electromagnetic field.

If the integral is extended over a finite volume, the surface integral on the right in Eq. (62) vanishes (the field at infinity is zero), and the law of energy conservation is written in the form

$$\frac{\partial}{\partial t}\int W\,dV = 0,$$

i.e.,

$$W = \text{const.}$$

6. Polarization of electromagnetic waves

The concept of polarization of a light wave was first introduced by Malus in his study of the properties of rays reflected from glass. A consistent treatment of polarization became possible only after the transversality of the light waves had been established.

For a quantitative description of polarization, we will use the complex notation of the electric and the magnetic waves (56). In general, the expression for the complex amplitudes $\mathscr{E}$ and $\mathscr{H}$ contains certain phase constants δ_i. Thus, the components of the complex amplitude $\mathscr{E}$ may be written in the form [1]

$$\mathscr{E}_x = a_1 e^{i\delta_1}, \quad \mathscr{E}_y = a_2 e^{i\delta_2}; \quad \mathscr{E}_z = a_3 e^{i\delta_3}. \tag{63}$$

Using Euler's identity

$$e^{i\phi} = \cos\phi + i\sin\phi,$$

we can write for the real components of the vector $\boldsymbol{E}$

$$
\begin{aligned}
E_x &= \operatorname{Re} a_1 e^{i(\omega t+\delta_1)} = a_1 \cos(\omega t + \delta_1); \\
E_y &= \operatorname{Re} a_2 e^{i(\omega t+\delta_1)} = a_2 \cos(\omega t + \delta_2); \\
E_z &= \operatorname{Re} a_3 e^{i(\omega t+\delta_3)} = a_3 \cos(\omega t + \delta_3).
\end{aligned} \tag{64}
$$

If the axis z points in the direction of wave propagation, $E_z = 0$ and $H_z = 0$ because of transversality, and only the x and y components of the vectors $\boldsymbol{E}$ and $\boldsymbol{H}$ are non-zero.

The problem of polarization boils down to the determination of the time dependence of the spatial position of the plane through the vector $\boldsymbol{E}$ (or $\boldsymbol{H}$) and the direction of wave propagation. Since the two vectors are perpendicular to the z axis at any time, we will consider the curve traced by the tip of the vector $\boldsymbol{E}$ or $\boldsymbol{H}$.

To eliminate the time factor from Eqs. (64), we multiply both sides of these equations by $\sin\delta_2$ and $\sin\delta_1$, respectively, expand the right-hand sides of the expressions for the components E_x and E_y, and subtract the second equation from the first:

$$
\frac{E_x}{a_1}\sin\delta_2 - \frac{E_y}{a_2}\sin\delta_1 = \cos\omega t \sin(\delta_2 - \delta_1) \tag{65}
$$

Similarly, multiplying by $\cos\delta_2$ and $\cos\delta_1$ and subtracting, we obtain

$$
\frac{E_x}{a_1}\cos\delta_2 - \frac{E_y}{a_2}\cos\delta_1 = \sin\omega t \sin(\delta_2 - \delta_1). \tag{66}
$$

Squaring (65) and (66) and adding up the results, we obtain

$$
\left(\frac{E_x}{a_1}\right)^2 + \left(\frac{E_y}{a_2}\right)^2 - 2\frac{E_x}{a_1}\frac{E_y}{a_2}\cos\delta = \sin^2\delta, \tag{67}
$$

where

$$
\delta = \delta_2 - \delta_1. \tag{68}
$$

Eq. (67) is an ellipse inscribed in a rectangle with sides $2a_1$ and $2a_2$ parallel to the coordinate axes. The end of the electric vector thus describes an ellipse in a plane perpendicular to the direction of wave propagation. In other words, a phase shift $\delta = \delta_2 - \delta_1$ between the field components leads to an elliptical polarization of the radiation.

Similarly, using (47) and (48), we can show that the end of the associated magnetic vector also describes an ellipse, which is inscribed in a rectangle of sides $2\sqrt{(\varepsilon/\mu)}a_1$ and $2\sqrt{(\varepsilon/\mu)}a_2$ parallel to the x and y axes.

In general, the axes of the two ellipses do not coincide with the coordinate axes. Let us now determine the axes of the ellipse and its orientation (e.g., the angle θ between the major axis and the direction x) for known amplitudes a_1, a_2 and a known phase difference δ. We introduce a new system of coordinates ξ, η which coincides with the ellipse axes. Then

$$E_\xi = E_x \cos\theta + E_y \sin\theta,$$
$$E_\eta = E_x \sin\theta + E_y \cos\theta, \tag{69}$$

and the equation of the ellipse takes the form

$$\left(\frac{E_\xi}{u}\right)^2 + \left(\frac{E_\eta}{w}\right)^2 = 1. \tag{70}$$

Inserting E_ξ and E_η in Eq. (70) and comparing the result with Eq. (67), we obtain the following relations between the parameters a_1, a_2, and δ in the coordinates x, y, on the one hand, and the parameters u, w, and θ in the coordinates ξ, η, on the other:

$$a_1 \sin\delta = \frac{uw}{\sqrt{w^2\cos^2\theta + u^2\sin^2\theta}};$$
$$a_2 \sin\delta = \frac{uw}{\sqrt{u^2\cos^2\theta + w^2\sin^2\theta}}; \tag{71}$$
$$\frac{a_1 a_2 \sin\delta}{\cos\delta} = \frac{uw}{\sin\theta\cos\theta\,(u^2 + w^2)}.$$

After simple manipulations, we finally obtain

$$a_1^2 + a_2^2 = u^2 + w^2;$$
$$\tan 2\theta = \frac{2a_1 a_2}{a_1^2 - a_2^2}\cos\delta = \tan 2\alpha \cos\delta, \tag{72}$$

where

$$\tan\alpha = \frac{a_2}{a_1};$$
$$\tan\beta = \frac{w}{u}; \tag{73}$$
$$\sin 2\beta = \pm \sin 2\alpha \sin\delta.$$

The double sign in the last equality corresponds to two different senses of rotation. Thus, if the parameters a_1, a_2, and δ are known, we can find the parameters u, w, and θ. Conversely, if the axes of the ellipse and its orientation, i.e., u, w, θ, are known, the amplitudes and the phase difference of the two field factors can be calculated.

Let us consider two highly important particular cases, when the elliptical polarization degenerates into linear or circular polarization. The case of linear polarization, as we see from Eq. (67), obtains when

$$\delta = \delta_2 - \delta_1 = \pm m\pi \quad (m = 0, 1, 2, \dots).$$

Eqs. (64) then give the following simple relation between the electric vector components:

$$\frac{E_y}{E_x} = (-1)^m \frac{a_2}{a_1}. \tag{74}$$

This expression also fixes the tangent of the angle θ between the direction of oscillation and the axis x. Note that the same result is obtained independently from Eq. (72).

Oscillations described by Eq. (74) are said to be linearly polarized. A simple rotation of the coordinate axes through an angle θ until one of the axes coincides with one of the components reduces the other component (E_y, say) to zero. Since the vectors $\boldsymbol{E}$ and $\boldsymbol{H}$ are perpendicular to each other, one of the components of the vector $\boldsymbol{H}$ also vanishes. For historical reasons, the plane through the wave vector and the electrical field vector is called the plane of oscillation of polarized light, and the plane through the magnetic vector and the field vector is called the *plane of polarization.*

Elliptical polarization degenerates into circular polarization if two further conditions are satisfied:

$$a_1 = a_2 = a;$$

$$\delta = \delta_2 - \delta_1 = \pm m \frac{\pi}{2} (m = 1, 3, 5, \dots).$$

Eq. (67) then reduces to an equation of a circle.

Two different cases of circular polarization are distinguished, depending on the sense of rotation of the vector $\boldsymbol{E}$. If an observer viewing the approaching wave sees the radius-vector rotating clockwise, the polarization is called right-hand; rotation in counterclockwise direction corresponds to left-hand polarization (Figure 11).

Figure 12 shows the effect of δ on the shape of the ellipse traced by the oscillating vector. For $\delta = 0$ and $\delta = \pi$ the ellipse degenerates into a straight line. For $\delta = \pi/2$ and $\delta = 3\pi/2$ we obtain right-hand and left-hand circular polarization, respectively. Intermediate phase differences correspond to different orientations of the ellipse.

7. Boundary conditions at the interface between two media

Maxwell's equations were formulated for regular points in space, i.e., those points around which the physical properties of the medium vary continuously. All the vec-

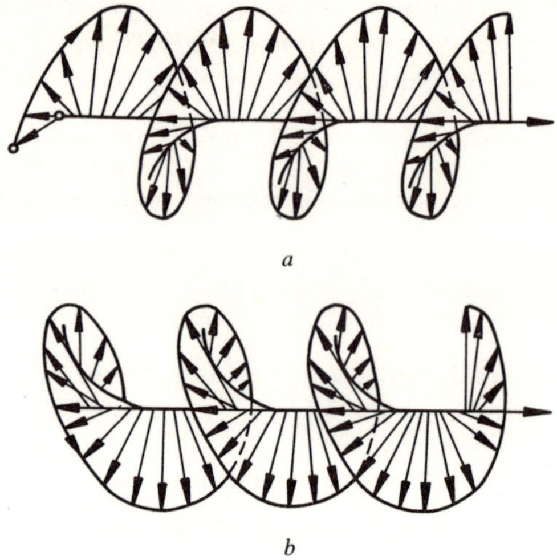

Figure 11
Spatial distribution of field intensity in a circularly polarized wave:
a right-hand polarization, *b* left-hand polarization.

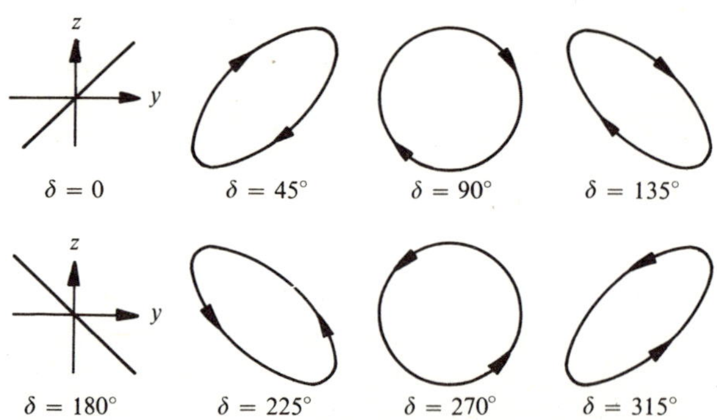

Figure 12
Various states of polarization as a function of the phase difference δ.

tors characterizing the electromagnetic wave should thus be finite, continuous, and differentiable. This naturally also applies to the constants ε, μ, and σ. However, when we cross an interface between two media with different ε, μ, and σ, the field vectors will change discontinuously, since they are related by material equations (41) which contain the constants ε, μ, and σ. Therefore, for a unique solution of Maxwell's equations with given initial conditions, we should also specify certain boundary conditions to be satisfied by the components of the electromagnetic vectors at the interface.

To derive these boundary conditions, we first assume that the interface S between medium 1 and medium 2 (Figure 13) is replaced by a thin transition layer where the parameters ε, μ, and σ vary rapidly, but nevertheless continuously. In this transitional layer, the field vectors and their derivatives are finite and continuous functions of position and time. In the following, we will mainly require the boundary conditions at the interface between a dielectric fiber and its casing, so that we may assume $\rho = 0$ and $\boldsymbol{j} = 0$.

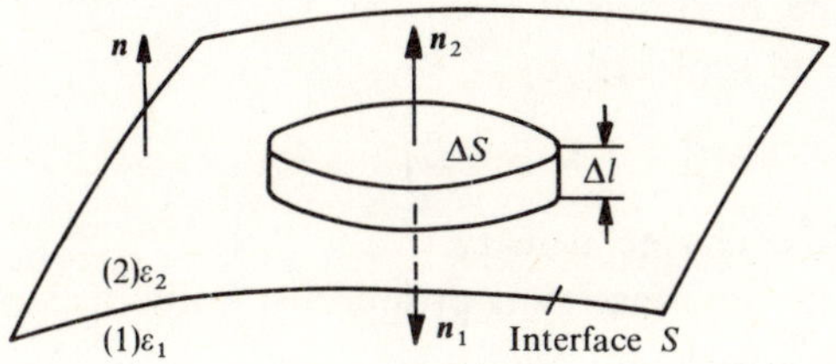

Figure 13

Derivation of the boundary conditions for the normal field component.

Consider a small cylindrical volume element ΔV in the transitional layer. The cylinder generatrices are normal to the surface S, and its bases coincide with the two boundary surfaces of the transitional layer, so that the height of the cylinder is equal to the layer thickness Δl. The direction of the positive normal n is chosen as shown in Figure 13. Medium 1 thus lies on the negative side of the surface S, and medium 2 on the positive side.

We integrate Eq. (39) for the electric displacement vector $\boldsymbol{D}$ over the cylindrical volume ΔV, using Gauss's theorem

$$\int_{\Delta V} \operatorname{div} \boldsymbol{D} \, dV = \oint_{\Delta S_{\text{tot}}} D_n \, dS.$$

The integrals are taken over the entire surface of the cylinder, including the side walls. Let us now gradually reduce the thickness of the transitional layer. In the limit ($\Delta l = 0$), we may ignore the part of the integral associated with the flux through the side walls and consider only the flux through the cylinder bases. Because of our particular choice of the positive normal, we may write

$$\int_{\Delta S_{\text{tot}}} (D_{2n} - D_{1n}) \, dS = 0.$$

i.e.,

$$D_{2n} = D_{1n}, \tag{75}$$

or in vector form

$$(\boldsymbol{D}_2 - \boldsymbol{D}_1)\, \boldsymbol{n} = 0. \tag{75a}$$

Similarly, from the condition div $\boldsymbol{B} = 0$ we have

$$B_{2n} = B_{1n} \tag{76}$$

or

$$(\boldsymbol{B}_2 - \boldsymbol{B}_1)\,\boldsymbol{n} = 0. \tag{76a}$$

Thus, in the absence of charges, the normal components of the vectors $\boldsymbol{D}$ and $\boldsymbol{B}$ are continuous at the interface.

If $\rho \neq 0$, Eq. (75) is replaced by

$$D_{2n} - D_{1n} = 4\pi\sigma,$$

where σ is the surface charge density.

Surface charge at the interface thus produces a finite discontinuity in the normal component of the displacement vector.

The behavior of the tangential components at the interface can be established from the first two Maxwell's equations. To this end, the cylinder (Figure 13) is replaced by a rectangle (Figure 14). Let the rectangle form a contour of length L and area S_0. A unit vector $\boldsymbol{\tau}$ defines the positive direction of motion along the top and the bottom sides of the rectangular contour. The vector $\boldsymbol{n}$, as before, corresponds to the positive normal to the surface S, $\boldsymbol{n}_0$ is a vector tangent to the surface at right angles to the vector $\boldsymbol{\tau}$.

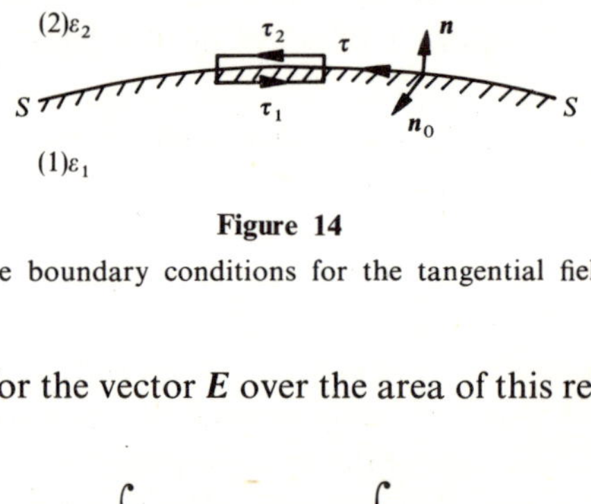

Figure 14

Derivation of the boundary conditions for the tangential field components.

Integrating Eq. (38) for the vector $\boldsymbol{E}$ over the area of this rectangle, we use Stokes' theorem

$$\int_{S_0} \operatorname{curl}_n \boldsymbol{a}\, dS = \oint_L a_l\, dl,$$

so that

$$\int_L E_l\, dl = -\int_{S_0} \left(\frac{\partial \boldsymbol{B}}{\partial t}\right)_{n_0} dS_0.$$

If the layer thickness Δl is allowed to shrink to zero, while the length of the rectangle remains fixed, the contribution from the two end sides of the rectangle becomes negligible. The left-hand side of the last equation reduces to integration over the two sides of the rectangle parallel to the interface. The right-hand side vanishes, since the field vectors and their derivatives are finite. As a result, we obtain

$$\int (E_{2t} - E_{1t})\, dt = 0,$$

i.e.,

$$E_{2t} = E_{1t} \tag{77}$$

or in vector form

$$\boldsymbol{n} \times (\boldsymbol{E}_2 - \boldsymbol{E}_1) = 0. \tag{77a}$$

Similarly, integrating Eq. (36) and remembering that the space current density $\boldsymbol{j}$ and the vector $\boldsymbol{D}$ are finite, we obtain a boundary condition for the tangential component of the magnetic vector

$$H_{2t} - H_{1t} = 0, \tag{78}$$

or in vector form

$$\boldsymbol{n} \times (\boldsymbol{H}_2 - \boldsymbol{H}_1) = 0. \tag{78a}$$

Eqs. (77) and (78) for the tangential components of the electric and magnetic vectors are the principal boundary conditions of our problem. Eqs. (75) and (76) describing the behavior of the vectors $\boldsymbol{D}$ and $\boldsymbol{B}$ at the interface can be derived from (77) and (78) [21].

Let us now derive the boundary condition at the dielectric–metal interface. The electromagnetic field virtually does not penetrate into the conductor [36, 23]. The field should therefore vanish when we cross the interface into an ideal conductor. Since condition (77) is quite general and therefore applies to this case also, we should have

$$E_{1t} = E_{2t} = 0. \tag{79}$$

In other words, the tangential component of the electric field vanishes at the surface of an ideal conductor ($\sigma \to \infty$).

8. Reflection and refraction of a plane wave at the interface between two dielectrics

The problem of light transmission through a plane interface between two dielectrics with dielectric constants ε_1 and ε_2 is completely solvable within the framework of the electromagnetic theory of light.

Consider a plane electromagnetic wave incident on the interface of two infinite homogeneous isotropic media (Figure 15). The directions of propagation of the

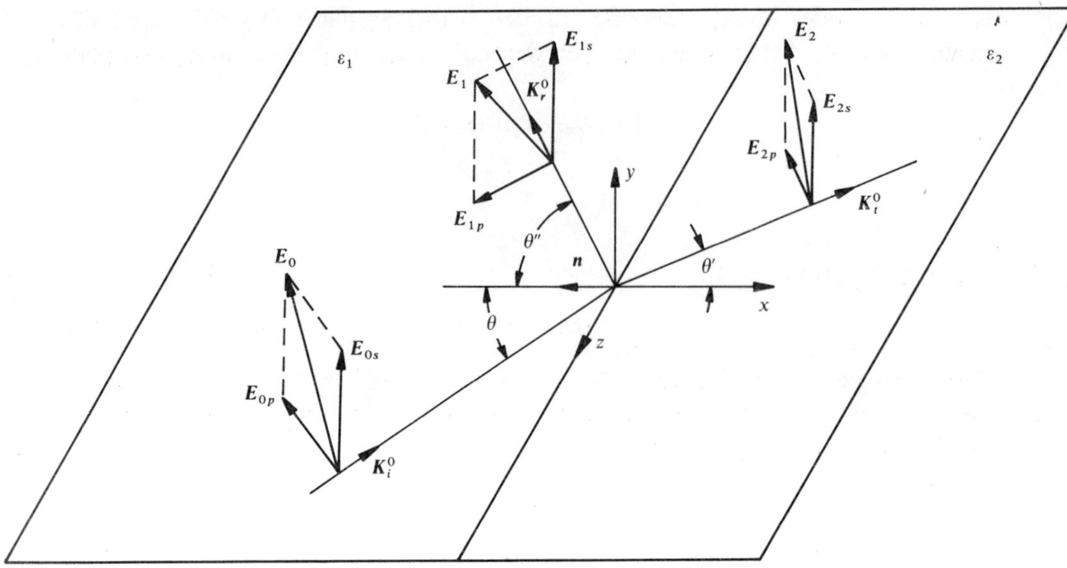

Figure 15

Reflection and refraction of a wave at an interface between two dielectrics.

incident, reflected, and refracted waves will be characterized by the unit vectors $\boldsymbol{k}_i^0$, $\boldsymbol{k}_r^0$, and $\boldsymbol{k}_t^0$, respectively. These vectors make the angles θ, θ'', and θ' with the unit vector $\boldsymbol{n}$, so that they point in the same direction as the wave vectors of the corresponding waves. In view of the above, the electric and the magnetic fields can be written in the form

$$\begin{aligned}
\boldsymbol{E}_i &= \boldsymbol{E}_0 \exp[i(\omega t - \boldsymbol{k}_i \boldsymbol{r})] = \boldsymbol{E}_0 \exp[2\pi i(vt - \boldsymbol{k}_i^0 \boldsymbol{r}/\lambda_1)] = \boldsymbol{E}_0 \exp[2\pi i v(t - \boldsymbol{k}_i^0 \boldsymbol{r}/v_1)];\\
\boldsymbol{H}_i &= \boldsymbol{H}_0 \exp[i(\omega t - \boldsymbol{k}_i \boldsymbol{r})] = \boldsymbol{H}_0 \exp[2\pi i(vt - \boldsymbol{k}_i^0 \boldsymbol{r}/\lambda_1)] = \boldsymbol{H}_0 \exp[2\pi i v(t - \boldsymbol{k}_i^0 \boldsymbol{r}/v_1)];\\
\boldsymbol{E}_r &= \boldsymbol{E}_1 \exp[i(\omega t - \boldsymbol{k}_r \boldsymbol{r})] = \boldsymbol{E}_1 \exp[2\pi i(vt - \boldsymbol{k}_r^0 \boldsymbol{r}/\lambda_1)] = \boldsymbol{E}_1 \exp[2\pi i v(t - \boldsymbol{k}_r^0 \boldsymbol{r}/v_1)];\\
\boldsymbol{H}_r &= \boldsymbol{H}_1 \exp[i(\omega t - \boldsymbol{k}_r \boldsymbol{r})] = \boldsymbol{H}_1 \exp[2\pi i(vt - \boldsymbol{k}_r^0 \boldsymbol{r}/\lambda_1)] = \boldsymbol{H}_1 \exp[2\pi i v(t - \boldsymbol{k}_r^0 \boldsymbol{r}/v_1)];\\
\boldsymbol{E}_t &= \boldsymbol{E}_2 \exp[i(\omega t - \boldsymbol{k}_t \boldsymbol{r})] = \boldsymbol{E}_2 \exp[2\pi i(vt - \boldsymbol{k}_t^0 \boldsymbol{r}/\lambda_2)] = \boldsymbol{E}_2 \exp[2\pi i v(t - \boldsymbol{k}_t^0 \boldsymbol{r}/v_2)];\\
\boldsymbol{H}_t &= \boldsymbol{H}_2 \exp[i(\omega t - \boldsymbol{k}_t \boldsymbol{r})] = \boldsymbol{H}_2 \exp[2\pi i(vt - \boldsymbol{k}_t^0 \boldsymbol{r}/\lambda_2)] = \boldsymbol{H}_2 \exp[2\pi i v(t - \boldsymbol{k}_t^0 \boldsymbol{r}/v_2)];
\end{aligned} \tag{80}$$

where $\boldsymbol{E}_0$, $\boldsymbol{H}_0$ are the amplitudes of the incident wave; $\boldsymbol{E}_1$, $\boldsymbol{H}_1$ the amplitudes of the reflected wave; $\boldsymbol{E}_2$, $\boldsymbol{H}_2$ the amplitudes of the refracted wave; $\boldsymbol{k}_i = (2\pi/\lambda_1)\,\boldsymbol{k}_i^0$, $\boldsymbol{k}_r = (2\pi/\lambda_1)\,\boldsymbol{k}_r^0$ and $\boldsymbol{k}_t = (2\pi/\lambda_2)\,\boldsymbol{k}_t^0$ are the wave vectors of the incident, reflected, and refracted waves; λ_1, v_1 and λ_2, v_2 are the wavelengths and the propagation velocities in the first and the second medium, respectively.

The magnetic vectors can be expressed in terms of the corresponding electric vectors. For a nonmagnetic dielectric, $\mu_1 = \mu_2 = 1$, and we have

$$H_i = \sqrt{\varepsilon_1}\,(k_i^0 \times E_i);$$

$$H_r = \sqrt{\varepsilon_1}\,(k_r^0 \times E_r); \tag{81}$$

$$H_t = \sqrt{\varepsilon_2}\,(k_t^0 \times E_t).$$

The electromagnetic field in the first medium is the sum of the incident and the reflected wave, and the field in the second medium is produced by the refracted wave. Therefore, the conditions of continuity of the tangential components, (77) and (78), can be written in the form [41]

$$n \times (E_i + E_r) = n \times E_t;$$

$$n \times (H_i + H_r) = n \times H_t.$$

Since both the amplitudes and the exponential factors of the tangential field components satisfy these equalities, we may write, using (80),

$$n \times (E_0 + E_1) = n \times E_2;$$

$$n \times (H_0 + H_1) = n \times H_2. \tag{82}$$

and

$$\frac{1}{v_1} k_i^0 r = \frac{1}{v_1} k_r^0 r;$$

$$\frac{1}{v_1} k_i^0 r = \frac{1}{v_2} k_t^0 r. \tag{83}$$

Eq. (83) signifies that the phases of the tangential field components are equal at the interface. This condition leads to the Decartes–Snell laws of refraction and reflection.

The Descartes–Snell laws

Following [41], we will consider an arbitrary vector r drawn from the point O on the interface between isotropic dielectrics (Figure 16). We decompose it into two components, one parallel and the other perpendicular to the interface. The magnitudes of the two components are clear from the figure. In general,

$$r = (nr)\,n - n \times (n \times r). \tag{84}$$

We are dealing with the boundary conditions at the interface, so that $(nr) = 0$,

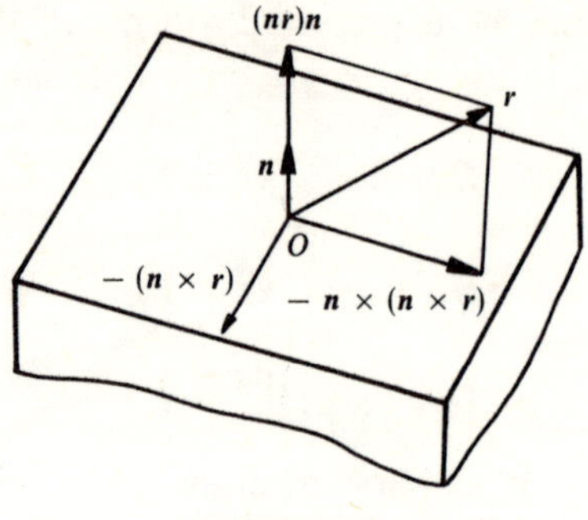

Figure 16
Derivation of the law of refraction.

since the vector $\boldsymbol{r}$ should be taken in the plane of the interface. Let us dot-multiply Eq. (84) by some vector $\boldsymbol{k}^0$ and use the standard vector identity

$$\boldsymbol{k}^0\boldsymbol{r} = \boldsymbol{k}^0\boldsymbol{n} \times (\boldsymbol{n} \times \boldsymbol{r}) = (\boldsymbol{k}^0 \times \boldsymbol{n})(\boldsymbol{n} \times \boldsymbol{r}).$$

This identity transforms the boundary condition (83) to the form

$$\begin{gathered}(\boldsymbol{k}_i^0 \times \boldsymbol{n} - \boldsymbol{k}_r^0 \times \boldsymbol{n})(\boldsymbol{n} \times \boldsymbol{r}) = 0; \\ \left(\boldsymbol{k}_i^0 \times \boldsymbol{n} - \frac{v_1}{v_2}\boldsymbol{k}_t^0 \times \boldsymbol{n}\right)(\boldsymbol{n} \times \boldsymbol{r}) = 0.\end{gathered} \tag{85}$$

The unit vectors $\boldsymbol{k}_i^0$, $\boldsymbol{k}_r^0$, and $\boldsymbol{k}_t^0$ make the angles θ, θ'', and θ' with the unit vector $\boldsymbol{n}$, so that

$$\boldsymbol{k}_i^0 \times \boldsymbol{n} = \sin\theta'' ;$$

$$\boldsymbol{k}_r^0 \times \boldsymbol{n} = \sin\theta'';$$

$$\boldsymbol{k}_t^0 \times \boldsymbol{n} = \sin\theta' .$$

Since $\boldsymbol{n} \times \boldsymbol{r}$ does not vanish, Eqs. (85) show that the vectors $\boldsymbol{k}_i^0$, $\boldsymbol{k}_r^0$, $\boldsymbol{k}_t^0$ and $\boldsymbol{n}$ are coplanar, i.e., they all lie in one plane (the plane of incidence). Moreover the first equation yields

$$\sin\theta = \sin\theta'', \tag{86}$$

and the second

$$\sin\theta = \frac{v_1}{v_2}\sin\theta'$$

or, alternatively,

$$\frac{\sin\theta}{\sin\theta'} = \frac{v_1}{v_2} = \sqrt{\frac{\varepsilon_2}{\varepsilon_1}} = \frac{n_2}{n_1} = n_{21} = n. \tag{87}$$

Eqs. (86) and (87) express the standard laws of wave reflection and refraction.

9. Fresnel equations

To arrive at a complete description of the behavior of electromagnetic waves at the interface between two dielectrics, we should also establish a relation between the amplitudes of the three waves. In general, natural light is incident on the interface, but Eqs. (47) and (48) show that the field vectors at any time are perpendicular to each other and to the direction of incidence, defined by the unit vector $\boldsymbol{k}_i^0$, so that

$$\boldsymbol{E}_0\boldsymbol{H}_0 = 0;$$

$$\boldsymbol{k}_i^0\boldsymbol{E}_0 = 0;$$

$$\boldsymbol{k}_i^0\boldsymbol{H}_0 = 0.$$

We decompose each of these vectors in both media into two components, so that the s component is normal to the plane of incidence and the p component is parallel to this plane (Figure 15). The solution of the general problem thus reduces to two particular cases.

Case 1. The incident light wave is polarized so that the electric vector is perpendicular to the plane of incidence (s component), and the positive direction of the vector corresponds to the direction from the observer into the drawing.

Since the unit vector $\boldsymbol{n}$ lies in the plane of incidence, the orthogonality conditions can be written in the form

$$\boldsymbol{n}\boldsymbol{E}_{0,s} = \boldsymbol{n}\boldsymbol{E}_{1,s} = \boldsymbol{n}\boldsymbol{E}_{2,s} = 0. \tag{88}$$

Thus, all the three electric vectors are parallel to the plane interface of the two media, and the boundary condition for the amplitudes of the tangential components (82) can be written in a simpler form

$$\boldsymbol{E}_{0,s} + \boldsymbol{E}_{1,s} = \boldsymbol{E}_{2,s}. \tag{89}$$

The second boundary condition (82) for the amplitudes of the tangential components of the magnetic vector can be expressed in terms of the electric field components, using (81):

$$\boldsymbol{n} \times (\boldsymbol{k}_i^0 \times \boldsymbol{E}_{0,s} + \boldsymbol{k}_r^0 \times \boldsymbol{E}_{1,s})\sqrt{\varepsilon_1} = \boldsymbol{n} \times (\boldsymbol{k}_t^0 \times \boldsymbol{E}_{2,s})\sqrt{\varepsilon_2}.$$

Using the vector identity $\boldsymbol{a} \times \boldsymbol{b} \times \boldsymbol{c} = \boldsymbol{b}(\boldsymbol{a}\boldsymbol{c}) - \boldsymbol{c}(\boldsymbol{a}\boldsymbol{b})$ and Eq. (88), and also remembering that $\boldsymbol{n}\boldsymbol{k}_i^0 = \cos\theta$, $\boldsymbol{n}\boldsymbol{k}_r^0 = -\cos\theta$, $\boldsymbol{n}\boldsymbol{k}_t^0 = \cos\theta'$, we obtain a second relation between the amplitudes of the s component of the electric vector,

$$\cos\theta\,(\boldsymbol{E}_{0,s} - \boldsymbol{E}_{1,s}) = \sqrt{\frac{\varepsilon_2}{\varepsilon_1}}\cos\theta'\,\boldsymbol{E}_{2,s}. \tag{90}$$

Solving Eqs. (89)–(90) for the amplitudes $E_{1,s}$ and $E_{2,s}$ of the reflected and the refracted wave, we obtain (with allowance for the sign of refraction)

$$E_{1,s} = -\frac{\sin(\theta - \theta')}{\sin(\theta + \theta')} E_{0,s};$$

$$E_{2,s} = \frac{2 \sin\theta' \cos\theta}{\sin(\theta + \theta')} E_{0,s}. \tag{91}$$

Case 2. The electric vector of the incident wave is parallel to the plane of incidence (p component). In this case, the magnetic vectors of all the three waves are perpendicular to the plane of incidence and the above scheme can be applied to the three vectors $H_{0,s}$, $H_{1,s}$, and $H_{2,s}$. As a result, we have

$$H_{0,s} + H_{1,s} = H_{2,s};$$

$$\cos\theta (H_{0,s} - H_{1,s}) = \sqrt{\frac{\varepsilon_1}{\varepsilon_2}} \cos\theta' H_{2,s}.$$

These equalities lead to the conditions

$$H_{1,s} = -\frac{\tan(\theta - \theta')}{\tan(\theta + \theta')} H_{0,s};$$

$$H_{2,s} = \frac{2 \sin\theta' \cos\theta}{\sin(\theta + \theta') \cos(\theta - \theta')} H_{0,s}. \tag{92}$$

To obtain the relations between the amplitudes of the electric vectors, we rewrite (47) in the form

$$E_{0,p} = -\frac{1}{\sqrt{\varepsilon_1}} (k_i^0 \times H_{0,s});$$

$$E_{1,p} = -\frac{1}{\sqrt{\varepsilon_1}} (k_r^0 \times H_{1,s}); \tag{93}$$

$$E_{2,p} = -\frac{1}{\sqrt{\varepsilon_2}} (k_t^0 \times H_{2,s}).$$

As a result, the boundary conditions for the electric field amplitudes take the form

$$\sqrt{\varepsilon_1} (E_{0,p} + E_{1,p}) = \sqrt{\varepsilon_2} E_{2,p};$$

$$\cos\theta (E_{0,p} - E_{1,p}) = \cos\theta' E_{2,p}. \tag{94}$$

Hence it follows that

$$
\begin{aligned}
E_{1,p} &= \frac{\tan(\theta - \theta')}{\tan(\theta + \theta')} E_{0,p};\\
E_{2,p} &= \frac{2\sin\theta'\cos\theta}{\sin(\theta + \theta')\cos(\theta - \theta')} E_{0,p}.
\end{aligned}
\tag{95}
$$

Some authors [31, 43] choose an opposite mutual orientation of the vectors $E_{0,p}$ and $E_{1,p}$, different from that used in most optics textbooks and original publications (and by us here, incidentally). As a result, the sign of the relation for $E_{1,p}$ is reversed. Without going into the relative merits of the two systems, we chose the traditional form of Eq. (95), since in the final account we are interested in the intensity of the refracted wave, which is independent of the direction of the vector $E_{1,p}$ and depends only on the square of its amplitude.

Similarly, using Eqs. (93), we obtain from (91) the expressions for the amplitudes of the magnetic vectors parallel to the plane of incidence,

$$
\begin{aligned}
H_{1,p} &= \frac{\sin(\theta - \theta')}{\sin(\theta + \theta')} H_{0,p};\\
H_{2,p} &= \frac{2\sin\theta'\cos\theta}{\sin(\theta + \theta')} H_{0,p}.
\end{aligned}
\tag{96}
$$

Eqs. (91), (92), (95), and (96) were first derived by Fresnel from his elastic theory of light. Using these relations, we can determine the amplitudes and the phases of the reflected and the refracted wave for incident light of any general polarization. Sometimes, the Fresnel relations are written in a somewhat different form,

$$E_{1,p} = \frac{\sqrt{\varepsilon_1}\cos\theta' - \sqrt{\varepsilon_2}\cos\theta}{\sqrt{\varepsilon_1}\cos\theta' + \sqrt{\varepsilon_2}\cos\theta} E_{0,p};$$

$$E_{1,s} = \frac{\sqrt{\varepsilon_1}\cos\theta - \sqrt{\varepsilon_2}\cos\theta'}{\sqrt{\varepsilon_1}\cos\theta + \sqrt{\varepsilon_2}\cos\theta'} E_{0,s};$$

$$E_{2,p} = \frac{2\sqrt{\varepsilon_1}\cos\theta}{\sqrt{\varepsilon_1}\cos\theta' + \sqrt{\varepsilon_2}\cos\theta} E_{0,p};$$

$$E_{1,p} = \frac{2\sqrt{\varepsilon_1}\cos\theta}{\sqrt{\varepsilon_1}\cos\theta + \sqrt{\varepsilon_2}\cos\theta'} E_{0,s}.$$

This notation is convenient in that it does not lead to ambiguities irrespective of the choice of the angle θ.

Let the real angles θ and θ' vary from 0 to $\pi/2$. Under this assumption, we will analyze the phase relations of all the three waves (for the electric field vectors only). Note that in accordance with our choice of the positive direction of the p and s components, the components $\boldsymbol{E}_{0,s}$, $\boldsymbol{E}_{1,s}$, and $\boldsymbol{E}_{2,s}$ and also $\boldsymbol{E}_{0,p}$ and $\boldsymbol{E}_{2,p}$ are in phase if they have the same sign and in counterphase if they have different signs. Conversely, $\boldsymbol{E}_{0,p}$ and $\boldsymbol{E}_{1,p}$ are in phase if their signs are different and in counterphase if their signs are alike. This follows from the different orientation of the wave vectors $\boldsymbol{k}_t^0$ and $\boldsymbol{k}_r^0$ relative to the normal $\boldsymbol{n}$ to the interface.

From Eqs. (91) and (95) we see that for any value of the angles θ and θ' in the interval chosen, the signs of the amplitudes of the two refracted waves ($\boldsymbol{E}_{2,s}$ and $\boldsymbol{E}_{2,p}$) are equal to the signs of the corresponding amplitudes of the incident wave ($\boldsymbol{E}_{0,s}$ and $\boldsymbol{E}_{0,p}$). This indicates that the phase of the refracted wave relative to the incident wave does not change at the interface.

A more complex situation is observed for reflected waves. In this case, the phase relations depend both on the angle of incidence θ (and the related θ') and on the ratio of the refractive indices of the two media.

If $n_2 > n_1$ $(\theta > \theta')$, the vector $\boldsymbol{E}_{1,s}$ will be in counterphase to the s component of the incident electric vector $\boldsymbol{E}_{0,s}$, irrespective of the angle $(\theta + \theta')$ (see Eq. (91)). Conversely, the phase of the p component of the reflected electric field vector, according to Eq. (95), depends on the sum of the angle of incidence and the angle of refraction: for $(\theta + \theta') > \pi/2$, $\boldsymbol{E}_{0,p}$ and $\boldsymbol{E}_{1,p}$ are in phase, and for $(\theta + \theta') < \pi/2$, $\boldsymbol{E}_{0,p}$ and $\boldsymbol{E}_{1,p}$ are in counterphase.

If $n_2 < n_1$ $(\theta < \theta')$, the amplitudes of the reflected waves reverse their sign, i.e., the phase of the s component of the vector $\boldsymbol{E}_{1,s}$ is analogous to the phase of $\boldsymbol{E}_{0,s}$, and the phase of the p component of the reflected wave does not change compared to the phase of $\boldsymbol{E}_{0,p}$ for incidence angles in the range $(\theta + \theta') < \pi/2$, while changing by π outside this range.

Let us now write the Fresnel relations for normal incidence $(\theta = \theta' = 0)$. Eqs. (91), (95) for the amplitudes of the reflected and the refracted wave are inapplicable to this case, since they assume an indeterminate form. Resolving the indeterminacy in the right-hand sides of Eqs. (91) and (95), we obtain

$$\boldsymbol{E}_{1,s} = \frac{n-1}{n+1}\boldsymbol{E}_{0,s};$$
$$\boldsymbol{E}_{2,s} = \frac{2}{n+1}\boldsymbol{E}_{0,s}. \tag{91a}$$

$$E_{1,p} = \frac{n-1}{n+1} E_{0,p};$$

$$E_{2,p} = \frac{2}{n+1} E_{0,p}. \tag{95a}$$

Similar relations can be derived between the amplitudes of the magnetic field vectors. These relations follow directly from the expressions on p. 31, where we should set $\cos\theta = \cos\theta' = 1$ and $\sqrt{\varepsilon_2/\varepsilon_1} = n_{21} = n$.

In normal incidence, the position of the plane of incidence is indeterminate and we cannot distinguish between the s and p components. Eqs. (91a) and (95a) thus quite properly show that the amplitudes

$$E_{1,s}, E_{1,p} \quad \text{and} \quad E_{2,s}, E_{2,p}$$

have equal magnitudes.

Experimental measurements give the intensity of the light wave, and not its amplitude. The intensity, in particular, is defined as the time-average energy flux $\bar{S}$ associated with the real part of the complex Poynting vector. According to the law of energy conservation, the normal component of the energy flux should remain continuous accross the interface, i.e.,

$$n(\bar{S}_i + \bar{S}_r) = n\bar{S}_t,$$

where

$$\bar{S}_i = \tfrac{1}{2}(E_i \times H_i) = \frac{c\sqrt{\varepsilon_1}}{8\pi} E_0^2 k_i^0;$$

$$\bar{S}_r = \frac{c\sqrt{\varepsilon_1}}{8\pi} E_1^2 k_r^0;$$

$$\bar{S}_t = \frac{c\sqrt{\varepsilon_2}}{8\pi} E_2^2 k_t^0.$$

In other words,

$$\sqrt{\varepsilon_1}\, E_0^2 \cos\theta + \sqrt{\varepsilon_1}\, E_1^2 \cos\theta = \sqrt{\varepsilon_2}\, E_2^2 \cos\theta'.$$

We define the reflection coefficient R as the ratio of the intensity of the reflected luminous flux to the intensity of the incident flux

$$R = \frac{n\bar{S}_r}{n\bar{S}_i} = \frac{E_1^2}{E_0^2}, \tag{97}$$

and the transmission coefficient T as the intensity ratio of the transmitted and the incident flux

$$T = \frac{n\bar{S}_t}{n\bar{S}_i} = \frac{E_2^2\sqrt{\varepsilon_2}\cos\theta'}{E_0^2\sqrt{\varepsilon_1}\cos\theta}. \tag{98}$$

The law of energy conservation then may be written in the form

$$R + T = 1.$$

Inserting for the amplitudes in (97) and (98) their expressions from the Fresnel equations, we obtain for the two components

$$R_p = \frac{\tan^2(\theta - \theta')}{\tan^2(\theta + \theta')}; \quad T_p = \frac{\sin 2\theta \sin 2\theta'}{\sin^2(\theta + \theta')\cos^2(\theta - \theta')};$$
$$R_s = \frac{\sin^2(\theta - \theta')}{\sin^2(\theta + \theta')}; \quad T_p = \frac{\sin 2\theta \sin 2\theta'}{\sin^2(\theta + \theta')}. \tag{99}$$

Eqs. (99) show that the expressions for the reflection and transmission coefficients do not change when θ and θ' are interchanged, i.e., when the ray direction is reversed.

In case of normal incidence, there is no distinction between the s and the p components, and in accordance with (91a) and (95a) the reflection and transmission coefficients are expressed in the form

$$R = \left(\frac{n-1}{n+1}\right)^2,$$
$$T = \frac{4\pi}{(n+1)^2}. \tag{100}$$

10. Polarization effects in reflection and refraction

Analysis of the Fresnel equations provides a complete answer to the topic of polarization of the reflected and refracted light.

For natural incident light, the intensities of the s and p components are equal, i.e.,

$$I_{0,s} = I_{0,p}.$$

Seeing that $I_{0,s} = I_{0,p} = \frac{1}{2}I_i$, the total reflected intensity is

$$I_r = I_{1,s} + I_{1,p} = \tfrac{1}{2}I_i\left[\frac{\sin^2(\theta - \theta')}{\sin^2(\theta + \theta')} + \frac{\tan^2(\theta - \theta')}{\tan^2(\theta + \theta')}\right]. \tag{101}$$

Since the reflection coefficient of the s component is always greater than the reflection coefficient of the p component (except for the extreme cases $\theta = 0, \frac{1}{2}\pi$), the light wave reflected from the interface of two dielectrics is invariably polarized in the plane of incidence ($E_{1,s} > E_{1,p}$). The degree of plarization is

$$P = \frac{I_{1,s} - I_{1,p}}{I_{1,s} + I_{1,p}}.$$

According to (101), the intensity of the s component of the reflected wave steadily increases with the increasing angle of incidence θ. The intensity $I_{1,p}$, on the other hand, at first decreases, dropping to zero for $\theta + \theta' = \frac{1}{2}\pi$, and then also increases to a maximum of $\frac{1}{2}I_i$ for $\theta = \frac{1}{2}\pi$ (grazing incidence). The corresponding intensity curves for ordinary glass ($n = 1.52$) are plotted in Figure 17. Thus, for $\theta + \theta' = \frac{1}{2}\pi$, the reflected light is totally polarized ($I_{1,p} = 0$ and $P = 1$), with the electric vector at right angles to the plane of incidence. In this case, the polarization of the transmitted light also reaches its maximum ($I_{2,p} > I_{2,s}$). Nevertheless, for transmitted light $|P| < 1$, since neither component is completely suppressed at any incidence angle, except $\theta = \frac{1}{2}\pi$. For the above condition, $\theta + \theta' = \frac{1}{2}\pi$, we have [24]

$$\frac{I_{2,s}}{I_{2,p}} = \frac{4n^2}{(1 + n^2)^2};$$

$$P = \frac{4n^2 - (1 + n^2)^2}{4n^2 + (1 + n^2)^2}.$$

For $n = 1.5$ (air-glass interface), $|P| = 0.08$, i.e., the transmitted light is weakly polarized.

The angle θ_0 at which the reflected light is linearly polarized is known as the angle of total polarization or *Brewster angle*. Seeing that in this case

$$\sin \theta' = \sin (\tfrac{1}{2}\pi - \theta_0) = \cos \theta_0,$$

we obtain from the law of refraction

$$\tan \theta_0 = \sqrt{\frac{\varepsilon_2}{\varepsilon_1}} = \frac{n_2}{n_1} = n_{21} = n. \tag{102}$$

Let us consider the case of linearly polarized incident light, with the electric field vector of amplitude $\mathbf{E}_0$ oscillating at an angle α to the p direction. Then,

$$\mathbf{E}_{0,p} = \mathbf{E}_0 \cos \alpha, \quad \mathbf{E}_{0,s} = \mathbf{E}_0 \sin \alpha,$$

and the corresponding intensities are

$$I_{0,p} = E_0^2 \cos^2 \alpha, \quad I_{0,s} = E_0^2 \sin^2 \alpha.$$

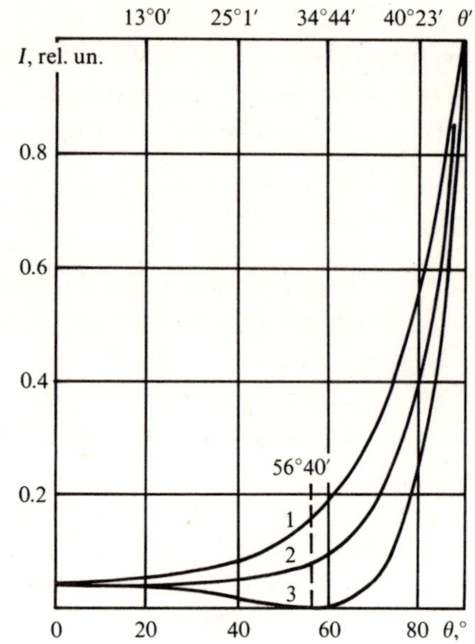

Figure 17

Intensity of reflected light vs. the angle of incidence for glass with $n = 1.52$:

1) *s*-component, 2) *p*-component, 3) natural light.

The total reflected intensity is

$$I_{1,\alpha} = I_{1,p} + I_{1,s} = \left(\frac{I_{1,p}}{I_{0,p}}\cos^2\alpha + \frac{I_{1,s}}{I_{0,s}}\sin^2\alpha\right)E_0^2,$$

so that

$$R_\alpha = \frac{I_{1,\alpha}}{E_0^2} = R_p\cos^2\alpha + R_s\sin^2\alpha, \tag{103}$$

where R_p and R_s are defined by (99).

If $\alpha = \frac{1}{4}\pi$, we have

$$R_\alpha = \tfrac{1}{2}(R_p + R_s),$$

i.e., the same result as for natural light.

It is significant, however, that when the incident light is linearly polarized, the directions of polarization in the reflected and the refracted wave are turned through a certain angle relative to the polarization of the incident wave. Let the polarization azimuths of the incident, reflected, and refracted waves, reckoned from the plane of incidence, be α, β, and γ. Then

$$\tan\alpha = \frac{E_{0,s}}{E_{0,p}}; \quad \tan\beta = \frac{E_{1,s}}{E_{1,p}}; \quad \tan\gamma = \frac{E_{2,s}}{E_{2,p}}.$$

Hence, using Eqs. (91) and (95), we find

$$\tan\beta = -\frac{\cos(\theta - \theta')}{\cos(\theta + \theta')}\tan\alpha;$$
$$\tan\gamma = \cos(\theta - \theta')\tan\alpha. \tag{104}$$

Since the angles θ and θ' vary from 0 to $\frac{1}{2}\pi$, we obtain from the last relations

$$|\tan\beta| \geqq |\tan\alpha|,$$
$$|\tan\gamma| \leqq |\tan\alpha|.$$

Thus, the plane of oscillations of the electric vector in the reflected wave is turned away from the plane of incidence, and the same plane in the refracted wave is turned toward the plane of incidence. It is significant that for an incidence angle equal to Brewster's angle, the azimuth of the reflected wave is precisely $\frac{1}{2}\pi$. Indeed, since in this case $\theta + \theta' = \frac{1}{2}\pi$, Eq. (104) gives $\tan\beta = \infty$, i.e., $\beta = \frac{1}{2}\pi$ irrespective of α.

Thus, in case of Brewster angle incidence, the oscillations of the electric vector in the reflected wave should be perpendicular to the plane of incidence. Moreover, we have seen from the phase relations at the interface between two media that Brewster's angle corresponds to a change of 180° in the phase of the reflected $\boldsymbol{E}_{1,p}$ component. Experimental results show, however, that neither of these conditions is exactly satisfied. Instead of linearly polarized light, a certain ellipticity of polarization is noted, which indicates that the p component is only partly extinguished and that the phase difference between $\boldsymbol{E}_{1p}$ and $\boldsymbol{E}_{1s}$ is neither 0 nor $\frac{1}{2}\pi$. Thus, the phase of $\boldsymbol{E}_{1p}$ does not change abruptly when the angle of incidence crosses Brewster's angle, but rather varies in a certain narrow interval of incidence angles.

Drude [12] has originally assumed that these deviations from the Fresnel equations near Brewster's angle are associated with the excessive idealization of the concept of a sharp interface between two media. In practice, the interface is in fact a very thin transitional layer, with the dielectric constant varying monotonically from ε_1 to ε_2. The existence of this layer converts the reflection and transmission coefficients to complex numbers, and the phase difference correspondingly deviates from the values characteristic of a sharp interface (0 and π). These deviations are very small and difficult to measure experimentally. However, near the angle of total reflection θ_0 the correction to the Fresnel coefficient R_p markedly increases and can no longer be ignored. The corresponding formula takes the form [31]

$$R_p = \frac{\sqrt{\varepsilon_2}\cos\theta\,(1 + ic_1) - \sqrt{\varepsilon_1}\cos\theta'\,(1 + ic_2)}{\sqrt{\varepsilon_2}\cos\theta\,(1 + ic_1) + \sqrt{\varepsilon_1}\cos\theta'\,(1 + ic_2)},$$

where c_1 and c_2 are coefficients which depend on the refractive indices of the adjacent media and the angles θ, θ'.

Since the effect is noticeable only very near Brewster's angle, we may take

$$\theta = \theta_0 + \theta_1,$$

where $\theta_0 = \tan^{-1} n$.

Retaining only terms linear in θ_1, c_1, and c_2, we obtain

$$R_p \simeq \frac{1}{2}\sqrt{\frac{\varepsilon_1}{\varepsilon_2}}\left(\frac{\varepsilon_1}{\varepsilon_2} - \frac{\varepsilon_2}{\varepsilon_1}\right)\theta_1 + \frac{i}{2}(c_1 - c_2).$$

If a linearly polarized plane wave is incident at 45° to the interface, we obtain

$$\frac{R_p}{R_s} = \frac{i\pi d\eta}{\lambda}\,\frac{\sqrt{\varepsilon_1 + \varepsilon_2}}{\varepsilon_1 - \varepsilon_2}, \tag{105}$$

where d is the thickness of the transitional layer, and

$$\eta = \bar{\varepsilon} + \varepsilon_1\varepsilon_2\,(1/\bar{\varepsilon}) - \varepsilon_1 - \varepsilon_2,$$

where ε is the dielectric constant of the transitional layer averaged over its entire thickness.

Drude's theory on the whole is in agreement with experimental findings. Thus the ellipticity coefficient R_p/R_s of the reflected light generally decreases when the interface is cleaned; it is particularly small for a freshly cleaved face of a crystal or for the surface of a liquid which is continuously renewed by pouring. Having measured the ellipticity of the reflected light, we can estimate from (105) the thickness of the transitional layer, which amounts to a few hundredths of a wavelength.

Nevertheless, Drude's theory should be used with some reservation. For example, it has been established that the ellipticity of the light reflected from one of the faces of a glass prism varies and may even reverse its sign when the other face of the prism is polished. These phenomena still remain unexplained, although recent theoretical and experimental studies (see bibliography in [31]) have provided adequate interpretation of some properties of thin transitional layers.

11. Total internal reflection

In the previous section we analyzed the phase–amplitude relations of the incident and reflected waves as a function of the incidence angle θ. We specifically avoided the case when the law of refraction (87) does not give a real refraction angle θ'. No real refraction angle can be obtained when the light propagating from a high-density medium into a low-density one ($n_2 < n_1$ or $n_{21} < 1$) strikes the interface at an angle θ such that

$$\frac{\sin\theta}{n_{21}} > 1. \tag{106}$$

For these incidence angles, no refracted wave is formed, and the entire incident radiation is reflected back into the first medium, i.e., total internal reflection is observed. The angle defined by the equality $\sin\theta_0 = n_{21} = n$, at which the refracted wave glances along the interface, is called the angle of total internal reflection.

Inequality (106) shows that in the case of total reflection $\cos\theta'$ takes on the value

$$\cos\theta' = \pm i\sqrt{\frac{\sin^2\theta}{n^2} - 1} = \pm\frac{i}{n}u, \tag{107}$$

where

$$u = \sqrt{\sin^2\theta - n^2},$$

i.e., a complex refraction angle is obtained.

From purely physical considerations, we should take the sign minus in (107), since otherwise the amplitude in the second medium would increase without bound as the wave moved away from the interface [1].

We will show that in total reflection, the reflection coefficients of the p and s components of the incident wave are both unity. Writing (91) and (95) in the form

$$\boldsymbol{E}_{1,p} = \frac{\sin\theta\cos\theta - \sin\theta'\cos\theta'}{\sin\theta\cos\theta + \sin\theta'\cos\theta'}\boldsymbol{E}_{0,p};$$

$$\boldsymbol{E}_{1,s} = -\frac{\sin\theta\cos\theta' - \sin\theta'\cos\theta}{\sin\theta\cos\theta' + \sin\theta'\cos\theta}\boldsymbol{E}_{0,s},$$

we divide the numerator and the denominator by $\sin\theta'$ and insert the results in (107), to obtain

$$\boldsymbol{E}_{1,p} = \frac{n^2\cos\theta - iu}{n^2\cos\theta + iu}\boldsymbol{E}_{0,p};$$

$$\boldsymbol{E}_{1,s} = \frac{\cos\theta - iu}{\cos\theta + iu}\boldsymbol{E}_{0,s}. \tag{108}$$

Hence, using (97), we find that $R_p = R_s = 1$. Thus, the intensity of reflected light is equal to the intensity of the incident light both for the wave polarized in the plane of incidence and for the wave polarized in the perpendicular direction. The same result can be obtained by examining the time average of the normal component of the Poynting vector.

The above conclusion is not to be interpreted, however, as indicating a zero electromagnetic field in the second medium. Indeed, the scalar value of the electric vector of the transmitted wave, according to Eq. (80), can be expressed in the form (Figure 15)

$$E_t = E_2 \exp\left[i\left(\omega t - \mathbf{k}_t^0 \mathbf{r}\right)\right] = E_2 \exp\left[2\pi i\left(\nu t - \mathbf{k}_t^0 \mathbf{r}/\lambda_2\right)\right] =$$

$$= E_2 \exp\left\{2\pi i\left[\nu t - (x\cos\theta' + z\sin\theta')/\lambda_2\right]\right\}.$$

Taking the law of refraction in the form $\lambda_2 = \lambda_1/n$ and using Eq. (107), we rewrite the last equation in the form

$$E_t = E_2 \exp\left\{2\pi i\left[\nu t - (z\sin\theta)/\lambda_1\right]\right\} \exp\left[-(2\pi x/\lambda_1)\sqrt{\sin^2\theta - n^2}\right]. \quad (109)$$

This expression describes a disturbance which decays exponentially in the direction x, i.e., as we go deeper into the second medium. The first exponential factor indicates that the disturbance propagates parallel to the interface in the direction z with the phase velocity

$$v_2 = \frac{dz}{dt} = \frac{v_1}{\sin\theta},$$

which is a function of the angle of incidence on the interface. Such waves with an amplitude which is variable in space are called inhomogeneous. In general, the phase velocity of an inhomogeneous wave in the second medium may be less than that of an ordinary homogeneous refracted wave in the same medium, and for the case of total reflection, it is actually equal to the phase velocity in the first medium.

Since the planes of constant amplitude and constant phase do not coincide (the former is parallel to the interface and the latter is perpendicular to it), this surface wave is not transversal.

The penetration of light into the second medium under conditions of total reflection has been confirmed experimentally [20]. The most striking experiment was performed by Mandelstam [27]. A glass prism was immersed in a fluorescein solution (Figure 18). The light ray hit the glass–liquid interface at an angle greater than the total reflection angle. A small volume of the solution near the point of incidence of the ray showed a strong glow, which provided an immediate proof of the penetration of a light wave.

The first experiments which established the penetration of light into the second medium under conditions of total reflection were staged by Newton. However, the exact origin of the light waves in the second medium remained unclear for a long time. The starting point was provided by the work of Eichenwald [45]. Proceeding from Maxwell's equations, he demonstrated the existence of a refracted beam whose direction varied with time. In Eichenwald's words, "The energy of the incident ray as if plunges into the second medium, only to reemerge completely into the first medium in the form of reflected ray. However, the entrance and exit

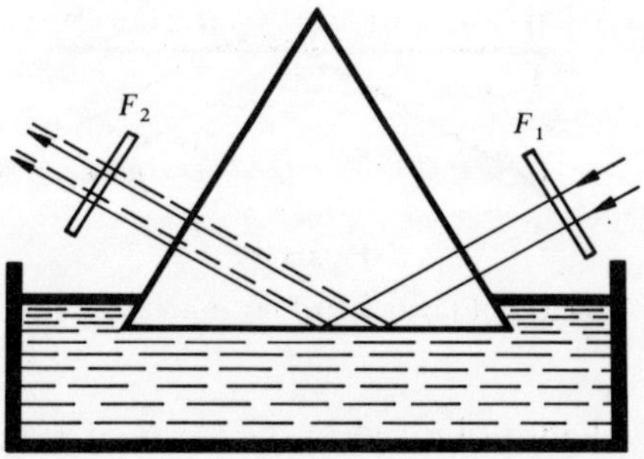

Figure 18
Schematic diagram of Mandelstam's experiment for observing the penetration of light into the low-density medium under conditions of total internal reflection (F_1 and F_2 are filters).

points of energy are in different places at different times, and for a fixed time, the energy enters the second medium at points separated from one another by half the distance between planes of equal phase."

The depth of field penetration into the second medium is very small. We see from (109) that a noticeable disturbance encompasses a layer whose thickness is of the order of one wavelength. Hence it follows that total reflection is better studied with electromagnetic waves of large λ, and this was indeed the course followed by Shaeffer and Gross, who used 15 cm waves and two paraffin prisms.

It follows from Eichenwald's theory that the point of entrance of the waves into the low-density medium and does not coincide with the point of emergence back into the high-density medium, and the displacement is determined by the wavelength and the incidence angle. Schematically, this is shown in Figure 19, where P is the incident ray, R is the ray corresponding to ordinary reflection at the interface at the point Q, and T is the actual ray associated with total reflection.

Total reflection thus causes a certain displacement of the ray. Special experiments [79] have shown that for natural light incident on a glass–air interface, the displacement D is expressed by the relation

$$D = Cn \frac{\lambda_1}{\sqrt{\sin^2 \theta - n^2}}, \tag{110}$$

where C is a constant.

Observations in polarized light gave the same value of the constant C, irrespective of the position of the plane of polarization. The last result contradicts the earlier data regarding the angular dependence of the depth of penetration of polarized light into the second medium. It has been established that a certain incidence angle θ^* exists such that for $\theta < \theta^*$ the light polarized in the plane of incidence penetrates to a smaller depth than the perpendicular component. Conversely, for $\theta > \theta^*$, the

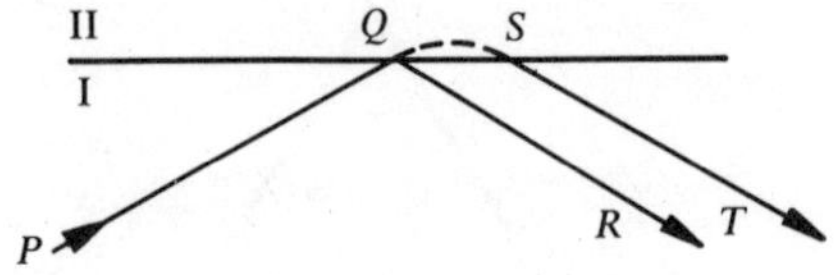

Figure 19
The displacement of the reflected ray in total internal reflection.

penetration depth of the light polarized in the plane of incidence is greater than the penetration depth of the perpendicular component. Similar results naturally can be expected for the displacement, since comparison of Eqs. (109) and (110) shows that the depth of penetration and the displacement are both determined by the same factor $\sqrt{\sin^2\theta - n^2/\lambda_1}$. Later polarization measurements [80] confirmed the dependence of displacement on the polarization of incident radiation.

The final theoretical solution of the problem of polarization effects in the case of total reflection and satisfactory interpretation of the experimental findings were advanced only recently by Fedorov [39] from an analysis of the inhomogeneous wave in the second medium. The mean energy density of the inhomogeneous wave was found to be

$$\overline{W}_{2,p} = \overline{W}_{2,s}A, \tag{111}$$

where $\overline{W}_{2,s}$ is the mean energy density in the second medium if the incident wave is polarized in the plane of incidence, $\overline{W}_{2,p}$ corresponds to polarization perpendicular to the plane of incidence.

The coefficient A is defined as follows:

$$A = \frac{n^2}{(n^2+1)\sin^2\theta - n^2}. \tag{112}$$

Since the penetration depth of a damped inhomogeneous wave into the second medium is determined by $\overline{W}_2$, we see from (111) that for a fixed incidence angle, the refracted wave is detected at various distances from the interface depending on the state of polarization of the incident radiation. Setting $A = 1$ in (112), we find the incident angle $\theta = \theta^*$ for which $\overline{W}_{2,p} = \overline{W}_{2,s}$ and the penetration depth into the second medium is independent of polarization (provided that $|E_{0,p}| = |E_{0,s}|$),

$$\sin\theta^* = n\sqrt{\frac{2}{1+n^2}}. \tag{113}$$

The angle θ^* for the case $n < 1$ $(n_2 < n_1)$ is always greater than the total reflection angle θ_0, i.e., it is one of the incidence angles satisfying the condition of total reflection $\sin\theta > n$. On the other hand, analyzing Eqs. (112) and (113), we find

$$A > 1 \text{ and } \overline{W}_{2,p} > \overline{W}_{2,s} \text{ for } \theta_0 < \theta < \theta^*;$$

$$A < 1 \text{ and } \overline{W}_{2,p} < \overline{W}_{2,s} \text{ for } \theta^* < \theta < \tfrac{1}{2}\pi.$$

This fits the above experimental data regarding the dependence of the penetration depth on the direction of polarization of the incident light.

The angle of incidence θ^* defined by (113) has still another important property. We will show that the maximum phase difference δ between the components $E_{1,p}$ and $E_{1,s}$ of the reflected wave corresponds to the angle θ^*. To this end, let us calculate the phase difference between the corresponding components of the incident and the reflected wave. We have shown before that for total reflection $R_p = R_s = 1$, i.e., $|E_{1,p}|^2 = |E_{0,p}|^2$ and $|E_{1,s}|^2 = |E_{0,s}|^2$, and we may therefore write

$$\frac{E_{1,p}}{E_{0,p}} = e^{i\delta_p} \quad \text{and} \quad \frac{E_{1,s}}{E_{0,s}} = e^{i\delta_s}.$$

By (108)

$$e^{i\delta_p} = \frac{n^2 \cos\theta - i\sqrt{\sin^2\theta - n^2}}{n^2 \cos\theta + i\sqrt{\sin^2\theta - n^2}}$$

and

$$e^{i\delta_s} = \frac{\cos\theta - i\sqrt{\sin^2\theta - n^2}}{\cos\theta + i\sqrt{\sin^2\theta - n^2}}$$

we find

$$\tan\tfrac{1}{2}\delta_p = \frac{\sqrt{\sin^2\theta - n^2}}{n^2\cos\theta}$$

and

$$\tan\tfrac{1}{2}\delta_s = \frac{\sqrt{\sin^2\theta - n^2}}{\cos\theta}. \tag{114}$$

Thus, the p and s components of the electric vector of the incident wave undergo different phase changes as a result of total reflection. From (114) we can find the angular dependence of the relative phase difference $\delta = \delta_p - \delta_s$ after reflection,

$$\tan\tfrac{1}{2}\delta = \tan\tfrac{1}{2}(\delta_p - \delta_s) = \frac{\tan\frac{1}{2}\delta_p - \tan\frac{1}{2}\delta_s}{1 + \tan\frac{1}{2}\delta_p \tan\frac{1}{2}\delta_s} = \frac{\cos\theta\sqrt{\sin^2\theta - n^2}}{\sin^2\theta}. \tag{115}$$

From this relation we see that $\delta = 0$ only for two values of the incidence angle, namely $\theta = \frac{1}{2}\pi$ (grazing incidence) and $\theta = \theta_0$ (total reflection). The phase difference δ is thus maximum for a certain incidence angle θ which lies between these limit values. Differentiating (115) with respect to θ and setting the result equal to zero, we find the condition of the maximum

$$\sin^2\theta = \frac{2n^2}{1 + n^2},$$

which coincides with Eq. (113).

A few words about total reflection for arbitrary polarization of the incident wave [39]. The effects observed in this case are fundamentally different from those obtained for linear polarization parallel or perpendicular to the plane of incidence. In this case, as opposed to the case of regular reflection at an interface between transparent dielectrics, the energy density W_2 and the Poynting vector $\boldsymbol{S}_2$ cannot be decomposed into p and s components. For elliptically polarized incident radiation, the mean energy flux in the second medium has a component perpendicular to the plane of incidence, which produces so-called lateral radiation pressure.

Finally, analysis shows that in case of circular polarization, not only the average energy flux through the interface is zero, but its instantaneous values are zero also. This apparently shows that the real properties of the incident light wave—its limited extent in space and in time—must be taken into consideration.

12. Coated optics

The light intensity reflected from an interface between two transparent dielectrics is determined by the refractive indices and the angle of incidence. It may reach a considerable fraction of the incident luminous flux. Thus, in a case of normal incidence, with the light beam crossing the air–glass interface only once, the reflection losses constitute 4 to 9% of the incident intensity, depending on the grade of the glass. The actual light losses at the end faces of a fiber, as we shall see in the next chapter, are much higher, due to the large aperture of the light beams. Moreover, fiber elements are generally joined with other optical systems, which also cause certain reflection losses, and the total loss of intensity may therefore reach a few tens of percent. On the other hand, reflected light penetrating into the optical system may create undesirable parasitic illumination. In fiber systems, this effect greatly reduces the contrast of the image.

Thus the reflection of light from interfaces between different media, and in particular from the end faces of fiber elements and the interfaces joining them to other optical elements, may constitute a serious problem. The losses can be reduced by using coated optics. The theoretical and experimental aspects of optical coating were developed in the USSR in the 1930s and the 1940s at the Vavilov Optical Institute [10].

In the simplest case, the optical surface is coated with a single thin dielectric film. The reflected intensity decreases because of destructive interference in the film.

Let the film thickness be d, its refractive index n_1, the refractive index of the coated element n_2, and the refractive index of the outer medium n_0. If the outer medium is air, we may take $n_0 = 1$ (Figure 20). By an appropriate choice of n_1, we can make equal the intensities of rays 1 and 2 reflected from the interfaces I and II. If now the film thickness d is such that the path difference between these two rays is equal to half the wavelength λ of the incident light, destructive interference will quench the two rays and the total reflected intensity will be zero. Elementary calculations give the following condition of zero reflected intensity for a single-film coat:

$$n_1 = \sqrt{n_2}, \tag{116}$$

provided that

$$\frac{n_2 - n_1}{n_1 - 1} > 0.$$

The layer thickness is determined by the relation

$$\Delta = 2dn_1 = \frac{\lambda}{2},$$

or in general

$$\Delta = (2m + 1)\frac{\lambda}{2}, \tag{117}$$

where $m = 0, 1, 2, \ldots$

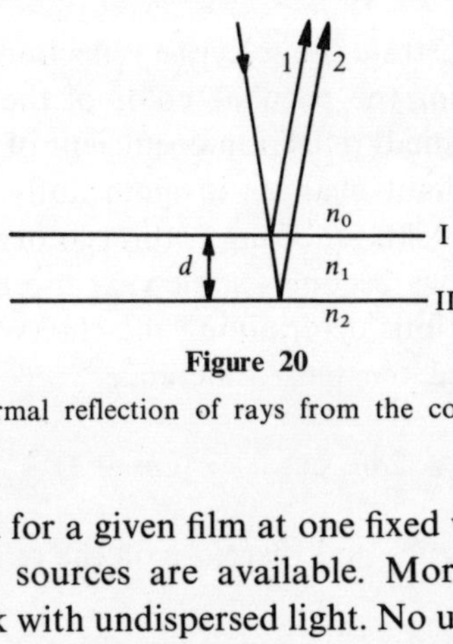

Figure 20
Normal reflection of rays from the coating.

This equality is satisfied for a given film at one fixed wavelength only. In practice, no strict monochromatic sources are available. Moreover, an optical system is generally expected to work with undispersed light. No universal coated optics can be devised for a wide range of wavelengths. For a given d, wavelengths which do not satisfy condition (117) show a certain finite reflection. The reflection increases with increasing deviation of the path difference between the interfering rays 1 and 2 from $(2m + 1)\lambda/2$. As an example, Figure 21 shows the theoretical curves of the reflection coefficient of glass ($n_2 = 1.52$) as a function of wavelength for an uncoated surface (curve 1), for a surface coated with a film of optical thickness $dn_1 = 1375$ Å (curve 2),

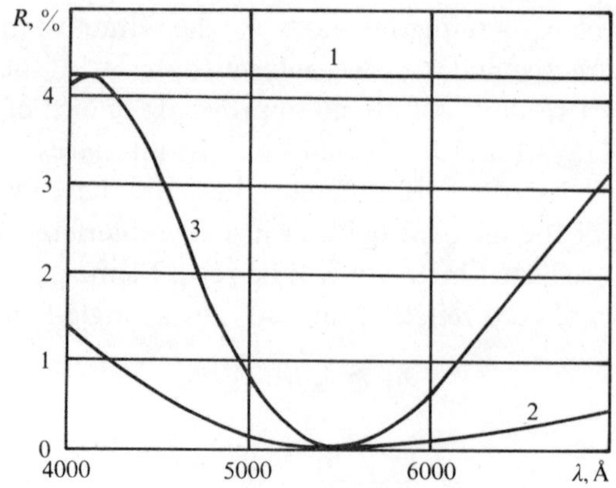

Figure 21

Reflection coefficient of glass vs. wavelength or different coating thicknesses.

and for a surface with a film of $dn_1 = 1375 + 5500/2$ Å (curve 3). The refractive index of the film in both cases meets condition (116). Curves 2 and 3 were calculated using the usual interference equation

$$I = I_1 + I_2 + 2\sqrt{I_1 I_2} \cos \frac{2\pi\Delta}{\lambda}.$$

These curves clearly demonstrate the selective reflection of light by coated optics, and this fact is responsible for the peculiar color of the coated optical elements. Nevertheless, because of the small reflection coefficient of glasses, the spectral composition of the transmitted light changes insignificantly. This factor, however, is nevertheless of importance in fields requiring faithful color reproduction of the image.

The analysis of coated optics becomes somewhat more complicated for oblique incidence of light. The conditions determining the choice of film thickness and the refractive index change, since the path difference

$$\Delta = 2dn_1 \cos \theta' = (2m + 1)\frac{\lambda}{2}$$

depends on the angle of incidence and therefore on the refraction angle θ' of the ray in the film.

In the case of oblique incidence, the maximum effect is attained at shorter wavelengths, i.e., the minimum of curves 2 and 3 in Figure 21 is shifted in accordance with the condition

$$\lambda_\theta = \lambda_0 \cos \theta',$$

where λ_θ and λ_0 are the wavelengths corresponding to zero reflection for normal and oblique incidence (Figure 22).

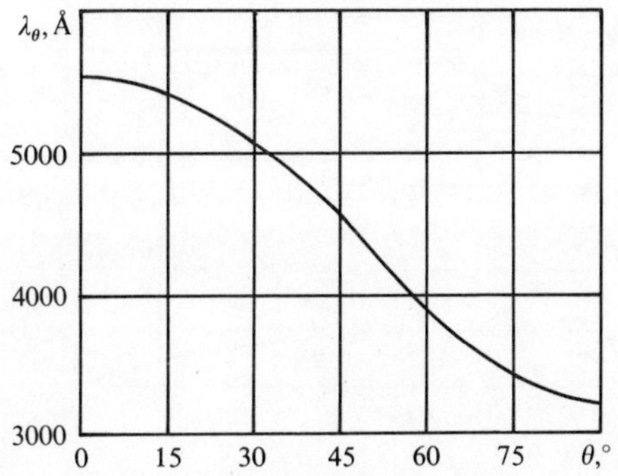

Figure 22

Wavelength of zero reflection vs. the incidence angle for fixed coating parameters.

We have so far assumed that condition (116) is observed in oblique incidence, although in reality this is not so. In this case, the difference in the reflection coefficients of the s and p components should be taken into consideration. The corresponding conditions of zero reflection take the following form:

for the s component,

$$n_1^2 - \sin^2\theta = \cos\theta\,(n_2^2 - \sin^2\theta)^{1/2}; \tag{118}$$

for the p component*,

$$\frac{n_1^2}{n_2}\cos\theta\,(n_2^2 - \sin^2\theta)^{1/2} = \frac{n_2}{n_1^2}(n_1^2 - \sin^2\theta). \tag{119}$$

For $\theta = 0$, both these relations reduce to (116). The curves of the optimum refractive index n_1 as a function of the incidence angle θ calculated from these relations for $n_2 = 1.52$ are shown in Figure 23 [10]. The left top corner shows part of the curves on a larger scale.

From Eqs. (118), (119) and the curves in Figure 23 we see that in case of oblique incidence the reflection cannot be reduced to zero for both components. Indeed, as the angle θ increases, the required value of the refractive index n_1 decreases for the s component and increases for the p component. Calculations show that the reflection coefficient of natural light from a coated glass surface is generally markedly different from zero for $\theta > 30°$, i.e., this effect is observed at larger incidence angles than the effect of the angular dependence of the zero-reflection wavelength.

Other shortcomings of single-layer coating include the shortage and frequent unavailability of coating materials which satisfy the optimum conditions. The re-

* We are dealing with the case $\theta \leq \tan^{-1} n$.

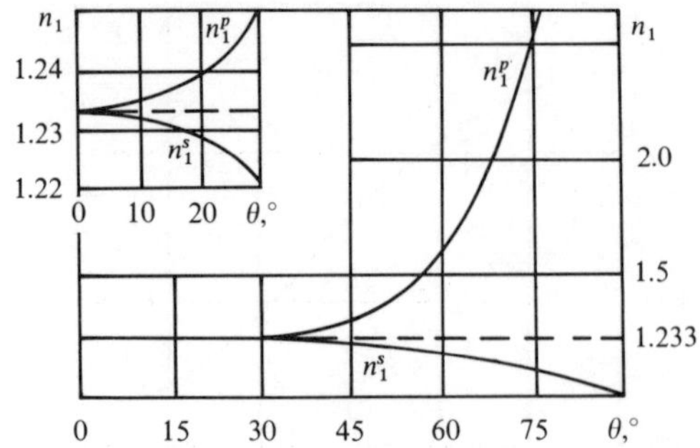

Figure 23

The optimum refractive index n_1 of the coating film vs. the angle of incidence θ for $n_2 = 1.52$.

fractive indices of common glasses are generally between 1.4 and 1.8. Therefore, in accordance with Eq. (116), the coating film should have a refractive index of 1.2–1.3. There are virtually no transparent solid materials with this low refractive index which can be used to obtain a mechanically sturdy surface layer. In the early days of coated optics, the refractive index of the film was artificially lowered by making it porous.

The advent of multilayer coats largely eliminated these difficulties. The first theoretical treatment of the interference properties of multicomponent systems was carried out by Vlasov in 1946 [9]. Subsequently, numerous studies were published on this subject.* Several methods are available for the theoretical treatment of multilayer coats. One of the approaches investigates the condition of interference of all the light beams emerging from the coat. Adequate information about every beam can be readily obtained by analyzing the reflection and refraction conditions at every interface of the multicomponent system. This approach, although very illuminating, is highly tedious and is mostly used in the theory of layered interferometers [31].

Another method calls for a derivation of recursion relations describing the reflection and the transmission of the system. In this case, one starts with the theory of the optical properties of a single-layer coat and examines the changes in these properties as additional layers are gradually added. The conditions of reflection and refraction at each added interface are written down using the coefficients T and R of all the underlying and overlying layers, as established in the preceding step. Calculations of this kind were carried out in [25] for the most general case of a system of entirely different layers in optical contact with one another.

Consider a multilayer system of m components, designated in Figure 24 by the numbers 2, 3, ..., $(m + 1)$, between two media designated 1 (on the side of the incident light beam) and $m + 2$ (on the side of the transmitted light beam). Let a linearly

* A bibliography can be found in [31].

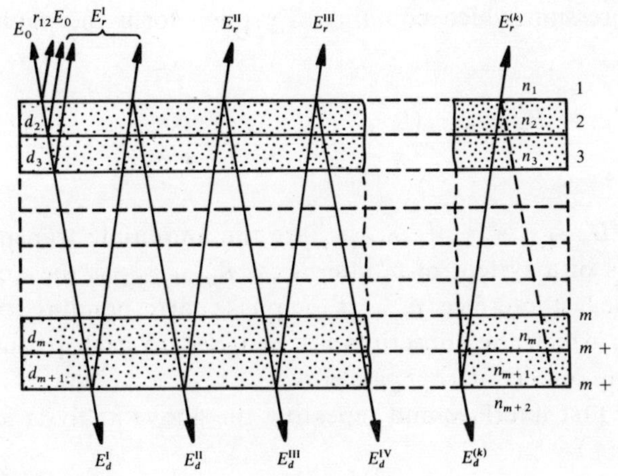

Figure 24

Rays in a multilayer coating.

polarized light wave with an electric vector of magnitude E_0 fall on this system. Separating the topmost interface from the entire system of $(m + 1)$ interfaces and examining the sequence of reflected wave amplitudes $E_r^{(\mathrm{I})}, E_r^{(\mathrm{II})}, \ldots, E_r^{(k)}$, we can readily show that the scalar magnitude E_{rm} of the electric vector of the reflected wave is representable as a geometrical progression with a sum

$$E_{rm} = \frac{R_{1,2} + \mathscr{R}_{2,m+2}e^{-i\delta_2}}{1 + R_{1,2}\mathscr{R}_{2,m+2}e^{-i\delta_2}} E_0, \tag{120}$$

where $R_{1,2}$ is the amplitude reflection coefficient for a beam traveling from medium 1 to medium 2; $\mathscr{R}_{2,m+2}$ is the amplitude reflection coefficient of the system of m interfaces for a wave propagating from 2 to $m + 2$; δ_2 is the phase acquired by the reflected on crossing the first layer, which is designated by the number 2.

For any layer, the corresponding phase can be written in the form

$$\delta_q = 2\pi k_q n_q d_q \cos\theta_q,$$

where $k_q = 2\pi/\lambda$, n_q, and d_q are, respectively, the wave number, the refractive index, and the thickness of the q-th layer.

Eq. (120) is applied in the following manner: first $\mathscr{R}_{2,m+2}$ is expressed in terms of $R_{2,3}$, $\mathscr{R}_{3,m+2}$, and δ_3 from an analog of (120) and the result is inserted in the expression for E_{rm}. Then, the other amplitude reflection coefficients $\mathscr{R}_{3,m+2}$, $\mathscr{R}_{4,m+2}$, etc., are successively found, until we obtain a final expression with parameters characterizing the $(m + 1)$-th layer and the last interface, i.e., $R_{m+1,m+2}$ and δ_{m+1}.

Eq. (120) contains the first layer constants in explicit form. If the first interface is replaced by the last interface, and the path of rays is analyzed as before, we obtain for E_{rm} an expression which contains in explicit form the parameters of the last layer, namely

$$E_{rm} = \frac{\mathscr{R}_{1,m+1} + R_{m+1,m+2}(D_{1,m+1}D_{m+1,1} - \mathscr{R}_{1,m+1}\mathscr{R}_{m+1,1})\, e^{-i\delta_{m+1}}}{1 - R_{m+1,m+2}\mathscr{R}_{m+1,1}e^{-i\delta_{m+1}}} E_0, \quad (121)$$

where $D_{1,m+1}$, $D_{m+1,1}$ $\mathscr{R}_{1,m+1}$, $\mathscr{R}_{m+1,1}$ are the amplitude transmission and reflection coefficients of a system of m interfaces; $R_{m+1,m+2}$ is the amplitude reflection coefficient of the last interface; δ_{m+1} is the phase corresponding to the layer $(m + 1)$.

The order in which the subscripts are assigned to each parameter corresponds to the direction of wave propagation.

Isolating the first interface and repeating the above analysis for the transmitted wave, we obtain

$$E_{dm} = \frac{T_{1,2}D_{2,m+2}e^{-i\delta_2/2}}{1 + R_{1,2}\mathscr{R}_{2,m+2}e^{-i\delta_2}} E_0. \quad (122)$$

If the analysis is carried out by isolating the last interface, we obtain for the transmitted wave

$$E_{dm} = \frac{D_{1,m+1}T_{m+1,m+2}e^{-i\delta_{m+1}/2}}{1 - R_{m+1,m+2}\mathscr{R}_{m+1,1}e^{-i\delta_{m+1}}} E_0. \quad (123)$$

In the above expressions, $T_{1,2}$ and $T_{m+1,m+2}$ are the amplitude transmission coefficients of the first and the last interface; $D_{1,m+1}$ and $D_{2,m+2}$ are the amplitude transmission coefficients of the m interfaces with the light propagating from medium 1 to $(m + 1)$ and from 2 to $(m + 2)$, respectively.

We have derived from (120)–(123) the overall reflection and transmission of the entire system, expressed in terms of the optical constants of the first and the last layer. Other recursion relations can be obtained along the same lines, each using in explicit form the optical parameters and the thickness of one of the layers. In practice, Eqs. (120) and (122) are the most useful, as the other relations are more complicated even for systems with few layers.

General relations including in explicit form the optical constants and the thicknesses of all the layers are naturally more convenient than recursion relations in practical applications [25]. Using such expressions, the calculations of a multilayer system can be completed much faster. General expressions follow, in particular, from an analysis of several layers and subsequent generalization of the results.

Consider a four-layer system. Expressing as before the amplitude reflection coefficients $\mathscr{R}_{2,6}$, $\mathscr{R}_{3,6}$, $\mathscr{R}_{4,6}$, $R_{5,6}$ in terms of the reflection coefficients of the individual interfaces, we can represent the scalar magnitude of the electric vector of the reflected wave in the form

$$E_{r4} = \frac{A}{B} E_0, \quad (124)$$

where

$$A = R_{1,2} + R_{2,3}e^{-i\delta_2} + R_{3,4}e^{-i(\delta_2+\delta_3)} + R_{4,5}e^{-i(\delta_2+\delta_3+\delta_4)} +$$
$$+ R_{5,6}e^{-i(\delta_2+\delta_3+\delta_4+\delta_5)} + R_{1,2}R_{2,3}R_{3,4}e^{-i\delta_3} +$$
$$+ R_{1,2}R_{2,3}R_{4,5}e^{-i(\delta_3+\delta_4)} + R_{1,2}R_{2,3}R_{5,6}e^{-i(\delta_3+\delta_4+\delta_5)} +$$
$$+ R_{1,2}R_{3,4}R_{4,5}e^{-i\delta_4} + R_{1,2}R_{3,4}R_{5,6}e^{-i(\delta_4+\delta_5)} +$$
$$+ R_{1,2}R_{4,5}R_{5,6}e^{-i\delta_5} + R_{2,3}R_{3,4}R_{4,5}e^{-i(\delta_2+\delta_4)} +$$
$$+ R_{2,3}R_{3,4}R_{5,6}e^{-i(\delta_2+\delta_4+\delta_5)} + R_{2,3}R_{4,5}R_{5,6}e^{-i(\delta_2+\delta_5)} +$$
$$+ R_{3,4}R_{4,5}R_{5,6}e^{-i(\delta_2+\delta_3+\delta_5)} + R_{1,2}R_{2,3}R_{3,4}R_{4,5}R_{5,6}e^{-i(\delta_3+\delta_5)};$$

$$B = 1 + R_{1,2}R_{2,3}e^{-i\delta_2} + R_{1,2}R_{3,4}e^{-i(\delta_2+\delta_3)} + R_{1,2}R_{4,5}e^{-i(\delta_2+\delta_3+\delta_4)} +$$
$$+ R_{1,2}R_{5,6}e^{-i(\delta_2+\delta_3+\delta_4+\delta_5)} + R_{2,3}R_{3,4}e^{-i\delta_2} + R_{2,3}R_{4,5}e^{-i(\delta_3+\delta_4)} +$$
$$+ R_{2,3}R_{5,6}e^{-i(\delta_3+\delta_4+\delta_5)} + R_{3,4}R_{4,5}e^{-i\delta_2} + R_{3,4}R_{5,6}e^{-i(\delta_4+\delta_5)} +$$
$$+ R_{4,5}R_{5,6}e^{-i\delta_5} + R_{1,2}R_{2,3}R_{3,4}R_{4,5}e^{-i(\delta_2+\delta_4)} +$$
$$+ R_{1,2}R_{2,3}R_{3,4}R_{5,6}e^{-(\delta_2+\delta_4+\delta_5)} + R_{1,2}R_{2,3}R_{4,5}R_{5,6}e^{-i(\delta_2+\delta_5)} +$$
$$+ R_{1,2}R_{3,4}R_{4,5}R_{5,6}e^{-i(\delta_2+\delta_3+\delta_5)} + R_{2,3}R_{3,4}R_{4,5}R_{5,6}e^{-i(\delta_3+\delta_5)}.$$

To simplify the notation, all $R_{q,q+1}$ and $T_{q,q+1}$ will be designated as R_q and T_q, respectively. The incident wave can be conveniently assigned zero phase ($\delta_1 = 0$), but for general application we will assume that $\delta_1 \neq 0$. The first five terms of the numerator A in (124) thus take the form

$$R_{1,2} = R_1 e^{-i\delta_1};$$
$$R_{2,3}e^{-i\delta_2} = R_2 e^{-i(\delta_1+\delta_2)};$$
$$R_{3,4}e^{-i(\delta_2+\delta_3)} = R_3 e^{-i(\delta_1+\delta_2+\delta_3)};$$
$$R_{4,5}e^{-i(\delta_2+\delta_3+\delta_4)} = R_4 e^{-i(\delta_1+\delta_2+\delta_2+\delta_4)};$$
$$R_{5,6}e^{-i(\delta_2+\delta_3+\delta_4+\delta_5)} = R_5 e^{-i(\delta_1+\delta_2+\delta_3+\delta_4+\delta_5)}.$$

Combining these five elements, we obtain all the terms in the numerator and the denominator. For example, any triple product can be obtained from the expression

$$\prod_{q=j_1,j_2,j_3} R_q \exp\left[(-1)^{f(q)}\Delta_{rq}\right]; \quad \Delta_{rq} = -i(\delta_1 + \delta_2 + \ldots + \delta_q), \tag{125}$$

where $f(q = j_l) = l - 1$ specifies the position of the factor R_q in the product.

In this notation, the common exponential describing the phase is formed by multiplying the exponentials of all the R_q, and the sign of the argument of each exponential is determined by the function $f(q)$. It is therefore highly significant to arrange all the R_q in each product in the correct order (i.e., in the order of increasing subscripts from left to right). In our four-layer case, we should take

$$1 \leqq j_1 < j_2 < j_3 \leqq 5.$$

For m layers, this condition takes the form

$$1 \leqq j_1 < f_2 < \ldots < j_v \leqq (m+1),$$

where v is the number of factors in the product.

Applying notation (125) to a system of m layers, adding up all the terms which contain the same number of elements, and combining like groups of terms, we obtain a general expression for the electric field strength of a reflected wave:

$$E_{rm} = \frac{\sum\limits_{v=1,3,\ldots,\binom{m}{m+1}} \sum\limits_{1 \leqq j_1 < j_2 < \ldots < j_v \leqq (m+1)} \prod\limits_{q=j_1,\ldots,j_v} R_q \exp[(-1)^{f(q)}\Delta_{rq}]}{1 + \sum\limits_{\mu=2,4,\ldots,\binom{m+1}{m}} \sum\limits_{1 \leqq j_1 < j_2 < \ldots < j_\mu \leqq (m+1)} \prod\limits_{q=j_1,\ldots,j_\mu} R_q \exp[(-1)^{f_1(q)}\Delta_{rq}]} E_0, \quad (126)$$

where the function $f_1(q)$ is defined as $f_1(q = j_l) = l$.

The maximum value of v for odd m is m and for even m it is $(m + 1)$. For μ in the denominator, the reverse rule applies. The parenthetical symbol under the first $\sum$ in the numerator and the denominator gives the corresponding maximum values of the two subscripts.

A simpler expression is obtained for a transmitted wave:

$$E_{dm} = \frac{\prod\limits_{q=1,2,\ldots,(m+1)} T_q \exp(\tfrac{1}{2}\Delta)}{1 + \sum\limits_{\mu=2,4,\ldots,\binom{m+1}{m}} \sum\limits_{1 \leqq j_1 < j_2 < \ldots < j_\mu \leqq (m+1)} \prod R_q \exp[(-1)^{f_1(q)}\Delta_{rq}]} E_0. \quad (127)$$

For normal incidence and layers of optical thickness $\lambda/4$, we have: for even m,

$$E_{rm} = \frac{n_1 \prod\limits_{v=3,5\ldots,m+1} n_v^2 - n_{m+2} \prod\limits_{\mu=2,4,\ldots,m} n_\mu^2}{n_1 \prod\limits_{v=3,5,\ldots,m+1} n_v^2 + n_{m+2} \prod\limits_{\mu=2,4,\ldots,m} n_\mu^2} E_0; \quad (128)$$

for odd m,

$$E_{rm} = \frac{n_1 n_{m+2} \prod\limits_{v=3,5,\ldots,m} n_v^2 - \prod\limits_{\mu=2,4,\ldots,m+1} n_\mu^2}{n_1 n_{m+2} \prod\limits_{v=3,5,\ldots,m} n_v^2 + \prod\limits_{\mu=2,4,\ldots,m+1} n_\mu^2} E_0. \quad (129)$$

Using Eqs. (128) and (129) we can write the condition of zero reflection losses for a surface coated with a four-layer film. The magnitude of the electric field vector of the reflected wave E_{rm} is set equal to zero. For even m, we have

$$n_1 \prod_{\nu=3,5,\ldots,m+1} n_\nu^2 - n_{m+2} \prod_{\mu=2,4,\ldots,m} n_\mu^2 = 0$$

or

$$\frac{\prod_{\nu=3,5,\ldots,m+1} n_\nu}{\prod_{\mu=2,4,\ldots,m} n_\mu} = \sqrt{\frac{n_{m+2}}{n_1}}. \tag{130}$$

For odd m,

$$\frac{\prod_{\mu=2,4,\ldots,m+1} n_\mu}{\prod_{\nu=3,5,\ldots,m} n_\nu} = \sqrt{n_1 n_{m+2}}\,. \tag{131}$$

If we take $n_1 = 1$ and $m = 1$, Eq. (131) is replaced by the previous condition (116) for the refractive index of a single-layer coat.

Thus, in the case of an m-component system (even one with fixed optical thicknesses of the layers), we have to deal with $2m$ parameters which can be formed into various combinations to satisfy the zero-reflection conditions.

In practice, an alternating stack of layers of two different materials is of considerable interest. The condition of zero reflection for such a system is particularly simple, if we write n_3 for all the n_ν, and n_2 for n_μ. For an even m, we have

$$\frac{n_3}{n_2} = \sqrt[m]{\frac{n_{m+2}}{n_1}},$$

and for an odd m,

$$\frac{n_2^{m+1}}{n_3^{m-1}} = n_1 n_{m+2}.$$

Simple considerations show that if numerous layers are used, their refractive indices may be fairly close to one another. This greatly facilitates the choice of coating materials. Another conclusion is that an even number of layers is somewhat more advantageous than an odd number for optical coatings.

In order to ensure achromatic optics, systems using films of various optical thicknesses have to be used [17, 47]. If achromatism is not relevant, the optical thickness of all the layers is generally taken equal to a multiple of $\lambda/4$ [43, 119, 157].

2 Propagation of Light in Thick Fibers

1. Requirements from fibers and fiber bundles

The theoretical framework of fiber optics is based on two significant assumptions:

(1) light is transmitted along the individual fibers as a result of total internal reflection from the fiber surfaces;

2) the fibers collected into a bundle transmit light independently of one another.

The first assumption shows that in every individual fiber, information is transmitted only by those light rays which hit the entrance end at certain angles. This, however, is not enough to ensure high light-transfer efficiency. The technology of fiber manufacture should meet exacting standards, since otherwise volume inhomogeneities and surface imperfections will cause absorption and scattering of light and thus affect the brightness and the quality of the image. A light ray undergoes a multitude of reflections even when propagating along a relatively short fiber. Calculations show that in a fiber about 50μ in diameter, there are upwards of 13,000 reflections per 1 m of fiber length [128]. The minimum losses in each reflection entirely depend on surface quality. On the other hand, fibers should be manufactured from optical materials which, in addition to a certain refractive index, also have a very low absorption coefficient. Otherwise, the absorption losses will impose severe limitations on the length of light pipes, an obviously undesirable effect.

The second of the two basic assumptions above is very difficult to satisfy in practice. The early experiments revealed a cross-coupling between the individual fibers, which became very prominent when they were placed at distances of the order of the λ from one another. This cross-coupling causes a definite leakage of light from every fiber into its neighbors. The effective contact area in a dense bundle of thin fibers is quite considerable, and the leakage constitutes a serious problem.

This effect is associated with the penetration of light into the low-density medium in the event of total internal reflection. The penetration depth is small, reaching at most 2λ. Leakage is therefore significant only in sufficiently dense fiber bundles.

To prevent leakage of light, the fibers are coated with special materials, in particular metal coatings [159, 160]. These coatings provide a high reflection coefficient, but their absorption is nevertheless high enough to produce an adverse effect after numerous reflections. As a result, the light-transfer efficiency of the coated fiber is markedly reduced. Metallic coatings were therefore soon replaced by dielectric coatings. Short fibers, transferring light from a nearby extended source, are still being coated in metal.

The widely used glass fibers are often enclosed in a glass sheath of lower refractive index. The optimum sheath thickness is approximately equal to the wavelength of transferred radiation. Further increase of skin thickness is inadvisable, as the resolving power of the system suffers.

These sheaths ensure adequate optical insulation, but they are incapable of eliminating the adverse effect of the scattered radiation which is produced when light hits the entrance end of the fiber at a supercritical angle. If the incidence angle is greater than the critical value, the condition of total internal reflection breaks down and light partly penetrates into the adjacent fibers, then into the next fiber, and so on, until it reaches the outer surface of the bundle or leaves through the exit end of some other fiber. This light wandering in the bundle is responsible for the bright background in the plane of the transmitted image.

Scattered light can be suppressed by coating the fiber in a second absorbing coat.

The propagation of light in fibers whose diameter is substantially greater than the light wavelength is generally considered within the framework of geometrical optics. In the visible spectrum, this approach is valid for fibers a few microns in diameter. As the fiber cross section is further reduced, the diffractional losses increase and the image quality deteriorates. In this case, the individual fibers should be considered as dielectric waveguides propagating certain electromagnetic modes only. In the present chapter, we will consider the geometrical theory of light propagation in fibers. The topics discussed include the effect of fiber shape and dimensions and of the properties of the core and the insulation material on the optical parameters of the individual fibers and of the fiber bundle as a whole.

2. Numerical aperture of a bundle

Meridional rays

The numerical aperture is defined as $A = n \sin\theta$, where θ is the angle subtended by the radius of the entrance pupil of the optical system at the point where the optical axis meets the object plane, and n is the refractive index of the medium in the object space [24]. If no immersion liquids are used, we have $n = 1$ (air). The numerical aperture is one of the fundamental characteristics of an optical system, since it directly determines the resolving power of the system. In most cases, the numerical aperture of optical fibers is greater than the average numerical aperture of conventional lens systems. It is this property of fibers that makes them particularly attractive in practical uses.

Light pipes in the form of fiber bundles not only have a high numerical aperture, but also a wide angle of view, which is difficult to attain in ordinary lens systems. The angle of view of the bundles is somewhat limited by the technical difficulties of packing optical fibers into the very thick bundles needed for collecting information over large surfaces.

The numerical aperture of a fiber, like that of a lens, characterizes the largest solid angle of the incoming pencil of rays. However, in fiber optics, the pencil apex lies on the end face of the fiber, and the maximum angle is determined by the condition of total internal reflection. In lens optics, on the other hand, the apex of the pencil is in the object plane, and the solid angle of the pencil is limited by the aperture of the entrance pupil and its distance from the object.

Let us consider meridional rays, i.e., rays propagating in diametrical planes and cutting the fiber axis (Figure 25). The numerical aperture for these rays can be calculated without much difficulty [94, 146]. Using the laws of refraction and total internal reflection for a straight cylindrical fiber, we have (Figure 26*a*)

$$A = n_0 \sin \theta_{cr} = (n_1^2 - n_0^2)^{1/2}, \tag{132}$$

where n_1 and n_0 are, respectively, the refractive indices of the fiber material and the surrounding medium; θ_{cr} is the maximum angle of incidence on the entrance end which still ensures total internal reflection from the cylindrical surface inside the fiber. All the light rays with incidence angles $\theta < \theta_{cr}$ will be transferred by the fiber.

If the fiber is held in air, we have

$$A = (n_1^2 - 1)^{1/2}. \tag{132a}$$

For single- and double-coated fibers (Figure 26, *b* and *c*), Eq. (132) is replaced with

$$A = n_0 \sin \theta_{cr} = (n_1^2 - n_2^2)^{1/2}, \tag{133}$$

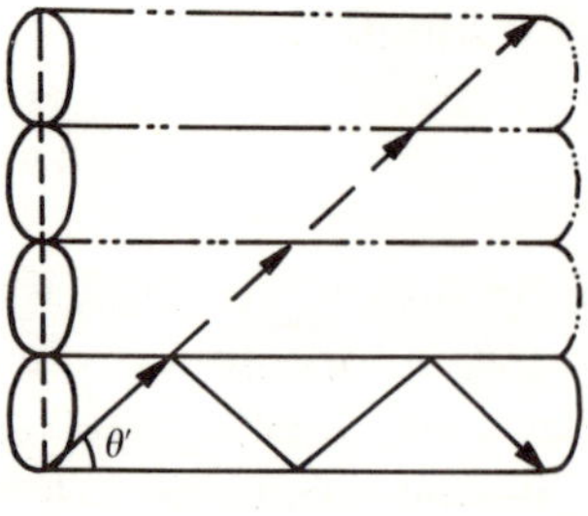

Figure 25

The path of a meridional ray in a straight cylindrical fiber.

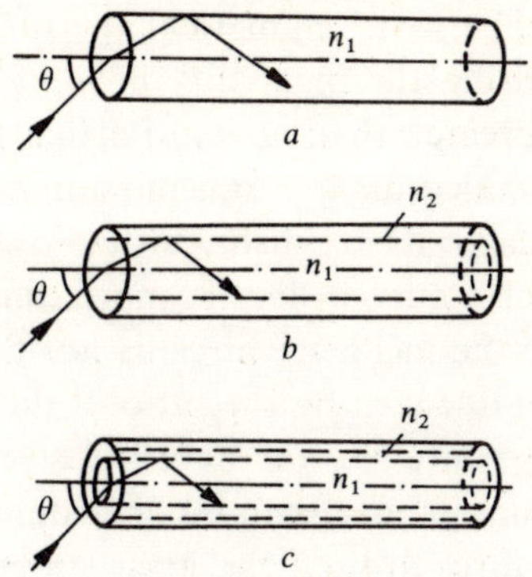

Figure 26

Different fibers:

a uncoated, *b* with a single coating, *c* a double-coat fiber.

where n_2 is the refractive index of the first coat, the one from which total internal reflection occurs.

Comparison of Eqs. (132) and (133) shows that a transparent coat on the fiber may reduce the numerical aperture; this reduction is more pronounced as the difference between the refractive indices of the core and the coat becomes smaller. Fiber cores are therefore generally manufactured from high-refraction heavy flints, whereas the outer envelope is made from light crowns [7, 89].

The dependence of the numerical aperture on the refractive index of the core (n_1) and the envelope (n_2) is plotted in Figure 27. We see that A increases with the increase in the ratio n_1/n_2.

If the thickness of the first envelope is greater than 2λ, the refractive index of the second envelope does not affect the numerical aperture of the fiber, and it may be manufactured from any absorbing material. Currently available materials are ade-

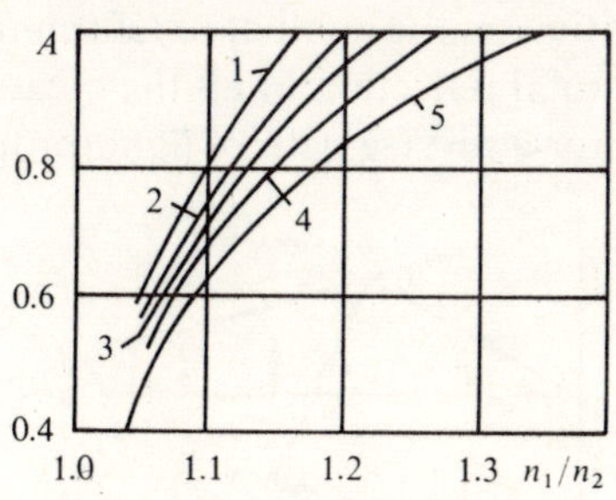

Figure 27

Numerical aperture as a function of the refractive indices of the core (n_1) and the insulating envelope (n_2):

1) $n_1 = 1.9$; 2) $n_1 = 1.8$; 3) $n_1 = 1.7$; 4) $n_1 = 1.6$; 5) $n_1 = 1.5$.

quate for the production of fibers with numerical apertures reaching unity. By using immersion liquids and ensuring the inequality $n_1^2 - n_2^2 \geqq 1$, we can obtain fibers with numerical apertures greater than unity. Particularly high values of A are characteristic of systems intended for the transmission of infrared light [146], since the corresponding optical materials are distinguished by high refractive indices.

The numerical aperture of a curved fiber is more difficult to calculate. Analysis shows [93] that in this case the incidence angle is not the only relevant factor: the numerical aperture is also influenced by the ratio of the radius of curvature of the fiber to its diameter. The aperture and the light-transfer efficiency remain virtually constant for radii of curvature greater than 60 fiber diameters. A greater curvature (attainable only at temperatures around the glass softening point) has an adverse effect on the total internal reflection, and some of the light escapes through the lateral surfaces.

If both ends of the fiber are immersed in the same medium (air, say), the entrance and exit apertures are equal. In some devices, however, the two ends are immersed in different media, and the appertures are correspondingly different. This arrangement can be applied to concentrate or disperse the light from the source, as desired.

A conical fiber has a similar effect [84, 85, 93]. The path of meridional rays in such a fiber with a conical angle of 2α is no different from the path of light between two plane mirrors set at the same angle. We see from Figure 28 that the reflection angle in a conical fiber is variable. Since any two successive angles of incidence differ by 2α, we obtain for the last reflection

$$\theta_\eta = 90^0 - [\theta' - \alpha(2\eta - 1)] = 90^0 - \left[\frac{n_0}{n_1}\sin^{-1}\theta - \alpha(2\eta - 1)\right], \tag{134}$$

where θ is the angle of incidence on the entrance face, η is the number of internal reflections.

Since η is generally very large, the term $\alpha(2\eta - 1)$ is significant even for small α, and the condition of total internal reflection may break down. This means that, in without adequate information on the number of reflections η, we cannot determine the angle θ_{cr} which ensures total reflection in all the η cases. This is so only for the case depicted in Figure 28. For a reverse path of light (propagating from the narrow

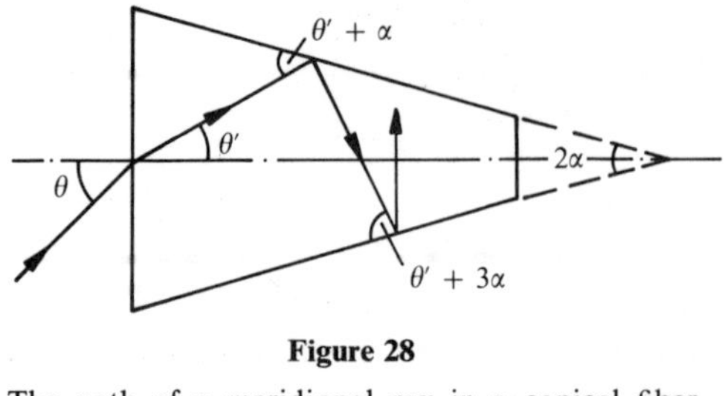

Figure 28
The path of a meridional ray in a conical fiber.

to the wide end of the fiber), θ_{cr} can be determined if we know the condition of total internal reflection for the first incidence on the lateral surface. Indeed, because of the reversibility of ray paths, the reflection angle will increase by 2α after each successive reflection. For a sufficiently small α, the numerical aperture of the narrow face coincides with that calculated from (133) for a cylindrical fiber.

To assess the critical angle θ_{cr} for the wide end, let us consider a limiting case of a hollow cone with ideally reflecting walls (Figure 29). A ray entering through the wide end at a sufficiently large angle (Figure 29*a*, dashed line) may reverse its direction as a result of multiple reflections and re-emerge through the same end. Thus, in distinction from a cylindrical fiber with an ideally reflecting surface, which transfers all the light rays to the exit end, a conical fiber will transmit only those rays whose incidence angle does not exceed a certain critical value θ_{cr}. This critical angle can be found by a simple geometrical construction [93] illustrated in Figure

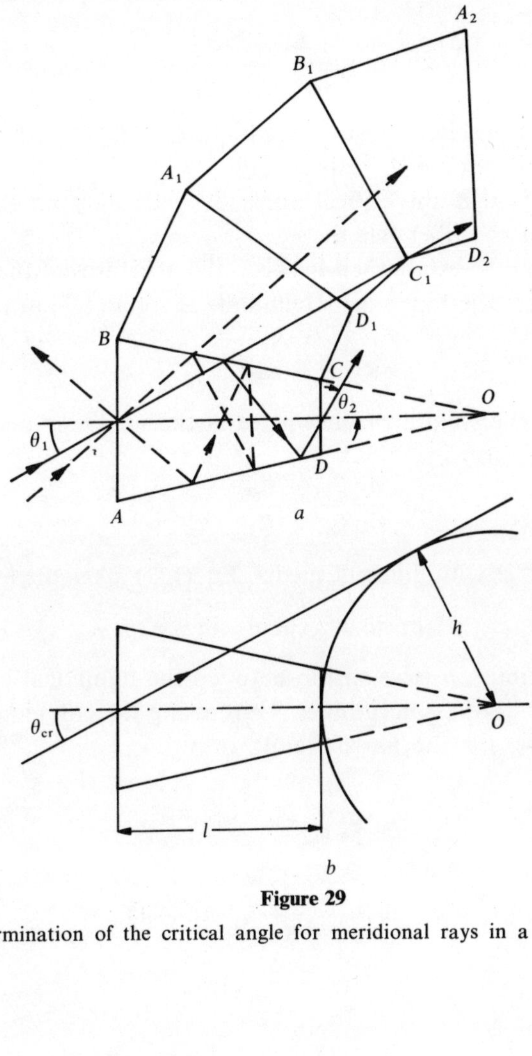

Figure 29

Determination of the critical angle for meridional rays in a conical fiber.

29*a*, where A_1D_1, B_1C_1, A_2D_2 are the virtual images of the sides AD and BC of the cone, obtained by successive reflections of the two sides in each other. The continuation of any incoming meridional ray meets the lateral surfaces of the auxiliary cones BA_1D_1C, $A_1B_1C_1D_1$ at an angle which complements to 90° the angles of incidence on the surface of the cone $ABCD$ in successive reflections. The path of the ray in the auxiliary cones is geometrically identical to its path in the principal cone. Therefore, if the continuation of the incoming ray crosses the polygonal line $DCD_1C_1\ldots$, the ray is transferred by the conical fiber; otherwise, the ray reverses its direction and re-emerges through the wide end.

Since the cross section of the exit end is small, the polygonal line $DCD_1C_1\ldots$, for small α can be replaced with fair accuracy by a circle centered at the cone apex O (Figure 29*b*). This simplification reduces the problem of finding θ_{cr} to a determination of the angle of incidence of a ray whose continuation is tangent to the circle. It is readily seen that

$$\sin\theta_{cr} = \frac{h}{h+l} = \frac{R_2}{R_1}, \tag{135}$$

where h is the radius of the circle, l is the length of the cone, R_1 and R_2 are the radii of the entrance and the exit end, respectively.

It follows from (135) that the critical angle depends only on the ratio R_2/R_1 and is independent of the fiber length.

A similar relation can be established between the apertures of the incoming and outgoing beams. Indeed, the Lagrange–Helmholtz theorem (27) in this case can be written in the form

$$\pi R_1^2\omega_1 = \pi R_2^2\omega_2, \tag{136}$$

where ω_1 and ω_2 are the solid aperture angles of the incoming and the outgoing light beams. By Eq. (136),

$$\frac{\sin\theta_1}{\sin\theta_2} = \frac{R_2}{R_1}. \tag{137}$$

If the ends are in contact with different media, Eq. (137) takes the form

$$R_1n_1\sin\theta_1 = R_2n_2\sin\theta_2. \tag{138}$$

These equalities establish a relationship between the numerical apertures of the wide and the narrow end of a conical fiber. Thus, using Eq. (137) and remembering that, by (133), we have for the narrow end

$$A_2 = n_0\sin\theta_2 = \sqrt{n_1^2 - n_2^2},$$

we obtain

$$A_1 = n_0\sin\theta_1 = \frac{R_2}{R_1}\sqrt{n_1^2 - n_2^2}\,. \tag{139}$$

where n_0 is the refractive index of the medium, n_1 and n_2 are the refractive indices of the core and the insulating envelope, respectively.

For $R_1 = R_2$, Eq. (139) reduces to the previous formula for the numerical aperture of a cylindrical fiber, and $A_1 = A_2$.

The relationship between the numerical apertures of the entrance and the exit faces of a conical fiber can be presented in the form of a nomogram (Figure 30). Here, to each ray of the first cone corresponds a certain ray of the second cone. The only information required is the ratio R_2/R_1 [123]. If $\sqrt{n_1^2 - n_2^2} > 1$, we have $A_1 > R_2/R_1$, i.e., the angle of incidence θ is greater than the critical angle θ_{cr}, and part of the light flux will reverse its direction after a number of reflections. This will

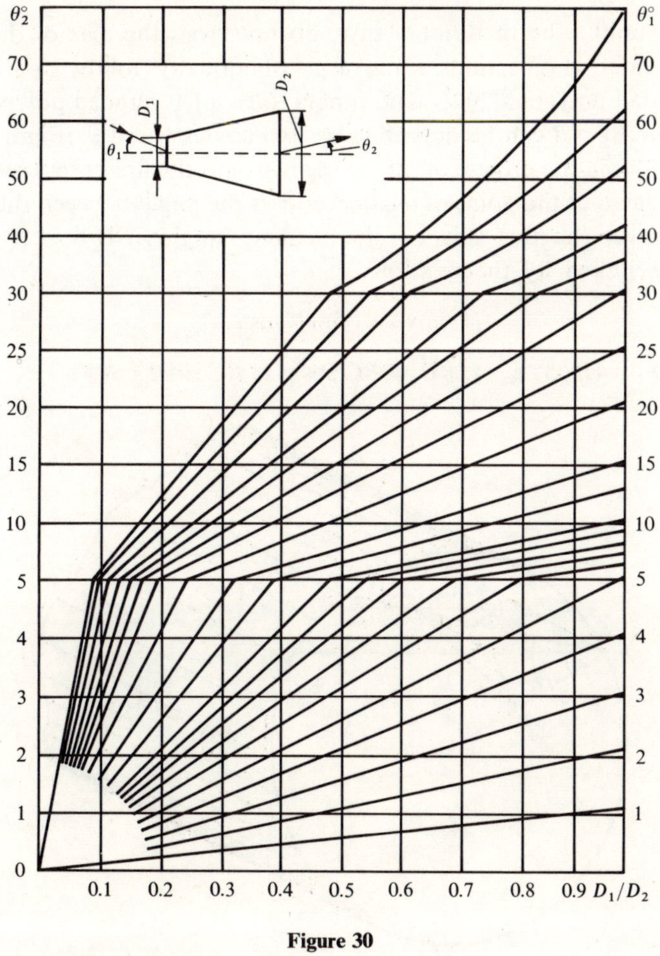

Figure 30

A nomogram interrelating the apertures of the incident and the emerging beams for a conical fiber (D_1 and D_2 are the diameters of the narrow and the wide ends of the fiber).

not happen if θ_{cr} satisfies the condition

$$A_{cr} = n_0 \sin\theta_{cr} = \begin{cases} \dfrac{R_2}{R_1}\sqrt{n_1^2 - n_2^2} & \text{for } n_1^2 - n_2^2 \leqq 1, \\ \dfrac{R_2}{R_1} & \text{for } n_1^2 - n_2^2 > 1, \end{cases} \tag{140}$$

which can be readily ensured by an appropriate choice of the core and envelope material.

Oblique rays

Oblique rays, unlike the meridional rays, do not cross the axis of the fiber. As a result of total internal reflections, some of the oblique rays follow a path in the form of a right-handed polygonal helix, and others trace a left-handed polygonal helix.

Any oblique ray AB can be described by two angles γ and θ' (Figure 31). Here γ is the angle between the projection BC of the ray onto the fiber cross section and the normal OB erected at the point of incidence; θ' is the angle between the ray and the line AC parallel to the fiber axis. Let the incidence angle ABO be α. Then the three angles can be linked by the relation

$$\cos\alpha = \sin\theta' \cos\gamma. \tag{141}$$

Indeed, $BD = AB\cos\alpha$, or $BD = BC\cos\gamma = BC\sin\theta'\cos\gamma$.

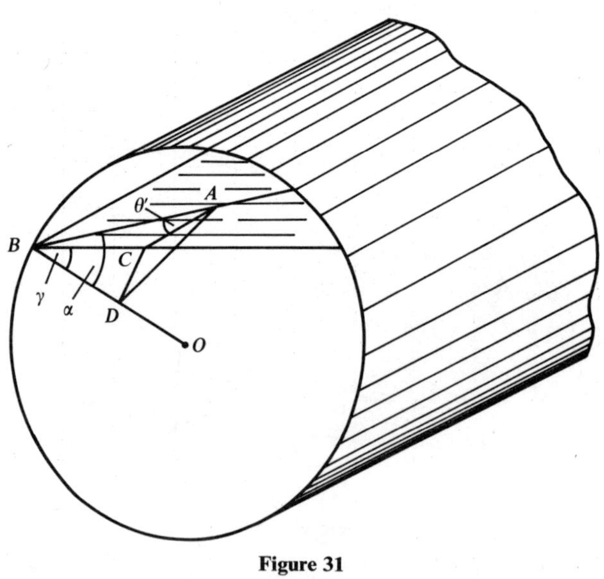

Figure 31

The path of an oblique ray in a cylindrical fiber.

Comparing the two expressions, we obtain (141), which is valid for any ray at any point of the cylindrical fiber. Since α is constant for a given ray, Eq. (141) shows that the angles θ' and γ are also fixed for any given ray, and this equality completely describes the path of the ray in the fiber.

The condition of total internal reflection for oblique rays

$$\sin\alpha \geqq \frac{n_2}{n_1}$$

combined with Eq. (141) lead to a certain relationship between the angles θ' and γ and the refractive indices of the fiber (n_1) and the envelope (n_2):

$$n_1 \sin\theta' \cos\gamma \leqq \sqrt{n_1^2 - n_2^2}. \tag{142}$$

If $\gamma_{\min} \to 0$, Eq. (141) reduces to the corresponding expression for meridional rays.

Using Eq. (142), we can estimate the minimum value of the angle γ for a fixed θ' at which the condition of total internal reflection is still satisfied. According to (142), the fiber will transfer oblique rays which make with the axis an angle θ' much greater than the critical angle of meridional rays (see (133)). In the case of oblique rays, θ' is limited by the condition of refraction at the entrance end. For grazing incidence ($\theta = 90°$), θ' goes to its limiting value

$$\theta'_{\text{lim}} = \sin^{-1}\frac{n_0}{n_1},$$

which is greater than the critical value θ'_{cr} for meridional rays. Clearly, for $0 \leqq \theta' \leqq \theta'_{\text{cr}}$, γ may take any value in the interval $0 \leqq \gamma \leqq \frac{1}{2}\pi$, where $\gamma = 0$ corresponds to a meridional ray and $\gamma = \frac{1}{2}\pi$ to a ray grazing along the cylindrical surface. In the latter case, we should naturally take $\theta' = 0$. For $\theta'_{\text{cr}} \leqq \theta' \leqq \theta'_{\text{lim}}$, a further restriction which follows from (142) is imposed on γ:

$$\gamma_{\min} \leqq \gamma \leqq \tfrac{1}{2}\pi; \quad \cos\gamma_{\min} = \frac{\sin\theta'_{\text{cr}}}{\sin\theta'}. \tag{143}$$

The curve of $\gamma_{\min}$ vs. θ' for fibers with different core refractive indices n_1 is plotted in Figure 32. In every individual case, the limiting value θ'_{lim} is determined by the equality

$$\theta'_{\text{lim}} = \sin^{-1}\frac{n_0}{n_1}.$$

The dependence of the critical angle θ'_{cr} for total internal reflection on the distance of the oblique ray from the fiber axis is readily established from Eq. (142), if we remember (Figure 31) that $\sin\gamma = 2x/D$ (x is the distance from the fiber axis to the plane through the ray, and D is the fiber diameter). The angle θ'_{cr} is obtained from

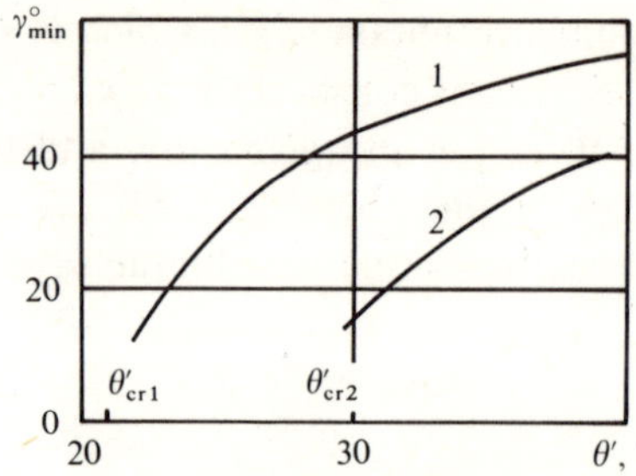

Figure 32

γ_{min} vs. θ' for fibers with $n_2 = 1.4$: 1) $n_1 = 1.5$; 2) $n_1 = 1.6$.

the equation

$$\sin\theta'_{cr}\sqrt{1-\left(\frac{2x}{D}\right)^2} = \frac{1}{n_1}\sqrt{n_1^2-n_2^2}\,. \tag{144}$$

θ'_{cr} as a function x/D for the particular case $n_1 = 1.5$ and $n_2 = 1.4$ is plotted in Figure 33. We see that already for $x/D = 0.43$, θ'_{cr} reaches its limiting value $\theta'_{lim} = \sin^{-1}(n_0/n_1) = 42°$. As the incident beam is further displaced toward the periphery of the fiber, the condition of total internal reflection from the lateral surface does not change.

The specific features of the angular distribution of oblique rays in a straight cylindrical fiber are clearly seen from the distribution of illumination over the exit end at angles greater than the meridional numerical aperture. A certain minimum permissible angle γ_{min} exists for which total internal reflection is still possible. As a result, for sufficiently large inclination angles of the rays, light propagates only in the peripheral part of the fiber [5, 131]. A similar effect was observed experimentally. Photomicrographs of the distribution of illumination over the exit end of a fiber for various propagation angles of oblique rays are shown in Figure 34. The dark region corresponding to the angles $\gamma < \gamma_{min}$ is clearly visible; it increases with the increase in the angle θ'.

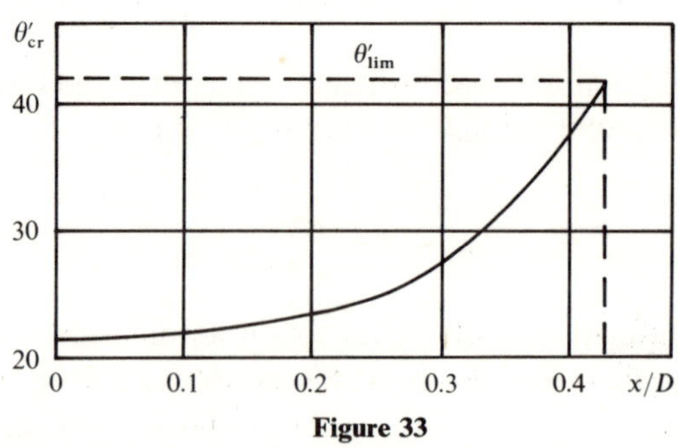

Figure 33

The critical angle θ'_{cr} vs. the distance of the ray from the fiber axis.

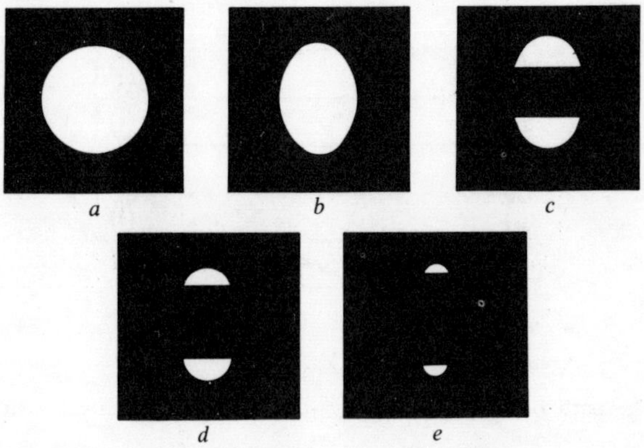

Figure 34

Photomicrograph of the exit end of a fiber taken for various angles θ':
a $\theta' = 0$; *b* $\theta' = 39°$; *c* $\theta' = 42°$; *d* $\theta' = 50°$; *e* $\theta' = 65°$.

3. Light-transfer efficiency of fibers for meridional rays

The light-transfer efficiency of a fiber depends on the transparency of the core and the envelope, the quality of the reflecting surface, the degree of light leakage through the envelope, Fresnel losses at the two fiber ends, etc. The great variety of the relevant factors accounts for the noticeable discrepancies in the calculations of different authors.

When calculating the total losses in a bundle of fibers, the internal and the end-face losses are generally treated separately. The attenuation of light through absorption is described with fair accuracy by an exponential formula

$$I = I_0 e^{-KL}, \tag{145}$$

where I_0 and I are, respectively, the intensity of the incoming and the outgoing light flux; K is the absorption coefficient, L is the effective path of the ray, which depends on the angle of incidence, the wavelength, the refractive index, and the fiber geometry.

The effective path of a meridional ray in a straight cylindrical fiber is determined without much difficulty. Let the ray be incident at an angle θ to the entrance end (Figure 35). The path length inside the fiber is independent of the fiber diameter. Indeed, we have from Figure 35

$$L = l \sec \theta' = \frac{l}{[1 - (\sin \theta / n_1)^2]^{1/2}}, \tag{146}$$

where n_1 is the refractive index of the fiber material.

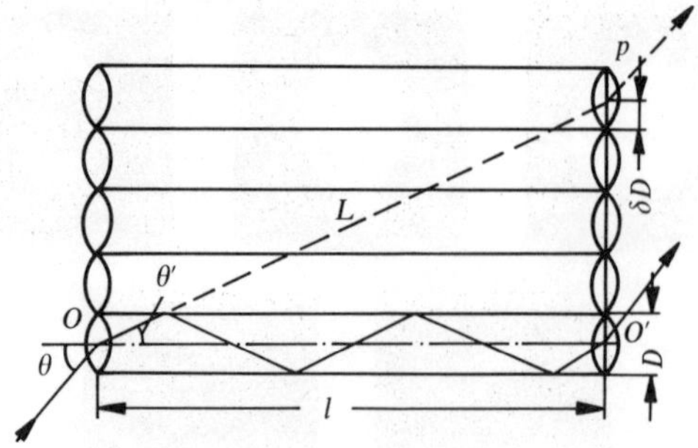

Figure 35

The path of a meridional ray in a straight cylindrical fiber.

After an even number of reflections, the emergent ray is parallel to the incident ray, whereas after an odd number of reflections, the ray emerges in a symmetrical direction relative to the optical axis.

The number of reflections $\eta(\theta')$ is also readily found from Figure 35:

$$l \tan \theta' = PO' = [\eta(\theta') - \tfrac{1}{2}]\, D + \delta D,$$

where $0 \leqq \delta \leqq 1$.

Since $\eta(\theta')$ may take on integral values only, we obtain

$$\eta(\theta') \cong \frac{l \tan \theta'}{D}. \tag{147}$$

Unlike L, the number of reflections $\eta(\theta')$ is a function of the fiber diameter. The number of reflections, as could have been expected, increases with the increase in fiber length and the decrease in its diameter. Thus, for $D = 25\mu$, $l = 25$ mm, $n_1 = 1.5$ and $\theta = 10°$, the number of reflections is 116, whereas for $\theta = 5°$ it is only 58. We see from (146), that even for these large η, the effective path length $L$is not substantially greater than the geometrical length l. For example, for $n_1 = 1.5$ and $\theta = 40°$, L is about $1.1\,l$.

Inserting for L in (145) its expression from (146), we obtain the following relationship between the light-transfer efficiency of a fiber and its principal parameters (ignoring end-face losses):

$$T = \frac{I}{I_0} = \exp\left\{\frac{-Kl}{[1 - (\sin\theta/n_1)^2]^{1/2}}\right\} = \exp[-Kl \sec\theta']. \tag{148}$$

For $0 \leqq \theta \leqq 90°$, the light transfer varies from its maximum value $T_{max} = \exp(-Kl)$ to the minimum value

$$T_{min} = \exp\left[\frac{-n_1 Kl}{(n_1^2 - 1)^{1/2}}\right].$$

The case $\theta = 90°$ is possible for a meridional ray only when $(n_1^2 - n_2^2)^{1/2} \geqq 1$, where n_2 is the refractive index of the fiber envelope. The curves of light transfer vs. the optical density Kl for axial ($\theta = 0°$) and grazing ($\theta = 90°$) incidence and for various refractive indices of a bare cylindrical fiber [93] are shown in Figure 36. In accordance with the previous theoretical considerations, the curves reveal a steady decrease in the difference $T_{\theta=0°} - T_{\theta=90°}$ with increasing refractive index.

The above method for the calculation of T is highly simplified and does not reflect the true state of things. In general, the expression for the light-transfer efficiency should be integrated over all the possible incidence angles. Recently, several techniques have been developed for the calculation of real light-transfer efficiencies of fiber systems. Thus the calculation of T carried out in [11, 57] takes into consideration not only the direct light, but also the light after one, two, three, ..., and m reflections in the fiber. In this case (again ignoring the end losses)

$$T = \sum_{m=0}^{\infty} T_m = \sum_{m=0}^{\infty} R^m \int_{\theta'_{m-1}}^{\theta'_m} e^{-Kl/\cos\theta'} \sin\theta' \, d\theta', \tag{149}$$

where T_m is the luminous flux at the exit end due to all the rays undergoing m reflections inside the fiber; R is the reflection coefficient; θ'_{m-1} and θ'_m are the maximum angles for which the ray undergoes respectively $m - 1$ and m reflections from the fiber surface.

Eq. (149) is highly general. Using this expression, we can find the light-transfer efficiency of metallized fibers, as well as of fibers working on the principle of total internal reflection. For a fiber with total internal reflection (seeing that the maximum number of total internal reflections η_{max} is given by (147), we insert θ'_{cr} for θ' and truncate the series in (149) so that

$$T = \sum_{m=0}^{m=\eta_{max}} T_m. \tag{150}$$

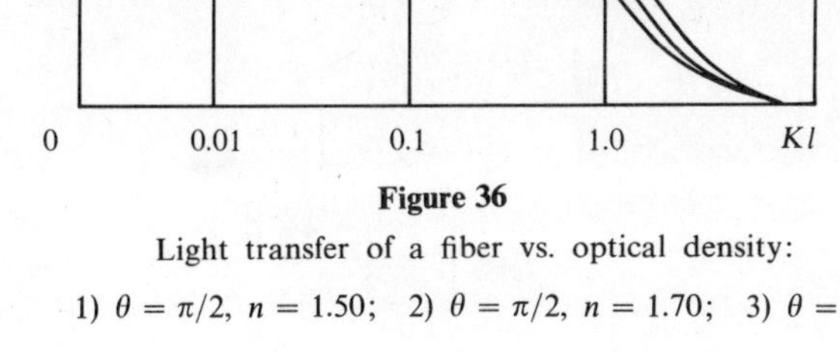

Figure 36

Light transfer of a fiber vs. optical density:

1) $\theta = \pi/2$, $n = 1.50$; 2) $\theta = \pi/2$, $n = 1.70$; 3) $\theta = 0$.

Moreover, in the case of total internal reflection, we may set $R = 1$, and then

$$T = \int_0^{\theta'_{cr}} \sin\theta' e^{-Kl/\cos\theta'}\, d\theta'. \qquad (151)$$

Replacing Kl with β and using $\cos\theta'$ as the new variable of integration, we obtain

$$T = e^{-\beta} - \frac{1}{n_1} e^{-n_1\beta} + \beta\,[\mathrm{Ei}(-\beta) - \mathrm{Ei}(-n_1\beta)], \qquad (152)$$

where Ei is a tabulated integral exponential [46].

The light-transfer efficiency of a light pipe illuminated by a Lambert-law radiator is determined in the same way. The result is

$$T_m = 2R^m \int_{\theta'_{m-1}}^{\theta'_m} \cos\theta' \sin\theta'\, e^{-Kl/\cos\theta'}\, d\theta'. \qquad (153)$$

For a light pipe with total internal reflection ($R = \rho = 1$)

$$T_m = 2 \int_0^{\theta'_{cr}} \cos\theta' \sin\theta'\, e^{-Kl/\cos\theta'}\, d\theta'. \qquad (154)$$

As before, we find

$$T = \left\{ e^{\beta}(1-\beta) - \frac{e^{-n_1\beta}}{n_1}\left(\frac{1}{n_1} - \beta\right) + \beta^2\,[\mathrm{Ei}(-n_1\beta) - \mathrm{Ei}(-\beta)] \right\}. \qquad (155)$$

Eqs. (152)–(155) adequately describe the transfer of meridional rays by straight cylindrical fibers. To obtain more accurate results, end losses and reflection losses must be taken into consideration. Further calculations for meridional and oblique rays will be found in [5, 8, 70, 92, 131, 132, 163].

Consider a straight insulated fiber of length l and diameter D, and a pencil of light rays with angular distribution $F(\theta)$ incident on the entrance end of the fiber. Assuming that the luminous flux is transferred by meridional rays only, we isolate from the incident cone those rays which hit the entrance end at angles between θ and $\theta + d\theta$. The flux in the corresponding solid angle thus can be written in the form

$$d\Phi = 2\pi F(\theta) \sin\theta d\theta. \qquad (156)$$

A certain fraction $R(\theta)$ of this flux is lost at the air–glass interface. This loss can be found from the Fresnel equations. Considering the s and the p components, we find

$$R(\theta) = \frac{1}{2}\left\{ \left[\frac{\cos\theta - (n_1^2 - \sin^2\theta)^{1/2}}{\cos\theta + (n_1^2 - \sin^2\theta)^{1/2}} \right]^2 + \right.$$

$$\left. + \left[\frac{-n_1^2\cos\theta + (n_1^2 - \sin^2\theta)^{1/2}}{n_1^2\cos\theta + (n_1^2 - \sin^2\theta)^{1/2}} \right]^2 \right\}. \qquad (157)$$

The loss in each total internal reflection can be allowed for by introducing an internal reflection coefficient $\rho \neq 1$, which depends on the angle of incidence on the fiber surface and therefore also on the refraction angle θ' at the entrance end. If the ray undergoes $\eta(\theta')$ internal reflections, the transparency of the fiber is $\rho^{\eta(\theta')}$ times the incident intensity (only the total reflection losses are considered).

Absorption losses inside the fiber are characterized by the quantity $1 - \exp(-Kl/\cos\theta')$, and the Fresnel losses at the glass–air interface, i.e., at the exit end of the fiber, are expressed by a reflection coefficient $R(\theta)$ from Eq. (157). For a conical fiber, the Fresnel losses are different at the two ends, and this difference should be taken into consideration.

Integration over all the possible angles gives for the total light-transfer coefficient

$$T = 2\pi A \int_0^{\theta'_{cr}} F(\theta')\,[1 - R(\theta')]^2 \rho^{\eta(\theta')} \exp\left(\frac{-Kl}{\cos\theta'}\right) \sin\theta'\, d\theta', \tag{158}$$

where A is a normalization factor defined by the equality

$$A 2\pi \int_0^{\theta_{cr}} F(\theta) \sin\theta d\theta = 1.$$

Eq. (158) characterizes the total transfer of meridional rays through a straight cylindrical fiber. Since optical fibers are generally manufactured from high-quality optical glass with very low values of K, whereas the number of internal reflections $\eta(\theta')$ is high, the main lossy factor in (158) is $\rho^{\eta(\theta')}$, and its value is largely determined by the manufacturing precision and quality of the fiber and the insulating envelope.

The curves of light transfer of fibers of various length and diameters vs. the angle θ calculated from (158) are shown in Figure 37. The dependence of T on fiber length l is shown in Figure 38. The calculations were carried out assuming zero Fresnel losses at the fiber ends and taking $\rho = 0.99$.

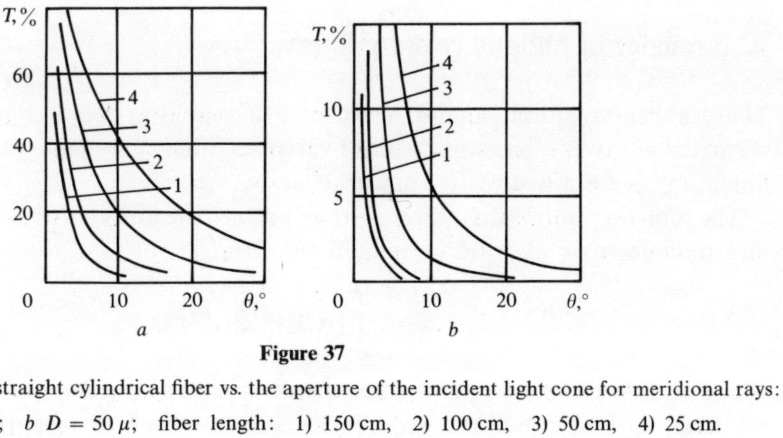

Figure 37

Light transfer of a straight cylindrical fiber vs. the aperture of the incident light cone for meridional rays: *a* $D = 200\,\mu$; *b* $D = 50\,\mu$; fiber length: 1) 150 cm, 2) 100 cm, 3) 50 cm, 4) 25 cm.

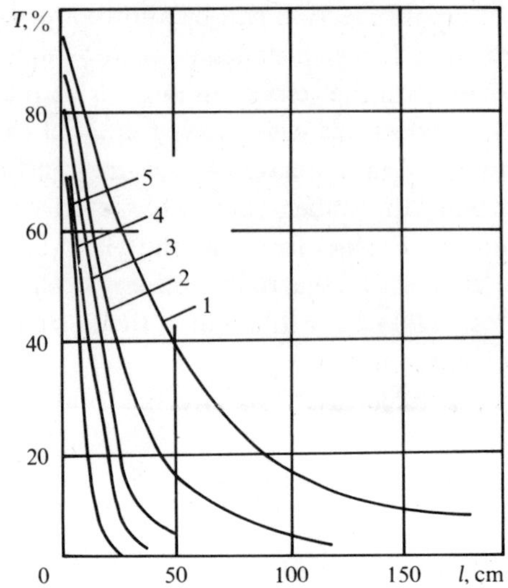

Figure 38

Light transfer vs. fiber length for meridional rays:

1) $D = 500\ \mu$; 2) $D = 200\ \mu$; 3) $D = 100\ \mu$; 4) $D = 50\ \mu$; 5) $D = 25\ \mu$.

Figures 37 and 38 show that T falls off rapidly with the decrease of fiber diameter and increase of fiber length. An increase of the angle θ has a similar effect. However, if T is calculated for $\rho = 1$, Eq. (158) does not give a strong dependence of T on the angle of incidence. This indicates that the main light losses are associated with the multiple reflections. Indeed, as the angle θ is increased, the number of reflections $\eta(\theta')$ increases much faster than the total path length.

4. Transfer of oblique rays by fibers

The problem is entirely analogous to that presented in Sec. 3, but the integration is carried out over a larger number of variables, since an oblique ray, unlike a meridional ray is described by two angles, θ' and γ.

The total flux transmitted through the fiber, i.e., the flux contributed by meridional and oblique rays, may be written in the form [131]

$$\Phi = \iint GF\,(\theta',\phi)\, dS\, d\Omega, \tag{159}$$

where $F(\theta',\phi)$ describes the angular distribution of the luminous intensity inside the fiber, dS is a cross sectional area element of the fiber near the entrance end,

$d\Omega$ is a solid angle element, θ' and ϕ are the coordinate angles, G is a function incorporating all the different light losses.

The function G may be expressed in the form

$$G = t^2(\theta')\,\rho^{\eta(\theta',\gamma)}\exp\left[-KL(\theta')\right], \tag{160}$$

where $t(\theta')$ is the light-transfer coefficient of the entrance and exit ends. If the entrance and exit ends are in contact with different media, their light-transfer coefficients are t_1 and t_2. The value of $t(\theta')$ can be found from the Fresnel equations,

$$t(\theta') = \frac{2\,(n_0^2 - n_1^2\sin\theta')^{1/2}}{\left[n_1\cos\theta' + (n_0^2 - n_1^2\sin^2\theta')^{1/2}\right]} + \frac{2n_0(n_0^2 - n_1^2\sin\theta')^{1/2}}{\left[n_0^2\cos\theta' + n_1\,(n_0^2 - n_1^2\sin^2\theta')^{1/2}\right]}. \tag{161}$$

In our case, the number of total internal reflections $\eta(\theta',\gamma)$ and the total path length are expressed by the equalities

$$\eta(\theta',\gamma) = \frac{l\tan\theta'}{D\cos\gamma}, \tag{162}$$

$$L(\theta') = \frac{l}{\cos\theta'}. \tag{163}$$

The final expression for the luminous flux transmitted by the fiber can be obtained from Eq. (159), remembering that the range of allowed angles γ depends on θ' (see Eq. (143)). Expressing the area element dS in terms of the angle γ and the fiber diameter D

$$dS = \tfrac{1}{4}D^2\cos^2\gamma\,d\gamma,$$

we find

$$\Phi = 2\pi D^2\int_{\theta'=0}^{\theta'=\theta'_{cr}}\int_{\gamma=0}^{\pi/2} t^2F(\theta')\,\rho^{l\tan\theta'/D\cos\gamma}\exp\left(-\frac{Kl}{\cos\theta'}\right)\times$$
$$\times\sin\theta'\cos^2\gamma\,d\gamma\,d\theta' + 2\pi D^2\int_{\theta'_{cr}}^{\theta'_{lim}}\int_{\gamma_{min}}^{\pi/2} t^2F(\theta')\,\rho^{l\tan\theta'/D\cos\gamma}\exp\left(-\frac{Kl}{\cos\theta'}\right)\times$$
$$\times\sin\theta'\cos^2\gamma\,d\gamma\,d\theta'. \tag{164}$$

In (164) it is implied that the incident flux intensity F is independent of the angle γ. This assumption somewhat lowers the generality of our treatment, since we automatically eliminate sources with an energy distribution dependent on γ; the great advantage of this approach, however, is that the numerical integration of (164) is greatly simplified. Equations for the calculation of the transfer of diffuse light by an ideally transparent straight fiber are given in [8].

Scintillation fibers have recently become quite popular in nuclear physics as particle detectors. The core is prepared from materials usually used in scintillation counters. Eq. (164) applies to this case also, provided the end-loss factor is taken to the first power; this corresponds to a source located inside the fiber. In the meridional approximation, the angles $\theta' > \theta'_{cr}$ are forbidden, and the second term in (164) vanishes, the result being an expression analogous to (158).

For $\rho = 1$, $K = 0$, and $t = 1$ (an ideally transparent light pipe with zero-reflection end faces and total reflection from the cylindrical surface), Eq. (164) can be integrated analytically.

In this case, the light transfer is entirely determined by the fiber geometry, the refractive index of the fiber material, and the angular distribution of the incident flux intensity. For a Lambert-law radiator, when $F(\theta) = F_0 \cos\theta$, we have for $\theta'_{lim} > \theta'_{cr}$

$$\Phi = \tfrac{1}{4}\pi^2 D^2 F_0 \sin^2\theta'_{lim} - \tfrac{1}{2}\pi D^2 F_0 \sin\theta'_{cr}(\sin^2\theta'_{lim} - \sin^2\theta'_{cr})^{1/2} - $$

$$- 2\pi D^2 \cos^{-1}\left(\frac{\sin\theta'_{cr}}{\sin\theta'_{lim}}\right)(\tfrac{1}{4}\sin^2\theta'_{lim} - \tfrac{1}{2}\sin^2\theta'_{cr}). \quad (165)$$

In the simpler case $\theta'_{lim} < \theta'_{cr}$, the second term in (165) is dropped and the total luminous flux transferred by the fiber is described by the equation

$$\Phi = \tfrac{1}{4}\pi^2 D^2 F_0 \sin^2\theta'_{lim}. \quad (166)$$

In practice, the numerical aperture of the fibers is only partly utilized. The optical properties of the systems then essentially depend on the illumination conditions. Thus for a Lambert-law radiator, the entire luminous flux transferred through the fiber is proportional to $(n_0 \sin\theta)^2$ (θ is the conical angle of the light cone hitting the entrance end; this angle is limited by the source geometry). A similar dependence is obtained in lens optics when a small Lambert-law radiator is used. In this case, the luminous flux transmitted by the system is also proportional to $(n_0 \sin\theta)^2$, but θ is determined by the size of the entrance pupil of the system.

The above properties of straight fibers are clearly apparent in a comparison of the transferred integrated flux $\Phi(\theta)$ as a function of the conical angle of the incident light cone for meridional and oblique rays. In the case of meridional rays, an increase in θ, $\theta \leqq \theta_{cr}$, raises $\Phi(\theta)$ in proportion to $\sin^2\theta$ (Figure 39*a*). As the angle θ is further increased ($\theta > \theta_{cr}$), the emergent luminous flux remains constant, since the condition of total internal reflection breaks down. However, if both terms are included in (164), the function $\Phi(\theta)$, although still proportional to $\sin^2\theta$ for $\theta < \theta_{cr}$, will continue increasing (at a somewhat slower rate) for $\theta > \theta_{cr}$ also. We see that the contribution from oblique rays for $\theta > \theta_{cr}$ may increase the transferred integrated flux by as much as 15–18% compared to the flux transferred for $\theta = \theta_{cr}$ [8, 132].

The luminous flux as a function of the aperture of the incident light cone for straight

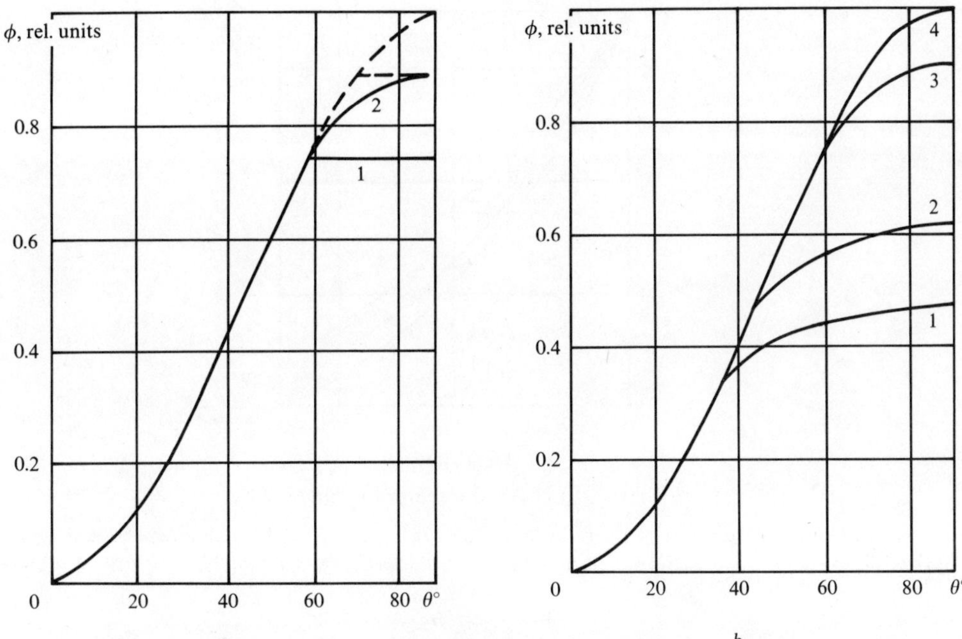

Figure 39

The emerging luminous flux vs. the aperture of the incident light cone:

a with allowance for meridional and oblique rays: 1) meridional rays allowed for, 2) oblique rays allowed for; *b* with allowance for all rays: 1) $n_1 = 1.62$, $n_2 = 1.52$; 2) $n_1 = 1.75$, $n_2 = 1.62$; 3) $n_1 = 1.75$, $n_2 = 1.52$; 4) $n_1 = 1.75$, $n_2 = 1.42$; the dashed curve corresponds to the sine-squared function.

fibers with various combinations of n_1 and n_2 (the refractive indices of the core and the insulating envelope) is plotted in Figure 39*b*. All the curves were calculated for $\rho = 1$, $K = 0$, and $t = 1$.

Eq. (164) has been verified experimentally in [131, 132]. The light-transfer values* obtained in these experiments for individual fibers with different refractive indices of the core and the envelope are plotted by the dots in Figure 40. The curves were calculated from (164) for $K = 1 \cdot 10^{-4}\ \text{cm}^{-1}$ and various values of ρ.

Eq. (164) also has been verified for a bundle of fibers each 70 μ in diameter (the diameter of the entire bundle was 3 mm). Although the individual fibers had an irregular cross section, the experimental results show an excellent fit with the theoretical curve (Figure 41) for $\rho = 0.9993 \pm 0.0001$. This fact indicates that slight deviations from a perfect circular cross section apparently do not affect the light-transfer properties.

---

* To find T, one should also measure the incident flux Φ_0, since $T = \Phi/\Phi_0$; the numerical value of Φ is expressed by Eq. (164).

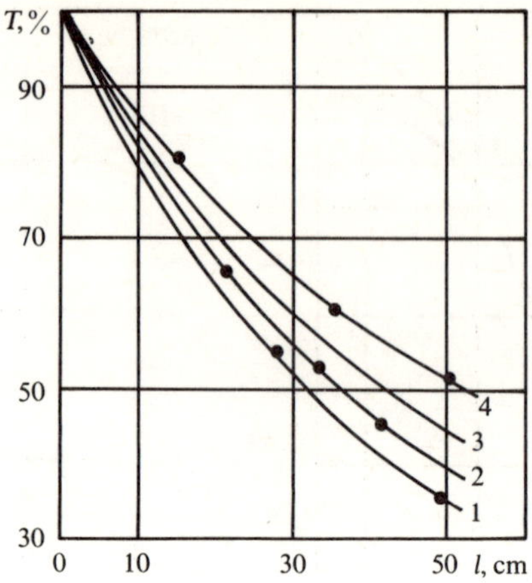

Figure 40

Light transfer of cylindrical fibers vs. fiber length:

curves 1 through 3 correspond to fibers with $n_1 = 1.62$, $n_2 = 1.52$, $D = 100\ \mu$, and $\rho = 0.9992$, 0.9993, 0.9994, respectively; curve 4 corresponds to a fiber with the parameters $n_1 = 1.75$, $n_2 = 1.52$, $D = 254\ \mu$, $\rho = 0.9992$.

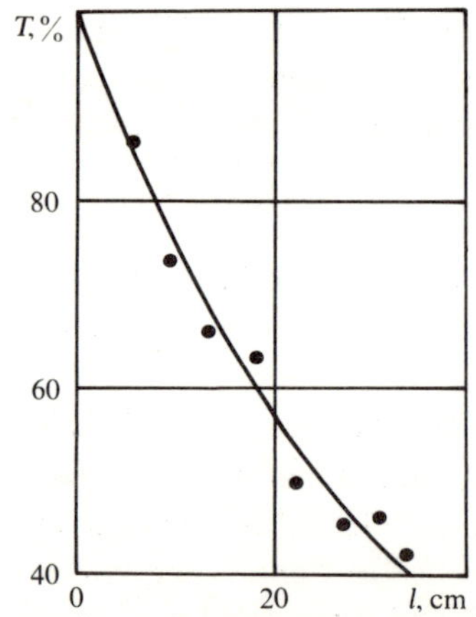

Figure 41

Light transfer of a fiber bundle 0.3 cm in diameter, consisting of fibers 70 μ in diameter, vs. length (dots mark the experimental findings).

Measurements with fibers from 1 to 1000 μ in diameter and from 5 to 150 mm long have shown [131] that in most cases the best fit between theoretical and experimental light-transfer values is obtained for an internal reflection coefficient.

However, the actual value of ρ may fluctuate between wide limits depending on the technology of fiber manufacture. Therefore the fiber drawing processes must meet very exacting precision standards.

Changes in the absorption coefficient K have a much less significant effect. For $10^{-4} \leqq K \leqq 10^{-3}$ cm^{-1}, the absorption in short fibers hardly affects the theoretical light-transfer curves. It may become noticeable only for very long bundles.

Using the experimental values of the parameters entering Eq. (164), we can calculate the outgoing integrated luminous flux as a function of the conical angle of the light cone incident on the entrance end of the fiber. The corresponding curves are plotted in Figure 42 for three fibers of different numerical apertures. The theory and the experiment again show an excellent agreement.

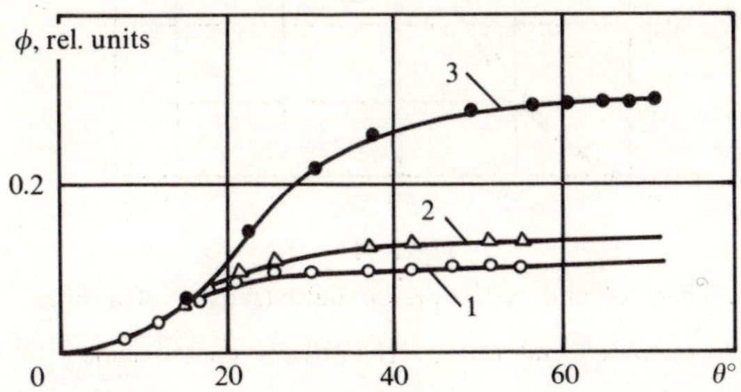

Figure 42

Theoretical and experimental curves plotting the emerging luminous flux as a function of the conical angle of the incident beam:

1) $A = 0.16$; 2) $A = 0.25$; 3) $A = 0.46$.

5. Fresnel losses at fiber ends

Losses associated with light reflection from the end faces of fibers and bundles are highly significant at large incidence angles (greater than Brewster's angle). This is illustrated by the curves (Figure 43) which plot the light-transfer efficiency of a fiber with optically coated and uncoated end faces for aperture angles $0° < \theta < 90°$. The calculations were carried out using Eqs. (161) and (164) for a fiber with a core refractive index of 1.75 and an envelope refractive index of 1.52 [132]. As the aperture angle of the incident light increases, the curves 1 and 2 diverge progressively. For $\theta = 90°$, the light transfer of an uncoated fiber is 0.8 of the light transfer T of a fiber with optically coated end faces. This corresponds to an aperture angle $\theta = 57°42'$ for the case shown by curve 2.

To calculate the end-reflection losses, we should first generalize the Fresnel formula for conical light beams [8]. For an aperture angle θ, the reflection coefficients are

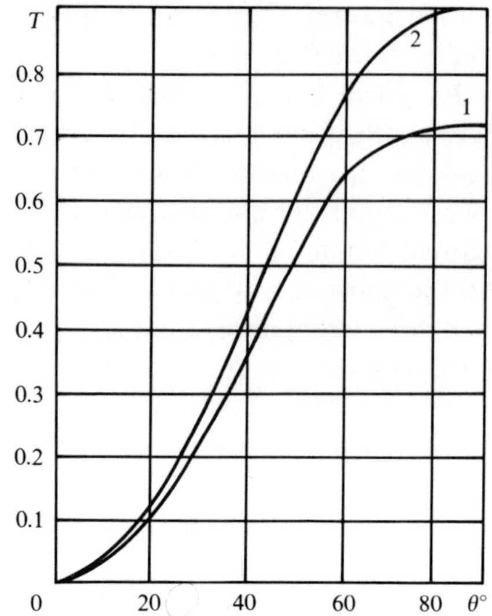

Figure 43

The effect of end reflection on light transfer of a fiber:

1) uncoated end faces; 2) optically coated ends.

$$R_s = \frac{2}{\sin^2 \theta} \int_0^{\theta} \frac{\sin^2 (\theta - \theta')}{\sin^2 (\theta + \theta')} \cos \theta \sin \theta \, d\theta,$$

$$R_p = \frac{2}{\sin^2 \theta} \int_0^{\theta} \frac{\tan^2 (\theta - \theta')}{\tan^2 (\theta + \theta')} \cos \theta \sin \theta \, d\theta, \tag{167}$$

where θ' is the aperture angle of the light cone inside the fiber.

Using the refraction law $n_0 \sin \theta = n_1 \sin \theta'$ (n_0 is the refractive index of the medium in contact with the fiber ends, n_1 is the refractive index of the fiber core) and setting $n_0 = 1$, we obtain after integration

$$R_s = \frac{(\sqrt{n_1^2 - \sin^2 \theta} + 3 \cos \theta)(n_1^2 - \sin^2 \theta)(n_1 + 1)(n_1 - 1)}{3 (\sqrt{n_1^2 - \sin^2 \theta} + \cos \theta)^3 \sin^2 \theta} - \frac{n_1^2 (n_1 - 1)(n_1 + 3)}{3 (n_1 + 1)^2 \sin^2 \theta}, \tag{168}$$

$$R_p = 1 - \frac{4}{n_1^2 \sin^2 \theta} (n_1 - \cos \theta \sqrt{n_1^2 - \sin^2 \theta}) + \frac{8 n_1^5 [n_1^2 \cos^2 \theta + (n_1 - 1) \sin^2 \theta - n_1 \cos \theta \sqrt{n_1^2 - \sin^2 \theta}]}{(n_1^4 - 1)^2 [n_1^2 - (n_1^2 + 1) \sin^2 \theta] \sin^2 \theta} +$$

$$+\frac{8n_1^6\sin^2\theta - 12n_1^5 + 12n_1^4\cos\theta\sqrt{n_1^2-\sin^2\theta} + 4n_1 - 4\cos\theta\sqrt{n_1^2-\sin^2\theta}}{n_1^2(n_1^4-1)^2\sin^2\theta} +$$

$$+\frac{4(n_1^2-1)}{n_1^2\sin^2\theta}\ln\frac{\sqrt{n_1^2-\sin^2\theta}+\cos\theta}{n_1+1} - \frac{16n_1^4(n_1^4+1)}{(n_1^4-1)^2(n_1^2+1)\sin^2\theta}\times$$

$$\times\ln\frac{n_1^2\cos\theta+\sqrt{n_1^2-\sin^2\theta}}{(n_1+1)\sqrt{n_1^2-\sin^2\theta}} + \frac{2(n_1^2-1)(4n_1^4+3n_1^2+1)}{n_1^2(n_1^2+1)^3\sin^2\theta}\times$$

$$\times\ln\frac{n_1\sqrt{n_1^2-\sin^2\theta}-\cos\theta}{(n_1-1)\sqrt{n_1^2-\sin^2\theta}} + \frac{2(4n_1^4-3n_1^2+1)}{n_1^2(n_1^2-1)\sin^2\theta}\times$$

$$\times\ln\frac{n_1\sqrt{n_1^2-\sin^2\theta}+\cos\theta}{(n_1+1)\sqrt{n_1^2-\sin^2\theta}}.$$

In practice, the case $\theta = 90°$ is of considerable interest, when the complex expressions (168) are substantially simplified. Seeing that the reflection coefficient for natural light is

$$R = \frac{R_p + R_s}{2},$$

we find

$$R(\theta = 90°) = \frac{3n_1^8 - 10n_1^7 - 10n_1^6 + 4n_1^5 + 16n_1^4 - 10n_1^3 + 2n_1^2 + 1}{3(n_1^4-1)^2} +$$

$$+\frac{8n_1^4(n_1^4+1)}{(n_1^4-1)(n_1^2+1)}\ln n_1 - \frac{n_1^2(n_1^2-1)}{(n_1^2+1)^3}\ln\frac{n_1+1}{n_1-1}. \qquad (169)$$

The curves plotting the reflection coefficient calculated from (168) vs. the core refractive index n_1 for various aperture angles of the incident light cone are shown in Figure 44. We see that the refractive index is the decisive factor in reflection for small aperture angles. As θ increases, the curve is translated without change of attitude toward larger values of R.

Figure 45 shows the curves of light transfer for various aperture angles vs. the core refractive index of an ideally transparent fiber, calculated by Veinberg and Sattarov [8] for $n_2 = 1.48$. The calculations used an expression incorporating the contribution from oblique rays. Although the method is different from that used in [131], the analysis nevertheless shows that this expression coincides with Eq. (164) if we take $K = 0$ and $\rho = 1$.

We see from Figure 45 that T increases with the increase in the refractive index of

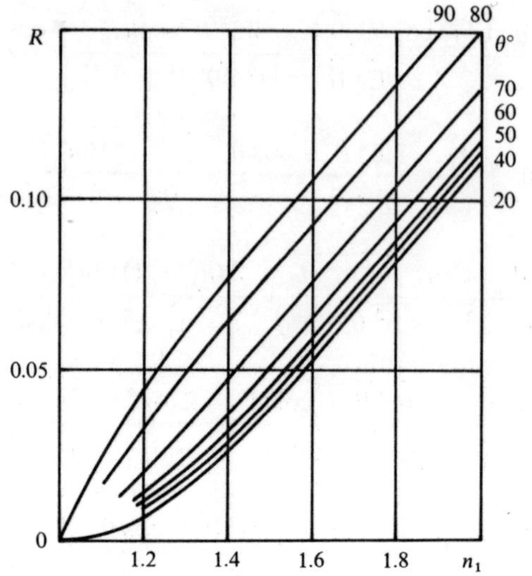

Figure 44

Integrated reflection coefficient of light cones of various apertures vs. the refractive index of the fiber.

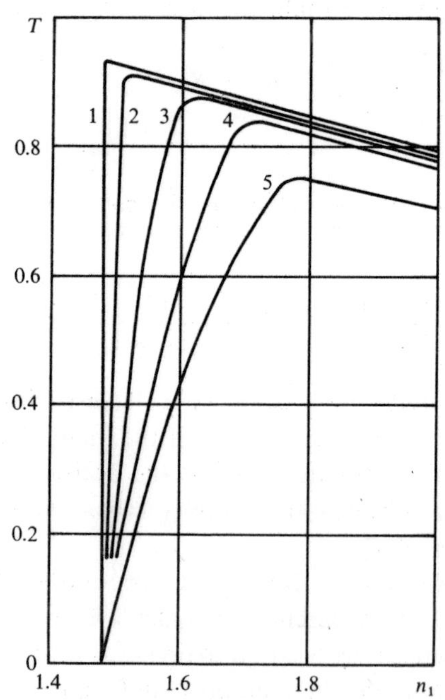

Figure 45

Light transfer vs. the refractive index of an ideally transparent fiber for light cones of various apertures θ:

1) $\theta = 0$; 2) $\theta = 20°$; 3) $\theta = 40°$; 4) $\theta = 60°$; 5) $\theta = 90°$.

the core, reaching a maximum for some n_1. Further increase of the core refractive index leads to decrease in T due to the increasing Fresnel losses at the fiber ends. For small aperture angles, T_{max} increases in absolute value and shifts toward smaller values of n_1. In the limiting case of normal incidence ($\theta = 0°$), T_{max} is attained for $n_1 = n_2$.

The above results correspond to ideal fibers ($K = 0$, $\rho = 1$). In real fibers, in the presence of light scattering, T_{max} are apparently shifted toward still higher values of n_1.

6. Angular distribution of radiation

An important characteristic of a fiber bundle is the angular distribution of the emerging luminous flux. This characteristic is particularly significant when the bundle is joined to various optical and electronic systems.

Assuming an axially symmetric light cone, we can derive the angular distribution of radiation inside the fiber $f(\theta')$ from the expression for the integrated luminous flux transferred by a straight cylindrical fiber [131, 132]. Since a solid angle element is given by

$$d\Omega' = \sin\theta'\, d\theta'\, d\phi,$$

we have

$$\Phi(\theta') = \int f(\theta')\, t \sin\theta'\, d\theta'\, d\phi, \tag{170}$$

where $\Phi(\theta')$ is the emerging luminous flux, expressed by Eq. (164); t is the light-transfer coefficient of the fiber end.

Comparison of Eqs. (164) and (170) gives

$$f(\theta') = D^2 tF(\theta') \exp\left(-\frac{Kl}{\cos\theta'}\right) \int_{\gamma=0}^{\pi/2} \rho^{l\tan\theta'/D\cos\gamma} \cos^2\gamma\, d\gamma +$$

$$+ D^2 tF(\theta') \exp\left(-\frac{Kl}{\cos\theta'}\right) \int_{\gamma_{min}}^{\pi/2} \rho^{l\tan\theta'/D\cos\gamma} \cos^2\gamma\, d\gamma, \tag{171}$$

where the notation is the same as in (164). To obtain the distribution function $f(\theta)$ of the outgoing radiation, we use the law of refraction and allow for the Fresnel losses. Thus,

$$f(\theta) \sin\theta\, d\theta\, d\phi = t f(\theta') \sin\theta'\, d\theta'\, d\phi. \tag{172}$$

The right-hand side of this expression corresponds to a luminous flux propagating inside a solid angle $d\Omega'$ in the fiber, and the left-hand side to radiation inside a solid angle beyond the exit end. After simple manipulations, we obtain

$$f(\theta) = t\,f(\theta')\,\frac{n_0\,(n_0^2 - n_1^2 \sin^2 \theta')^{1/2}}{n_1^2 \cos \theta'} =$$

$$= t\,f(\theta)\,\frac{n_0^2\,(n_1^2 - \sin^2 \theta)^{1/2}}{n_1^2\,(n_1^2 - n_0 \sin^2 \theta)^{1/2}}\,. \tag{173}$$

The angular distribution of the emerging radiation for two fibers of different numerical apertures is plotted in Figure 46 [132]. Both curves were calculated for a Lambert-law radiator. The two curves are entirely different for large emergence angles. This is readily understood on the basis of the relationship between the maximum aperture of the cone of transferred miridional rays and the numerical aperture of the fiber A. Curve 1 plotting the angular distribution of the radiation emerging from a fiber with $A < 1$ ($A = 0.66$) is made up of two sections. For $\theta < \theta_{cr}$ ($\theta_{cr} = 41.44°$), the emergent light flux is formed by both meridional and oblique rays. Any ray hitting the entrance end at some angle $\theta < \theta_{cr}$ leaves the fiber at the same angle. This ensures that the angular distribution of radiation over the exit end is identical to that at the entrance end. This naturally implies that the refractive indices of the two media in contact with the fiber ends are equal.

For $\theta > \theta_{cr}$, radiation is transferred by oblique rays only. Curve 1 therefore reveals a significant drop of $f(\theta)$.

An essentially different picture is observed for fibers with a numerical aperture greater than unity (curve 2). The angular distribution of light is transferred without any changes to the exit end.

The angular scattering function of light pipes is of some importance in connection

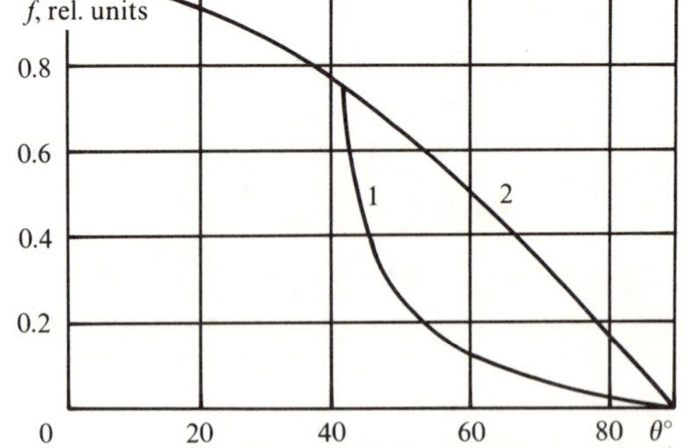

Figure 46

Angular distribution of the emerging light for two fibers with different numerical apertures: 1) $A < 1$, $n_1 = 1.75$, $n_2 = 1.62$; 2) $A > 1$, $n_1 = 1.75$, $n_2 = 1.42$.

with our problem. It was studied in [4], where the authors measured the scattering function of one light pipe by two different methods. According to the first method, a bundle of glass fibers was mounted in fixed position in front of a photometric objective and the entrance end was illuminated by a narrow parallel beam, sent at a certain angle (Figure 47*a*). The brightness of the exit end was compared with the brightness of a fixed frosted-glass plate.

According to the second method (Figure 47*b*), the bundle was attached to the light source and the two rotated together. The brightness of the exit end was measured as a function of the angle of observation. The results were identical in both cases. The experimental dependence of the brightness B on the angle θ obtained for two different bundles is shown in Figure 48. The general shape of the curves indicates that the scattering function is highly directional, and the directivity improves as the ratio D/l is increased. This is associated with the reduction in the number of internal reflections in the fiber, so that the effect of surface defects diminishes. The higher the quality of the fiber, the better is the directivity of the scattering function and the smaller is its dependence on fiber geometry.

7. Depolarization

When oblique rays are transferred through a straight cylindrical fiber, half of the rays trace a left-handed polygonal helix, and the other half a right-handed polygonal helix. Therefore, a plane-polarized wave entering the fiber will be partly depolarized

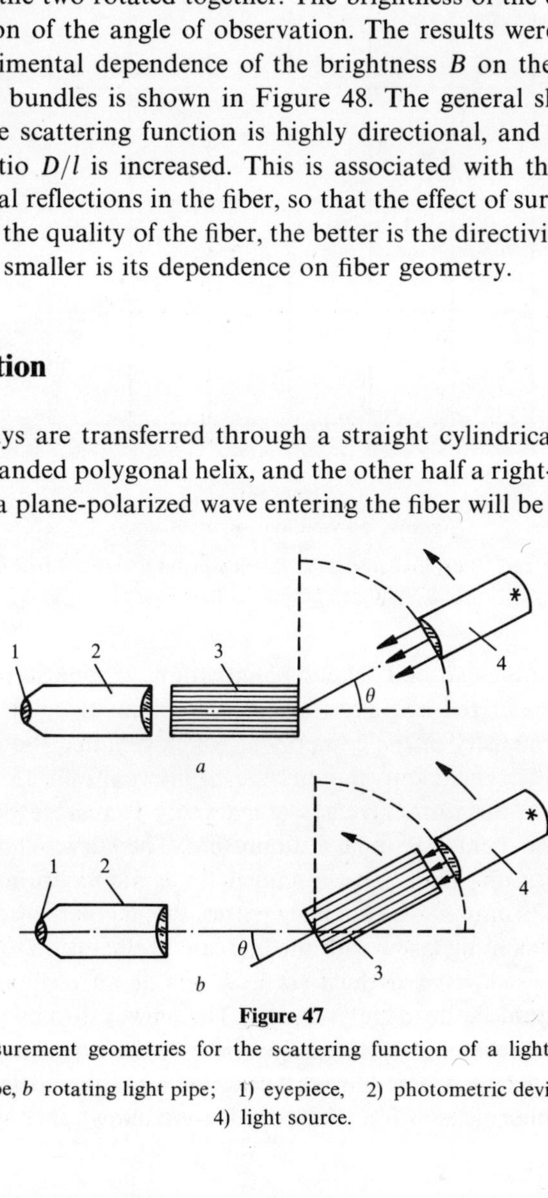

Figure 47

Measurement geometries for the scattering function of a light pipe:

a fixed light pipe, *b* rotating light pipe; 1) eyepiece, 2) photometric device, 3) light pipe, 4) light source.

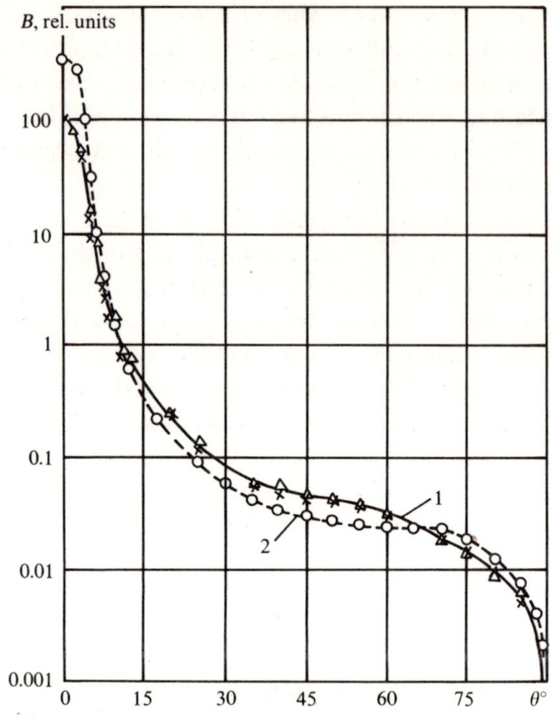

Figure 48

Angular distribution of brightness:

1) a bundle of fibers, each 100 μ in diameter; Δ illumination at $\theta = 0°$, $\times$ observation at $\theta = 0°$;
2) light pipe 2.7 mm in diameter.

upon emerging from the exit end. The depolarization Q, defined as the ratio of the outgoing intensity polarized at right angles to the plane of polarization of the incident light to the intensity of the component which retained the original plane of polarization, is highly sensitive to the number of internal reflections and therefore to the ratio D/l [4]. As this ratio increases, Q markedly decreases; the rate of decrease becomes faster as the incidence angle θ diminishes. The curves shown in Figure 49 confirm this proposition. They were obtained for a monochromatic radiation of $\lambda = 0.52\ \mu$ and $l = 72$ mm. As the angle increases, the depolarization also increases, the rate of change becoming faster with the increase in the ratio D/l. For a normally incident plane-polarized wave, a thick ($D \gg \lambda$) acts as an ordinary plane-parallel plate, and there should be no depolarization. The curves do not pass through the origin apparently because of imperfect adjustment of the experimental setup, slight divergence of the incident beam, and scattering from the cylindrical surface of the light pipe. The angular distribution of depolarization shown in Figure 49 is associ-

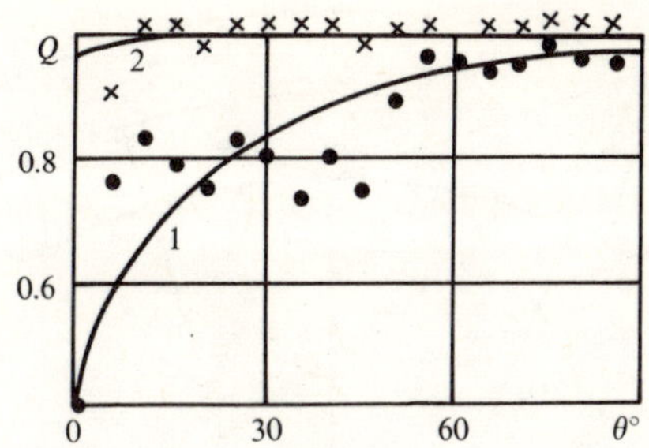

Figure 49

Angular distribution of depolarization:

1) light pipe 2.7 mm in diameter; 2) a fiber 100 μ in diameter.

ated with the dependence of the number of internal reflections on the angle θ (see (162)) and the decisive role of oblique rays for large incidence angles.

In experimental studies of depolarization, the possible departure of the real fiber shape from the ideal cylinder should be taken into consideration. This shape imperfection may possibly explain the dependence of the angular distribution of depolarization on the orientation of the electrical vector of the incidence light wave, observed in [4].

8. A curved cylindrical fiber

The rapidly increasing uses of flexible light pipes focus our attention on the optical properties of curved cylindrical fibers. Unfortunately, no detailed analysis of this problem has been attempted so far. Only Kapany [93] published some tentative results relating to curve fibers.

Consider a cylindrical fiber of diameter D curved along an arc with a radius of curvature R (Figure 50). In the general case, the apex of the incident light cone lies in the plane of the entrance end at a certain distance h away from the fiber axis. Therefore, simple geometrical considerations and the sine theorem lead to the equality

$$\sin i = \frac{R + h}{R + r} \cos \theta', \tag{174}$$

where $r = \frac{1}{2}D$ is the fiber radius, i is the angle of incidence of the light ray on the cylindrical surface, θ' is the angle of refraction at the entrance end.

The light path between two successive reflections is

$$AB = \frac{(R - r) \sin \alpha}{\sin i}. \tag{175}$$

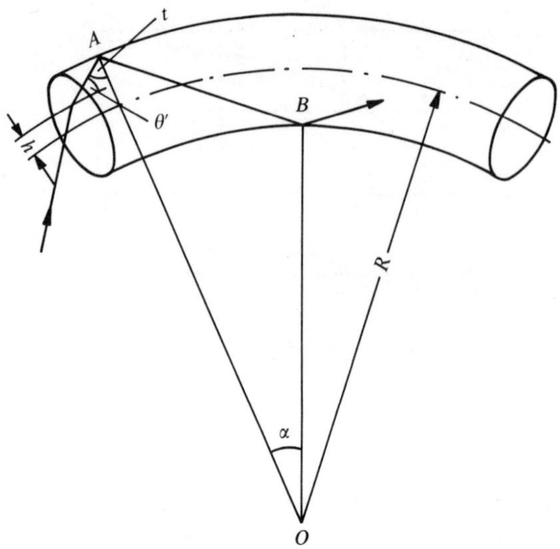

Figure 50

The path of rays in a curved cylindrical fiber.

The ratio of the total path length L of a meridional ray to the length of the curved fiber satisfies the inequality

$$\frac{L}{l} < \frac{\sin \alpha}{\alpha \sin i}.$$

Seeing that $\sin \alpha/\alpha \leqq 1$ and $\sin i \simeq \cos \theta'$, we find

$$L < l \sec \theta'. \tag{176}$$

A meridional ray in a curved fiber thus traverses a shorter distance than the same ray in a straight fiber of the same length.

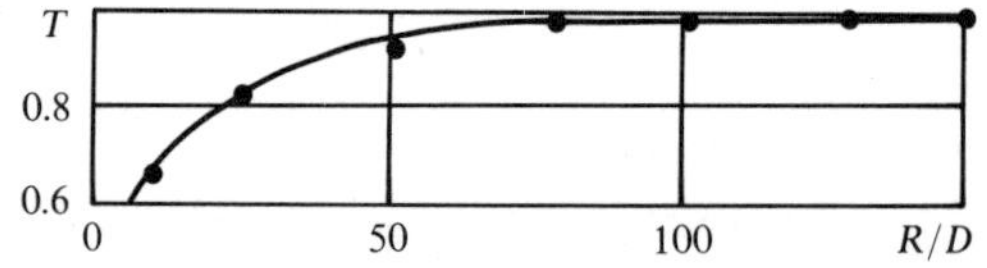

Figure 51

The effect of fiber curvature on light transfer.

Using Eq. (174), we can find the minimum permissible radius of curvature R for which the condition of total internal reflection is just satisfied. The results of [93] show that for $R > (3\text{–}4)\,D$ the light transfer for meridional rays should not differ significantly from the light transfer of a straight fiber. Nevertheless, the experimental curve of T vs. R/D obtained in [93] (Figure 51) reveals a considerable decrease of T starting with radii of curvature equal to 20 D. This indicates that Eqs. (174)–(176), although adequately rigorous for the meridional rays, cannot be used to describe the behavior of oblique rays in a curved fibers.

3 Propagation of Light in Thin Fibers

1. Fundamental equations

The optical properties of thin fibers should be considered within the framework of wave optics, since for fibers a few microns and less in diameter, diffraction effects begin to be felt. Such thin fibers therefore should be treated as dielectric waveguides propagating only certain allowed electromagnetic modes (a mode is a certain space and time distribution of the electromagnetic field). The propagated modes depend on the fiber parameters, the surrounding medium (insulating envelope), and the frequency of the incident light wave. Each mode satisfies Maxwell's equations and certain boundary conditions determined by the geometry and the optical characteristics of the fiber.

In general, electromagnetic modes are also sustained in fibers of relatively thick diameters, but because of the inevitable superposition of a tremendous number of such modes, the distinctive features of each individual mode are not as pronounced as in thin fibers.

Originally, the theory of guided electromagnetic waves was developed for metallic waveguides. The principal results, however, were found to apply to dielectric waveguides also [34]. In a metallic waveguide, the electromagnetic wave is reflected from the highly conducting walls, whereas in fibers the wave propagation is governed by the total internal reflection at the interface between dielectrics with different refractive indices. As a result, dielectric waveguides show a number of peculiar features associated with the boundary conditions at the interface [139, 149].

The electromagnetic field of a wave in a three-dimensional space will be described by the electric and the magnetic vector components E_1, E_2, E_3 and H_1, H_2, H_3, respectively. For convenience, x_3 is directed along the waveguide axis.

In general, an electromagnetic wave can be treated as a superposition of two fields with the components

$$\begin{cases} 0 & E_2, & 0 \\ H_1, & 0, & H_3 \end{cases} \tag{177}$$

and

$$\begin{cases} E_1, & 0, & E_3 \\ 0, & H_2, & 0 \end{cases} \tag{178}$$

The superposition of fields (177) and (178) gives a general field. Here, (177) describes a so-called transverse electric wave (TE wave), since the electric vector is perpendicular to the direction of propagation, and (178) a transverse magnetic field (TH wave). A general TE wave should have $E_3 = 0$, while other components do not necessarily vanish; similarly a TH wave is the one with $H_3 = 0$.

In so-called transverse electromagnetic waves (TEH waves), both E_3 and H_3 vanish simultaneously.

Analysis of wave processes shows [21] that all the three wave types can be sustained independently under appropriate conditions.

We will now consider the fundamental equations of guided waves in a general waveguide. Maxwell's equations describing an electromagnetic wave in the absence of extraneous currents can be written in the form

$$\frac{\partial E_\gamma}{\partial x_\beta} - \frac{\partial E_\beta}{\partial x_\gamma} = \frac{i\omega\mu}{c} H_\alpha, \tag{179}$$

$$\frac{\partial H_\gamma}{\partial x_\beta} - \frac{\partial H_\beta}{\partial x_\gamma} = \frac{\sigma - i\omega\varepsilon}{c} E_\alpha, \tag{180}$$

where ε and μ are the dielectric constant and the magnetic permeability of the medium, σ is the conductivity of the medium, and ω is the radian frequency.

Eqs. (179) and (180) constitute a set of six scalar equations for the six wave components. They hold true when the subscripts α, β, and γ are assigned the values 1, 2, and 3, or any cyclic permutation of this triad.

Eqs. (179) and (180) describe stationary fields. The time dependence is incorporated by multiplying the solutions by the exponential time factor $e^{-i\omega t}$.

Each of the components of E and H satisfies the Helmholtz equation of the form

$$\Delta w + k^2 w = 0,$$

where

$$k^2 = \frac{\omega^2 \varepsilon\mu + i\omega\mu\sigma}{c^2}. \tag{181}$$

The solution $w(x_1,x_2,x_3)$ of the Helmholtz equation is conveniently sought in the form of a product $u(x_1,x_2)\,v(x_3)$. Substituting, we find

$$\frac{1}{u}\left(\frac{\partial^2 u}{\partial x_1^2} + \frac{\partial^2 u}{\partial x_2^2}\right) + \frac{1}{v}\frac{\partial^2 v}{\partial x_3^2} + k^2 = 0. \tag{182}$$

Separation of variables now gives

$$\frac{\partial^2 u}{\partial x_1^2} + \frac{\partial^2 u}{\partial x_2^2} + (k^2 - \gamma^2)\, u = 0 \tag{183}$$

and

$$\frac{\partial^2 v}{\partial x_3^2} + \gamma^2 v = 0, \tag{184}$$

where γ^2 is some complex or real number.

A physically significant case is that when γ^2 is a positive real number. Otherwise, the solution $v(x_3)$ of (184) is an exponentially decaying function (in the direction x_3). For $\gamma^2 > 0$, the integral of Eq. (184) has the form

$$v\,(x_3) = C_1 e^{-i\gamma x_3} + C_2 e^{i\gamma x_3}.$$

It describes two waves propagating in opposite directions along the x_3 axis. Since these waves are identical, the final solution need only contain the factor $e^{i\gamma x_3}$, provided we remember that the direct wave is always mixed with a reflected wave.

The propagation constant γ may take on positive and negative real values. The solution of the Helmholtz equation is thus written in the general form

$$w\,(x_1, x_2, x_3) = Au(x_1, x_2).$$

Thus Eqs. (183) and (184) should be considered in conjunction with the equations describing the derivatives of the field components with respect to x_3,

$$\frac{\partial E_\alpha}{\partial x_3} = i\gamma E_\alpha \quad \text{and} \quad \frac{\partial H_\alpha}{\partial x_3} = i\gamma H_\alpha \quad (\alpha = 1, 2, 3). \tag{185}$$

Since we are dealing with wave propagation in dielectric fibers, we may take $\sigma = 0$. This group includes waveguides in the form of hollow or dielectric-filled conducting cylinders, since the nonzero conductivity of the cylinder walls affects only the boundary conditions, and not the general field equations inside the waveguide.

Using (185) and setting $\sigma = 0$, we obtain from (179) and (180) the following expressions for the transverse components of the electric and the magnetic vectors:

$$\begin{aligned}
(k^2 - \gamma^2)\, E_1 &= i\gamma \frac{\partial E_3}{\partial x_1} + \frac{i\omega\mu}{c}\frac{\partial H_3}{\partial x_2}\,;\\
(k^2 - \gamma^2)\, E_2 &= i\gamma \frac{\partial E_3}{\partial x_2} - \frac{i\omega\mu}{c}\frac{\partial H_3}{\partial x_1}\,;\\
(k^2 - \gamma^2) H_1 &= i\gamma \frac{\partial H_3}{\partial x_1} - \frac{i\omega\varepsilon}{c}\frac{\partial E_3}{\partial x_2}\,;
\end{aligned} \tag{186}$$

$$(k^2 - \gamma^2)H_2 = i\gamma \frac{\partial H_3}{\partial x_2} + \frac{i\omega\varepsilon}{c} \frac{\partial E_3}{\partial x_1}, \tag{186}$$

where

$$k^2 = \frac{\omega^2 \varepsilon\mu}{c^2}. \tag{187}$$

To obtain a complete description of the TE and TH waves for $k^2 \neq \gamma^2$, we only need the longitudinal field components from equations of the form (183). Simple differentiation of the longitudinal components will then yield the transverse components. The TEH waves constitute a case in themselves, however, since their longitudinal components cannot be found in this way (for TEH waves, $E_3 = H_3 = 0$). Eqs. (186) show that for TEH waves, nontrivial solutions exist only when $k^2 = \gamma^2$, i.e.,

$$\gamma = \omega \frac{\sqrt{\varepsilon\mu}}{c}. \tag{188}$$

The components of the TEH wave can be obtained from the two-dimensional Laplace equations which follow from Eq. (183) for $k^2 = \gamma^2$. For E_1 we have

$$\frac{\partial^2 E_1}{\partial x_1^2} + \frac{\partial^2 E_1}{\partial x_2^2} = 0. \tag{189}$$

The same equation can be obtained directly from the field equations (179) and (180), using (185) and (188).

The general relations suggest a number of conclusions regarding the character of wave processes in dielectric waveguides. Let us consider a TE wave, for example. Since $E_3 = 0$, Eqs. (186) take the form

$$\frac{(k^2 - \gamma^2)c}{i\omega\mu} E_1 = \frac{\partial H_3}{\partial x_2};$$

$$\frac{(k^2 - \gamma^2)c}{i\omega\mu} E_2 = -\frac{\partial H_3}{\partial x_1};$$

$$\frac{(k^2 - \gamma^2)}{i\gamma} H_1 = \frac{\partial H_3}{\partial x_1};$$

$$\frac{(k^2 - \gamma^2)}{i\gamma} H_2 = \frac{\partial H_3}{\partial x_2}.$$

These expressions lead to a relationship between the transverse components of the electric and the magnetic field of a TE wave:

$$E_1 = \frac{\omega\mu}{\gamma c} H_2 \quad \text{and} \quad E_2 = -\frac{\omega\mu}{\gamma c} H_1. \tag{190}$$

Similarly for a TH wave,

$$E_1 = \frac{\gamma c}{\omega\varepsilon} H_2 \quad \text{and} \quad E_2 = -\frac{c\gamma}{\omega\varepsilon} H_1. \tag{191}$$

Eqs. (190) and (191) show that the transverse electric and magnetic components of TE and TH waves are mutually perpendicular and proportional in magnitude to one another. At a given frequency, the proportionality is determined by the dielectric properties and the waveguide parameters (the proportionality coefficient contains the propagation constant γ). Since the transverse components are perpendicular, the vectors $\boldsymbol{E}$ and $\boldsymbol{H}$ as such are also perpendicular.

Further analysis of TE and TH waves is impossible without taking into consideration the boundary conditions. In light pipes, we need the boundary conditions at the interface between two dielectrics (the core and the envelope). A detailed analysis of this case will be given later on. At this stage, we will only consider some general properties of waveguides with a conducting surface. This category includes, in particular, cylindrical glass fibers with a metallized surface.

If the waveguide is an ideal conductor with a dielectric filling, the tangential electric field component vanishes on its surface [36]:

$$E_3 = 0. \tag{192}$$

Therefore TE and TH waves exist only when $k^2 - \gamma^2 > 0$. Using (187), we obtain

$$\gamma^2 < \frac{\omega^2}{c^2} \varepsilon\mu, \tag{193}$$

or, alternatively,

$$\gamma^2 = \frac{\omega^2}{v^2} - h^2, \tag{193a}$$

where $v^2 = c^2/\varepsilon\mu$ is the square of the propagation velocity of the electromagnetic wave in the dielectric filling of the waveguide; $h^2 = k^2 - \gamma^2$ is a real number.

Differential equations of the form (183) have nontrivial solutions only for certain real positive values of the parameter $(k^2 - \gamma^2)$, which constitute an infinitive discrete sequence of numbers $h_1, h_2, h_3, \ldots$. If h_1 is the least of these allowed values, we see from (193) that for frequencies ω below some critical value

$$\omega_0 = vh_1, \tag{194}$$

the propagation constant γ is imaginary, since $\gamma^2 < 0$. In this case, we are dealing with wave processes which are damped in the direction of propagation x_3. The existence of the critical frequency ω_0 (termed the *cutoff frequency*), below which only damped waves propagate, is one of the most significant characteristics of TE and TH waves.

Using (194), we write the propagation constant in the form

$$\gamma^2 = \frac{\omega^2}{v^2} - \frac{\omega_0^2}{v^2} = \frac{\omega^2}{v^2}\left(1 - \frac{\omega_0^2}{\omega^2}\right). \tag{195}$$

To obtain the time and position dependence of the TE and TH waves in the direction x_3, we introduce the factor

$$\exp\left[i(\gamma x_3 - \omega t)\right] = \exp\left[i\left(\frac{x_3\sqrt{\omega^2 - \omega_0^2}}{v} - \omega t\right)\right] =$$

$$= \exp\left[i\,\frac{2\pi}{\lambda}(x_3 - v_{\text{ph}}\,t)\right],$$

where

$$\lambda = \frac{2\pi v}{\sqrt{\omega^2 - \omega_0^2}} \tag{196}$$

is the wavelength, and

$$v_{\text{ph}} = \frac{v}{\sqrt{1 - \omega_0^2/\omega^2}} = \frac{c\omega}{\sqrt{\varepsilon\mu(\omega^2 - \omega_0^2)}} \tag{197}$$

is the phase velocity of the wave. We see from (197) that the phase velocity v_{ph} of TE and TH waves is higher than the velocity v of energy transfer in the dielectric medium. The dependence of v_{ph} on frequency points to the existence of dispersion: the wave propagating through the waveguide is not localized within a bounded volume, gradually spreading to occupy a progressively larger part of space.

Consider a TEH wave whose components are obtained from the solution of the two-dimensional equation (189). Inserting the propagation constant γ from (188) into exp $(i\gamma x_3)$ and multiplying by exp $(-i\omega t)$, we obtain a factor which describes the dependence of the amplitude of a TEH wave in a dielectric waveguide on time and positions along the x_3 axis:

$$\exp\left[i(\gamma x_3 - \omega t)\right] = \exp\left[i\,\frac{\omega\sqrt{\varepsilon\mu}}{c}\left(x_3 - \frac{c}{\sqrt{\varepsilon\mu}}t\right)\right] =$$

$$= \exp\left[ik(x_3 - v_{\text{ph}}t)\right]. \tag{198}$$

Unlike the phase velocity of TE and TH waves, the phase velocity of a TEH wave is equal to the velocity of its propagation in the dielectric filling. Since v_{ph} is independent of frequency, the TEH wave remains undispersed.

The transverse components of the TEH wave are linked by relations of the form

$$\sqrt{\varepsilon}\,E_1 = \sqrt{\mu}\,H_2 \quad \text{and} \quad \sqrt{\varepsilon}\,E_2 = -\sqrt{\mu}\,H_1. \tag{199}$$

They show that the electric and the magnetic vectors are perpendicular to one another and the magnitudes of these vectors are proportional, with a proportionality coefficient

$$\beta = \sqrt{\mu/\varepsilon}\,,$$

which depends neither on wave frequency nor on waveguide parameters, and is entirely determined by the dielectric properties.

TEH waves have no cutoff frequency, i.e., fields of this kind may exist at very large wavelengths (and even under d.c. conditions). Nevertheless, the modern theory of waveguides, concerned mainly with metallic waveguides, largely ignores these field modes [2, 42], because *TEH* waves do not propagate in a waveguide with conducting walls. Indeed, Eq. (189) shows that the field components of this wave are harmonic functions of x_1 and x_2 in any plane x_3 = const, i.e., at any time the field configuration coincides with the configuration of static electric and magnetic fields produced by the corresponding boundary conditions. Therefore, these field components can be written as derivatives of some potential with respect to the coordinates x_1 and x_2. The potential is also a harmonic function. Seeing that the potential is constant over the conducting waveguide surface, we conclude that it is constant inside the waveguide too. The electric and the magnetic field components, obtained by differentiation of this potential, therefore identically vanish inside the waveguide and there is no field.

2. Wave propagation in a dielectric waveguide

A fiber waveguide comprises a core of dielectric constant ε_1 encased in an envelope of lower dielectric constant ε_2. Both dielectrics are assumed to be perfect insulators with certain magnetic permeabilities. For simplicity, the envelope will be regarded as infinitely thick. This assumption is quite justified under normal circumstances.

Let us consider the waveguide effects in dielectric fibers in the shape of a circular cylinder of radius r_0. The field inside the waveguide can be computed for an arbitrary cross section [42]. The main distinction of dielectric waveguides, as compared to metallic waveguides, is that they are capable of sustaining mixed modes, in which both $\boldsymbol{E}$ and $\boldsymbol{H}$ have nonzero longitudinal components. The actual number of modes in these waveguides is generally higher than the number of *TE* and *TH* waves.

The problem will be considered in cylindrical coordinates r, ϕ, and z, with the axis Oz directed along the fiber axis.

The longitudinal components of the vectors $\boldsymbol{E}$ and $\boldsymbol{H}$ can be obtained as solutions of the Helmholtz equation for E_z and H_z. Eqs. (183) give for the field inside the fiber

$$\frac{1}{r}\frac{\partial}{\partial r}\left(r\frac{\partial E_z^{(1)}}{\partial r}\right)+\frac{1}{r^2}\frac{\partial^2 E_z^{(1)}}{\partial \phi^2}+(k_1^2-\gamma_1^2)\,E_z^{(1)}=0, \tag{200}$$

$$\frac{1}{r}\frac{\partial}{\partial r}\left(r\frac{\partial H_z^{(1)}}{\partial r}\right)+\frac{1}{r^2}\frac{\partial^2 H_z^{(1)}}{\partial\phi^2}+(k_1^2-\gamma_1^2)\,H_z^{(1)}=0. \tag{201}$$

For the field in the surrounding medium

$$\frac{1}{r}\frac{\partial}{\partial r}\left(r\frac{dE_z^{(2)}}{\partial r}\right)+\frac{1}{r^2}\frac{\partial^2 E_z^{(2)}}{\partial\phi^2}+(k_2^2-\gamma_2^2)\,E_z^{(2)}=0, \tag{202}$$

$$\frac{1}{r}\frac{\partial}{\partial r}\left(r\frac{\partial H_z^{(2)}}{\partial r}\right)+\frac{1}{r^2}\frac{\partial^2 H_z^{(2)}}{\partial\phi^2}+(k_2^2-\gamma_2^2)\,H_z^{(2)}=0, \tag{203}$$

where

$$k_1^2=\frac{\omega^2}{c^2}\varepsilon_1\mu \quad\text{and}\quad k_2^2=\frac{\omega^2}{c^2}\varepsilon_2\mu,$$

the subscript 1 corresponds to the waveguide, and the subscript 2 to the surrounding medium. As before, the transverse components can be found by simple differentiation. The corresponding equations in cylindrical coordinates are written as

$$(k_1^2-\gamma_1^2)\,E_r^{(1)}=i\gamma_1\frac{\partial E_z^{(1)}}{\partial r}+\frac{i\omega\mu}{c}\frac{1}{r}\frac{\partial H_z^{(1)}}{\partial\phi}; \tag{204}$$

$$(k_1^2-\gamma_1^2)\,E_\phi^{(1)}=i\gamma_1\frac{1}{r}\frac{\partial E_z^{(1)}}{\partial\phi}-\frac{i\omega\mu}{c}\frac{\partial H_z^{(1)}}{\partial r}; \tag{205}$$

$$(k_1^2-\gamma_1^2)\,H_r^{(1)}=i\gamma_1\frac{\partial H_z^{(1)}}{\partial r}-\frac{i\omega\varepsilon_1}{c}\frac{1}{r}\frac{dE_z^{(1)}}{\partial\phi}; \tag{206}$$

$$(k^2-\gamma_1^2)\,H_\phi^{(1)}=i\gamma_1\frac{1}{r}\frac{\partial H_z^{(1)}}{\partial\phi}+\frac{i\omega\varepsilon_1}{c}\frac{\partial E_z^{(1)}}{\partial r}. \tag{207}$$

Similar equations hold true for the transverse field components outside the cylinder ($r > r_0$). The corresponding equations are obtained by changing from subscript 1 to 2.

The solutions of Eqs. (200)–(203) give the dependence of the longitudinal components E_z and H_z on the coordinates r and ϕ, just as Eqs. (183) specified the dependence of the field components on the coordinates x_1 and x_2. The dependence of E_z and H_z on the coordinate z is established by equations of the form (184), i.e.,

$$\frac{\partial^2 E_z^{(1)}}{\partial z^2}+\gamma_1^2E_z^{(1)}=0,\qquad \frac{\partial^2 E_z^{(2)}}{\partial z^2}+\gamma_2^2E_z^{(2)}=0 \tag{208}$$

and

$$\frac{\partial^2 H_z^{(1)}}{\partial z^2}+\gamma_1^2H_z^{(1)}=0,\qquad \frac{\partial^2 H_z^{(2)}}{\partial z^2}+\gamma_2^2H_z^{(2)}=0, \tag{209}$$

whose solutions are exponential functions of the form exp $(i\gamma_1 z)$ and exp $(i\gamma_2 z)$.

Particular solutions of Eqs. (200)–(203) and (208)–(209) incorporating the dependence on the coordinate z can be written in the form

$$
\begin{aligned}
E_z^{(1)} &= A_n Z_n(h_1 r)\, e^{i\gamma_1 z} \cos(n\phi + \xi_n^{(1)});\\
H_z^{(1)} &= B_n Z_n(h_1 r)\, e^{i\gamma_1 z} \cos(n\phi + \zeta_n^{(1)});\\
E_z^{(2)} &= C_n Z_n(h_2' r)\, e^{i\gamma_2 z} \cos(n\phi + \xi_n^{(2)});\\
H_z^{(2)} &= D_n Z_n(h' r)\, e^{i\gamma_2 z} \cos(n\phi + \zeta_n^{(2)});\\
n &= 0, 1, 2, \ldots,
\end{aligned} \tag{210}
$$

where

$$h_1^2 = k_1^2 - \gamma_1^2, \quad h_2^2 = k_2^2 - \gamma_2^2; \tag{211}$$

A_n, B_n, C_n, D_n are some constant coefficients; $\xi_n^{(1)}$, $\xi_n^{(2)}$, $\zeta_n^{(1)}$ and $\zeta_n^{(2)}$ are phase constants; $Z_n(h_1 r)$ and $Z_n(h_2' r)$ are cylindrical functions. The other notation as before.

Since the field inside the waveguide should remain finite, $Z_n(h_1 r)$ is identified with Bessel's function $J(h_1 r)$. In the surrounding medium, the field should fall off rapidly with distance from the waveguide axis. The required radial dependence can be obtained using for $Z_n(h_2' r)$ Hankel's modified function of the first kind $K_n(h_2 r)$ which is real-valued for all real values of the argument h_2 ($ih_2 = h_2'$) (see Appendix 1), i.e.,

$$h_2^2 = \gamma_2^2 - k_2^2. \tag{212}$$

Before writing Eq. (210) for the field components in the final form, we will establish some simplifying relations between the components entering these equations. We recall that the tangential field components at the interface between two dielectrics are constant,

$$E_z^{(1)}\big|_{r=r_0} = E_z^{(2)}\big|_{r=r_0}, \quad E_\phi^{(1)}\big|_{r=r_0} = E_\phi^{(2)}\big|_{r=r_0}, \tag{213}$$

$$H_z^{(1)}\big|_{r=r_0} = H_z^{(2)}\big|_{r=r_0}, \quad H_\phi^{(1)}\big|_{r=r_0} = H_\phi^{(2)}\big|_{r=r_0}. \tag{214}$$

Moreover, the dependence of the electric and the magnetic field components on the coordinate z is expressed by exponential factors of the form $\exp(i\gamma_1 z)$ and $\exp(i\gamma_2 z)$, whence it follows that $\gamma_1 = \gamma_2$, i.e., the subscripts of the propagation constant can be dropped.

The same boundary condition leads to two further relations between the phase constants,

$$\cos(n\phi + \xi_n^{(1)}) = \cos(n\phi + \xi_n^{(2)})$$

and

$$\cos(n\phi + \zeta_n^{(1)}) = \cos(n\phi + \zeta_n^{(2)}).$$

Hence,

$$\xi_n^{(1)} = \xi_n^{(2)} = \xi_n \quad \text{and} \quad \zeta_n^{(1)} = \zeta_n^{(2)} = \zeta_n. \tag{215}$$

Eqs. (210) for the z components of the field inside the waveguide and in the encasing dielectric, incorporating the time dependence, are thus written in the form

$$E_z^{(1)} = A_n J_n(h_1 r) \cos(n\phi + \xi_n) \exp[i(\gamma z - \omega t)], \tag{216}$$

$$H_z^{(1)} = B_n J_n(h_1 r) \cos(n\phi + \zeta_n) \exp[i(\gamma z - \omega t)], \tag{217}$$

$$E_z^{(2)} = C_n K_n(h_2 r) \cos(n\phi + \xi_n) \exp[i(\gamma z - \omega t)], \tag{218}$$

$$H_z^{(2)} = D_n K_n(h_2 r) \cos(n\phi + \zeta_n) \exp[i(\gamma z - \phi t)]. \tag{219}$$

Using Eqs. (204)–(207), we can write the expressions for the transverse components. Inside the fiber ($r < r_0$), we have

$$E_r^{(1)} = \left[A_n \frac{i\gamma_1}{h_1} J'_n(h_1 r) \cos(n\phi + \xi_n) - B_n \frac{i\omega\mu n}{ch_1^2 r} J_n(h_1 r) \sin(n\phi + \zeta_n) \right] F, \tag{220}$$

$$E_\phi^{(1)} = -\left[A_n \frac{in\gamma}{h_1^2 r} J_n(h_1 r) \sin(n\phi + \xi_n) + \right.$$
$$\left. + B_n \frac{i\mu\omega}{ch_1} J'_n(h_1 r) \cos(n\phi + \zeta_n) \right] F, \tag{221}$$

$$H_r^{(1)} = \left[A_n \frac{ink_1^2 c}{h_1^2 r\mu\omega} J_n(h_1 r) \sin(n\phi + \xi_n) + \right.$$
$$\left. + B_n \frac{i\gamma}{h_1} J'_n(h_1 r) \cos(n\phi + \zeta_n) \right] F, \tag{222}$$

$$H_\phi^{(1)} = \left[A_n \frac{ik_1 c}{\mu\omega h_1} J'_n(h_1 r) \cos(n\phi + \xi_n) - \right.$$
$$\left. - B_n \frac{i\gamma n}{h_1^2 r} J_n(h_1 r) \sin(n\phi + \zeta_n) \right] F, \tag{223}$$

where $F = \exp[i(yz - \omega t)]$.

At any point of the dielectric casing ($r > r_0$), the field components are expressed by similar expressions where the components A_n and B_n have been replaced with C_n and D_n, respectively, and Bessel's function $J_n(h_1 r)$ is changed to Hankel's function $K_n(h_2 r)$ which goes to zero at infinity. The ϕ-component of the magnetic vector, for example, is written in the form

$$H_\phi^{(2)} = \left[C_n \frac{ik_2^2 c}{\mu\omega h_2} K'_n(h_2 r) \cos(n\phi + \xi_n) - \right.$$
$$\left. - D_n \frac{i\gamma n}{h_2^2 r} K_n(h_2 r) \sin(n\phi + \zeta_n) \right] F. \tag{224}$$

The coefficients A_n, B_n, C_n, D_n and the propagation constant γ are unknown at this stage. A relationship between these constants can be established using the boundary conditions (213) and (214).

The continuity of the tangential components of $\boldsymbol{E}$ leads to the following equalities:

$$A_n J_n(s_1) = C_n K_n(s_2), \tag{225}$$

$$A_n \frac{n\gamma}{s_1^2} J_n(s_1) \sin(n\phi + \xi_n) + B_n \frac{\mu\omega}{cs_1} J_n'(s_1) \cos(n\phi + \zeta_n) =$$

$$= -C_n \frac{n\gamma}{s_2^2} K_n(s_2) \sin(n\phi + \xi_n) - D_n \frac{\mu\omega}{cs_2} K_n'(s_2) \cos(n\phi + \zeta_n). \tag{226}$$

Similarly, for the components of the vector $\boldsymbol{H}$,

$$B_n J_n(s_1) = D_n K_n(s_2), \tag{227}$$

$$A_n \frac{k_1^2 c}{\mu\omega s_1} J_n'(s_1) \cos(n\phi + \xi_n) - B_n \frac{n\gamma}{s_1^2} J_n(s_1) \sin(n\phi + \zeta_n) =$$

$$= C_n \frac{k_2^2 c}{\mu\omega s_2} K_n'(s_2) \cos(n\phi + \xi_n) - D_n \frac{n\gamma}{s_2^2} K_n(s_2) \sin(n\phi + \zeta_n), \tag{228}$$

where $s_1 = h_1 r_0$, $s_2 = h_2 r_0$.

Eqs. (225)–(228) constitute a system of homogeneous linear equations for the coefficients A_n, B_n, C_n, and D_n. The system has nontrivial solutions only when its determinant vanishes. Expanding the determinant and writing

$$\eta_1 = \frac{J_n'(s_1)}{s_1 J_n(s_1)} \quad \text{and} \quad \eta_2 = \frac{K_n'(s_2)}{s_2 K_n(s_2)},$$

we obtain

$$\frac{(\eta_1 + \eta_2)(k_1^2\eta_1 + k_2^2\eta_2)}{n^2\gamma^2 (1/s_1^2 + 1/s_2^2)^2} = -\frac{\sin(n\phi + \xi_n)\sin(n\phi + \zeta_n)}{\cos(n\phi + \xi_n)\cos(n\phi + \zeta_n)}. \tag{229}$$

The left-hand side of (229) is independent of ϕ, and therefore the phase constants ξ_n and ζ_n entering the right-hand side of the equation should be such that the entire expression in the right-hand side of (229) is likewise independent of ϕ. This is so when

$$\zeta_n - \xi_n = \pm \tfrac{1}{2}\pi.$$

Then

$$\frac{(\eta_1 + \eta_2)(k_1^2\eta_1 + k_2^2\eta_2)}{n^2\gamma^2 (1/s_1^2 + 1/s_2^2)^2} = -\frac{\sin(n\phi + \xi_n)\sin(n\phi \pm \frac{1}{2}\pi + \xi_n)}{\cos(n\phi + \xi_n)\cos(n\phi \pm \frac{1}{2}\pi + \xi_n)} = 1,$$

i.e.,

$$(\eta_1 + \eta_2)(k_1^2\eta_1 + k_2^2\eta_2) = n^2\gamma^2 (1/s_1^2 + 1/s_2^2)^2. \tag{230}$$

The roots of this transcendental equation, together with (211), define the allowed values of the propagation constant γ, i.e., they fix the characteristic waves which may propagate along the cylindrical fiber.

Since Bessel's functions are transcendental and have an infinity of roots for any n (each of these roots can be assigned an index m), Eq. (220) also has an infinity of discrete roots γ_{nm}. Eq. (230) is quadratic in γ, and its roots therefore form a two-dimensional infinite manifold whose physical meaning will be established in what follows.

No general method can be proposed for solving Eq. (230), and yet in every particular case a careful analysis of the boundary conditions leads to certain general conclusions which greatly simplify the solution of the problem. In dielectric waveguides, to satisfy the boundary conditions, we always have to consider superpositions of electric and magnetic waves, except in the symmetrical case $n = 0$. Indeed, let us show that a dielectric waveguide will not propagate TH waves with $n \neq 0$. From the boundary condition (213) for the component E_ϕ, using Eq. (205) and the condition $H_z = 0$, we write

$$\frac{1}{h_1^2}\left.\frac{\partial E_z^{(1)}}{\partial \phi}\right|_{r=r_0} = \frac{1}{h_2'^2}\left.\frac{\partial E_z^{(2)}}{\partial \phi}\right|_{r=r_0}. \tag{231}$$

For $n = 0$, i.e., for TH waves whose amplitude depends on the coordinate r alone, Eq. (231) is automatically satisfied. If, however, $n \neq 0$, comparison of Eqs. (211) and (212) taking into consideration the particular form of the components $E_z^{(1)}$ and $E_z^{(2)}$ gives

$$\frac{1}{h_1^2} E_z^{(1)} = -\frac{1}{h_2^2} E_z^{(2)}.$$

Since (213) is satisfied, we obtain

$$h_1^2 = -h_2^2.$$

This equality, in conjunction with Eq. (187), is equivalent to the assertion

$$\varepsilon_1 = \varepsilon_2 \quad \text{for} \quad n \neq 0.$$

We have thus demonstrated in a general way that a waveguide of dielectric constant ε_1 encased in an envelope of a *different* dielectric constant ε_2 will propagate transverse magnetic waves in the axisymmetric case $n = 0$ only. TH waves with a field distribution which depends on the azimuthal angle ϕ do not exist in such waveguides.

It is similarly shown that the transverse electric field is independent of the angle ϕ, and is therefore sustained in the axisymmetric case $n = 0$ only. In other cases, the field is not transverse in relation to the electric or the magnetic vector. The field configuration is such that $E_z \neq 0$ and $H_z \neq 0$ simultaneously.

Note that if only the symmetric mode ($n = 0$) is excited and the field is independent of ϕ, the coefficients A_n, C_n and B_n, D_n are independent of one another. Indeed, Eqs. (225)–(228) show that in this case the transverse magnetic wave ($H_z = 0$) satisfies the boundary conditions

$$A_0 J_0(s_1) = C_0 K_0(s_2),$$

$$A_0 \frac{k_1^2}{s_1} J_0'(s_1) = C_0 \frac{k_2^2}{s_2} K_0'(s_2), \tag{232}$$

whereas for the transverse electric wave ($E_z = 0$)

$$B_0 \frac{1}{s_1} J_0'(s_1) = - D_0 \frac{1}{s_2} K_0'(s_2),$$

$$B_0 J_0(s_1) = D_0 K_0(s_2). \tag{233}$$

Eqs. (232) and (233) constitute two independent systems of linear homogeneous equations. Setting the respective determinants equal to zero and using the previous notation for η_1 and η_2, we find that the allowed transverse magnetic modes are defined by the roots γ_{0m} of the transcendental equation

$$k_1^2 \eta_2 = k_2^2 \eta_2, \tag{234}$$

and the transverse electric modes are obtained from the equation

$$\eta_1 = - \eta_2. \tag{235}$$

Let us return to the general case described by Eqs. (232). For $n \neq 0$, the waveguide sustains "hybrid" modes which are neither transverse electric nor transverse magnetic. These modes are generally divided into two types, depending on the relative contribution of E_z and H_z into the longitudinal field component. If E_z makes the major contribution, the result is an EH_{nm} mode; otherwise we have an HE_{nm} mode.

The field equations (216)–(223) can be written in a more compact form [149] using a new parameter

$$P = - \frac{\mu\omega}{h} \frac{B_n \cos(n\phi + \zeta_n)}{A_n \sin(n\phi + \xi_n)} = \frac{n(1/s_1^2 + 1/s_2^2)}{\eta_1 + \eta_2}. \tag{236}$$

P is proportional to the ratio between the longitudinal components of the magnetic and the electric vectors (ignoring the phase relation in ϕ). Using this parameter, we write

$$E_z^{(1)} = J_n(h_1 r) F_{nc},$$

$$E_r^{(1)} = \frac{i\gamma}{h_1}\left[J_n'(h_1 r) - \frac{n}{h_1 r} P J_n(h_1 r) \right] F_{nc}, \tag{237}$$

$$E_{\phi}^{(1)} = \frac{i\gamma}{h_1}\left[PJ'_n(h_1 r) - \frac{n}{h_1 r}J_n(h_1 r)\right]F_{ns},$$

$$H_z^{(1)} = -\frac{\gamma}{\mu\omega}PJ_n(h_1 r)\,F_{ns},$$

$$H_r^{(1)} = -\frac{ik_1^2 c}{\mu\omega h_1}\left[\frac{\gamma^2}{k_1^2}PJ'_n(h_1 r) - \frac{n}{h_1 r}J_n(h_1 r)\right]F_{ns},$$

$$H_{\phi}^{(1)} = \frac{ik_1^2 c}{\mu\omega h_1}\left[J'_n(h_1 r) - \frac{\gamma^2 n}{k_1^2 h_1 r}PJ_n(h_1 r)\right]F_{nc},$$

where

$$F_{nc} = A_n \cos(n\phi + \xi_n)\exp[i(\gamma z - \omega t)],$$

$$F_{ns} = A_n \sin(n\phi + \xi_n)\exp[i(\gamma z - \omega t)].$$

Thus the two-dimensional infinite manifold of solutions of Eq. (232) is a mathematical reflection of the reflection of the existence of two different types of solutions in dielectric waveguides, corresponding to *HE* and *EH* modes. In the symmetrical case $n = 0$, when the waves are transversal, the coefficient P is naturally equal to 0 (a *TH* wave) or 1 (a *TE* wave) [139].

3. Qualitative treatment of the mode characteristics in dielectric waveguides

Using simple physical and geometrical considerations, we will establish and analyze the principal properties of dielectric waveguides.

Undamped modes in a waveguide are sustained for real values of h_1 and h_2. Therefore, Eqs. (211) and (212) lead to the inequalities

$$k_1^2 \leqq \gamma^2 \leqq k_2^2. \tag{238}$$

Since

$$v_{\text{ph}} = \frac{\omega}{\gamma} = \frac{c}{\sqrt{\varepsilon\mu}} = \frac{c}{n},$$

inequalities (238) can be written in the form

$$\frac{c}{n_1} \leqq v_{\text{ph}} \leqq \frac{c}{n_2}. \tag{239}$$

This signifies that the phase velocity of the mode in a dielectric waveguide lies between the phase velocities in the core and the casing. The phase velocity is thus limited from above by the casing value

$$v_{ph} = \frac{c}{n_2}, \tag{240}$$

which corresponds to the case $\gamma = k_2$.

Eq. (240) defines the critical (cutoff) wavelength of the mode. In the geometrical optics of light pipes, Eq. (240) corresponds to the existence of the critical angle of total internal reflection.

Consider a plane wave with a normal vector $\boldsymbol{p}$ incident on the surface of the fiber at a certain angle α (Figure 52). Let Ω_1 and Ω_2 be two phase surfaces separated by λ/n_1 (this is the wavelength in the dielectric medium, λ being the wavelength in vacuum). The projection of λ onto the fiber axis is

$$\lambda_g = \frac{\lambda}{n_1 \sin \alpha}. \tag{241}$$

If α is equal to the critical angle, we may write $\lambda_g = \lambda/n_2$ (seeing that $n_1 \sin \alpha_{cr} = n_2$), and this coincides with (240). Thus the treatment of the optical properties of thin light pipes within the framework of wave optics recovers the basic results of geometrical optics, in addition to yielding a whole range of qualitatively new effects.

Interesting physical conclusions regarding the properties of thin fibers may be reached using a simple nomogram (Figure 53) [143]. The horizontal and vertical axes give c/λ and $1/\lambda_g$, respectively. The mode lines are schematically shown by curves asymptotically approaching the straight line with a slope factor of c/n_1 on one side and meeting the line with a slope factor of c/n_2 on the other. Since in the particular coordinates of the nomogram the slope of any line corresponds to a certain phase velocity $v_{ph} = v/\lambda_g$, the restriction of the mode curves between the lines c/n_1 and c/n_2 follows from (239). The intersection point of a mode curve with the line c/n_2 corresponds to the critical condition $v_{ph} = c/n_2$ or, equivalently, $\lambda_g = \lambda/n_2$. The only exceptional mode is HE_{11}. It has no cutoff frequency, and the corresponding curve in Figure 53 therefore does not meet the line c/n_2, asymptotically approaching this line as the frequency decreases.*

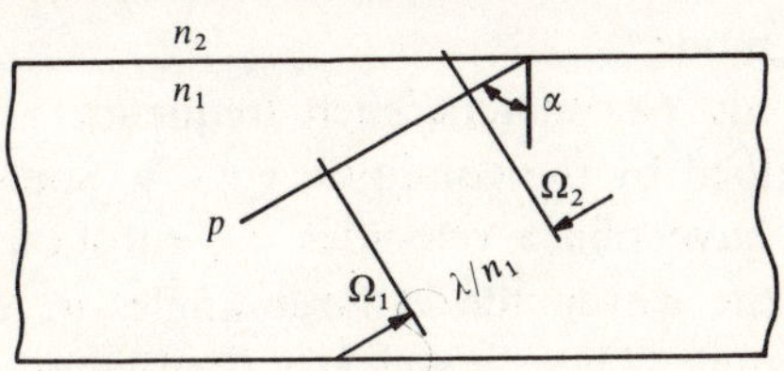

Figure 52

Propagation of a plane wave in a fiber.

* All the modes in Figure 53 are highly schematic. The actual curves have been calculated so far for two modes only, HE_{11} and TE_{01} [48, 140].

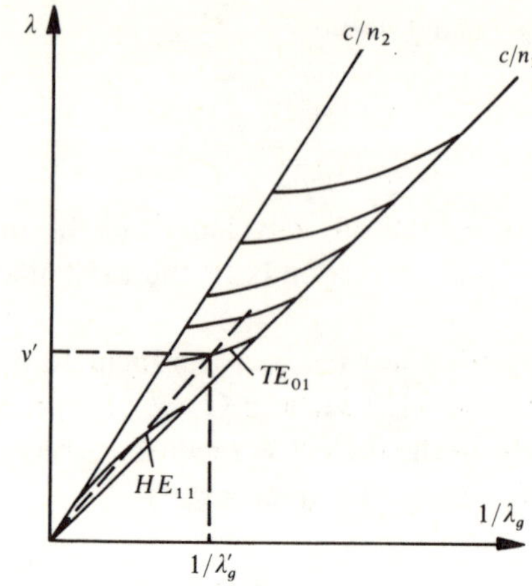

Figure 53

Schematic diagram showing c/λ as a function of $1/\lambda_g$ for modes with waveguide parameters n_1 and n_2.

As an example, let us consider the TE_{01} mode excited at a frequency ν' (Figure 53). The corresponding value of $1/\lambda'_g$ is readily obtained as the abscissa of the intersection point of the mode curve with the horizontal line at ν'. The slope of the line drawn from the origin to this intersection point is c/n'. It characterizes the phase velocity v'_{ph}, whereas the group velocity v'_{gr} corresponds to the slope factor of the tangent at that point.

For any mode (Figure 53) propagating along the waveguide axis, $v_{\text{ph}} = v_{\text{gr}} = c/n_1$. The figure also leads to a number of interesting qualitative conclusions.

1. For a fixed $1/\lambda_g$, the waveguide sustains a fixed number of modes (which, in general, depends on waveguide parameters) with different frequencies $\nu_1, \nu_2, \nu_3, \ldots, \nu_n$ and different phase and group velocities $v_{\text{ph}\,1}, v_{\text{ph}\,2}, v_{\text{ph}\,3}, \ldots, v_{\text{ph}\,n}$ and $v_{\text{gr}\,1}, v_{\text{gr}\,2}, \ldots, v_{\text{gr}\,n}$. Since $v = c/\lambda$, Eq. (241) leads to a discrete sequence of angles α, each corresponding to one mode only. The existence of these discrete angles is a natural consequence of the discreteness of the propagation constant γ, each of its values corresponding to a certain mode.

2. For given waveguide parameters, each frequency ν corresponds to a fixed number of modes described by the condition $\nu_{\text{cr}} < \nu$. Some of the modes, with ν_{cr} sufficiently close to ν, have phase velocities v_{ph} substantially greater than c/n_1. They propagate along the waveguide at large angles θ', which nevertheless does not exceed the critical angle. These are clearly the highest order modes among all those sustained in the given waveguide. There are always some modes with lower critical frequencies whose phase velocities are very close to c/n_1. These modes should propagate at small angles to the waveguide axis.

Figure 53 thus shows that at a fixed frequency ν, the mode with a high critical frequency will propagate at a large angle to the waveguide axis. This mode corresponds to a small value of $1/\lambda_g$, and hence to a small angle of incidence on the waveguide surface (see Eq. (241)).

3. The dependence of the number of modes and of their angular distribution on the refractive indices of the core and the casing is a topic of special interest.

If the refractive index n_1 of the core is given, the slope of the line c/n_1 (Figure 54) is known. Let us analyze the changes in the mode distribution over the wavelength λ_1 as a function of the refractive index n_2 of the casing. For a fixed $n_2 = n_2'$ (Figure 54) and some frequency ν', the waveguide sustains a combination of modes which are determined by the intersection points of the horizontal line ν' with the mode curves inside the sector between the lines c/n_2 and c/n_1. The mode tentatively identified as $(p + 2)$ makes the largest angle with the axis. The phase velocity of this mode is determined by $\tan \gamma_{p+2}$. Modes of lower orders $p + 1$, p, $p - 1, \ldots$ propagate at smaller angles and have lower phase velocities. Since the mode curves asymptotically approach the line c/n_1, only the modes of lowest order (i.e., those with the lowest indices) can be regarded as axial.

The slope of the line c/n_2 increases as the casing refractive index n_2 decreases. As a result, for some $n_2'' < n_2'$, a new mode $(p + 3)$ may be sustained by the waveguide, apart from the previous modes, whose critical frequency will be $\nu_{p+3} > \nu_{p+2}$.

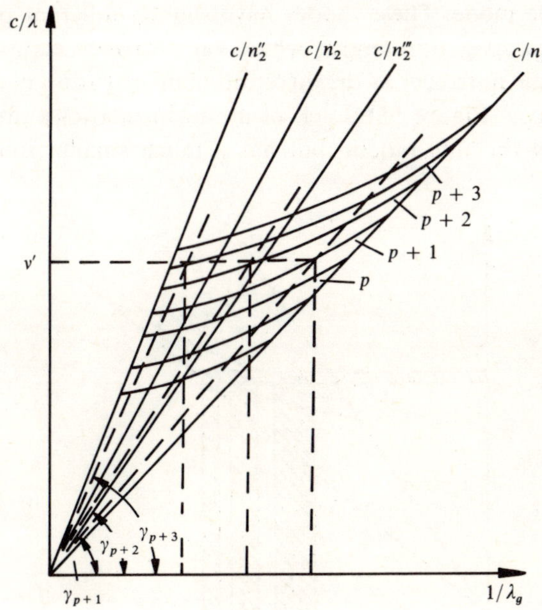

Figure 54

Curves showing the number of modes and their angular distribution for various refractive indices of the core and the casing.

This mode has the highest phase velocity $v_{ph} = \tan \gamma_{p+3}$ and correspondingly makes the largest angle θ'_{p+3} with the waveguide axis.

In geometrical optics, this multiplication of side modes corresponds to the increase in the aperture of the light cone transferred by the fiber following the increase in the critical angle for total internal reflection with a decrease in the casing refractive index.

Conversely, as n_2 is increased (Figure 54), the total number of modes diminishes. Thus, if $n_2 = n_2'''$, the modes $(p + 3)$ and $(p + 2)$ are missing, since their phase velocities exceed the allowed maximum $v_{ph} = c/n_2$, i.e., they propagate at super critical angles. In the limit $n_2 \to n_1$ the waveguide will sustain only modes propagating at infinitesimal angles to the axis (low-order modes).

4. Let us consider the effect of wavelength on the mode configuration in the waveguide. For fixed waveguide parameters, a decrease in λ leads to an increase in the number of sustained modes (the corresponding ordinate in Figure 54 increases). Since the horizontal lines c/λ_2 now cross higher mode curves, new high-order modes appear. At the same time, the points of intersection with lines corresponding to previously existing modes (Figure 54) are shifted toward the line c/n_1, which indicates that the modes approach the waveguide axis.

5. Another important question is that of wave monochromaticity. Even a small frequency spread is enough to produce a whole combination of modes of one type, instead of a single mode. These modes have slightly different phase velocities and slightly different propagation angles, which vary between certain limits. The limits of variation become narrower as the direction of propagation of the particular mode approaches the axis (Figure 55). Lack of monochromaticity thus causes a definite angular spread of the side modes, but has a much smaller influence on the axial modes.

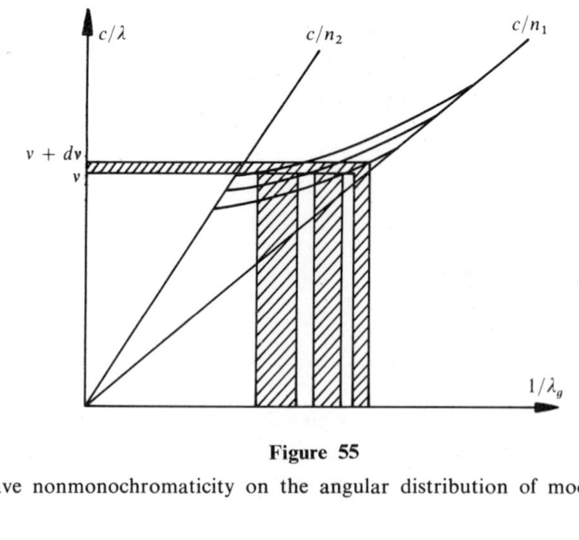

Figure 55

The effect of wave nonmonochromaticity on the angular distribution of modes in the waveguide.

4. Solution of the fundamental waveguide equation

The fundamental waveguide equation (230) cannot be solved in a general way because of its transcendental character. We will therefore proceed with a graphical analysis of this equation and derive solutions for some limiting cases.

1. Let us consider the case $n = 0$, when Eq. (230) may be written in the form

$$(\eta_1 + \eta_2)(k_1^2\eta_1 + k_2^2\eta_2) = 0. \tag{242}$$

Since

$$\eta_1 = \frac{J_n'(s_1)}{s_1 J_n(s_1)} \quad \text{and} \quad \eta_2 = \frac{K_n'(s_2)}{s_2 K_n(s_2)},$$

one of the factors in (242) must vanish. Both factors cannot vanish simultaneously, since $k_1 \neq k_2$. Thus, for $n = 0$, the allowed values of the propagation constant γ are determined by one of the following two equations:

$$\eta_1 + \eta_2 = 0, \tag{243}$$

$$k_1^2\eta_1 + k_2^2\eta_2 = 0. \tag{244}$$

The set of homogeneous equations (225)–(228) then breaks up into two independent systems with determinants (243) and (244) which do not vanish simultaneously. Therefore, at least one pair of coefficients A_n, C_n or B_n, D_n vanish. In the former case, $E_z = 0$, and in the latter $H_z = 0$. In other words, for $n = 0$, the waveguide may sustain only TH_{0m} or TE_{0m} waves.

The roots γ of Eqs. (243) and (244) can be found graphically (Figure 56). The values of the parameter u laid off along the horizontal axis are determined by the waveguide parameters. Indeed, we have seen that $k_1^2 \leqq \gamma^2 \leqq k_2^2$. Therefore, the range of values of $s_1 = h_1 r_0$ is given by

$$0 \leqq s_1 \leqq r_0\sqrt{k_1^2 - k_2^2}. \tag{245}$$

The inequality $0 \leqq s_1$ corresponds to axial modes with $\lambda \ll \lambda_{cr}$, and the inequality $s_1 \leqq r_0\sqrt{k_1^2 - k_2^2}$ to modes propagating at a large angle to the axis ($\lambda \simeq \lambda_{cr}$). By (211) and (212), the limits for the parameter $s_2 = h_2 r_0$ are the same as above, with the only difference that the maximum value of s_1 corresponds to $s_2 = 0$ and, conversely, for $s_1 = 0$, $s_2 = r_0\sqrt{k_1^2 - k_2^2}$. As a result, the waveguide is conveniently characterized by a certain parameter R, such that $R^2 = s_1^2 + s_2^2$. After simple manipulations, we find

$$R = \frac{\pi D}{\lambda}(n_1^2 - n_2^2). \tag{246}$$

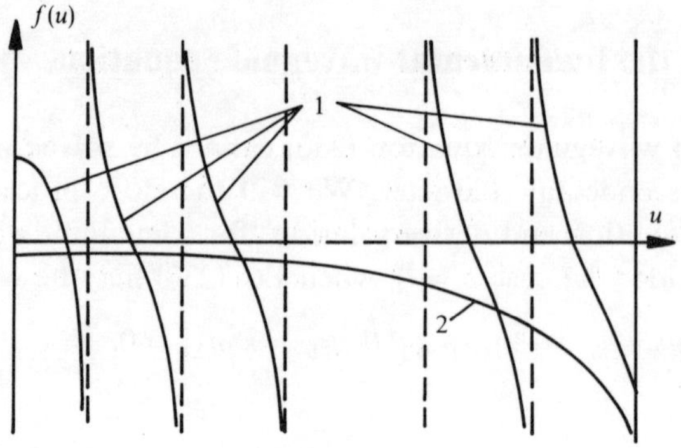

Figure 56

Solution of the fundamental waveguide equation for $n = 0$: curves 1 correspond to the functions η_1, curves 2 to η_2.

Since for sufficiently large s_2 (small s_1), curve 2 in Figure 56 corresponding to the function η_2 can be approximated with the function $1/s_2$, the intersection points of curves 1 and 2 (and thus the solution of Eq. (234)) are determined by the roots of the equation $\eta_1 = 0$. Thus the values of s_1 corresponding to transverse electric and transverse magnetic modes for wavelengths far from the critical value are determined by the roots of Bessel's function $J_n(s_1)$. We see from the figure that the solution $s_1 = 0$ is unfeasible for TE and TH waves, which means that pure axial modes do not exist.

2. Let us consider the case $n \neq 0$. Hybrid HE and EH modes are sustained in this case. To clarify the point, we return to Eq. (232) whose graphical solution is shown in Figure 57. Curves 1 plot the left-hand side of the equation, and curve 2 corresponds to the right-hand side for a fixed n. We see from the general trend of the curves that an asymptotic solution corresponding to an axial mode exists for $s_1 \to 0$.

Since the parameter R characterizes the range of variation of s_1 and s_2, the hybrid modes ($n \neq 0$) clearly exist when R is made infinitesimal (either by reducing the difference $n_1 - n_2$ of the refractive indices or by contracting the core diameter D), when no TE and TH waves are observed.

The two intersection points of curve 2 with each of the V-shaped branches of curve 1 correspond to the existence of two modes with identical indices (HE_{nm} and EH_{nm} modes).

We see from Figure 57 that the number of side modes diminishes as the difference in the refractive indices of the core and the casing becomes smaller. Indeed, for $\varepsilon_1 \to \varepsilon_2$, the range of allowed s_1 values, (245), contracts. In the limit of equal refractive indices, $k_1 = k_2$ and $s_1 = 0$. In this case, the waveguide may only sustain axial modes. This problem was considered in detail by Fox and Li [76] in connection with resonant modes in a maser interferometer.

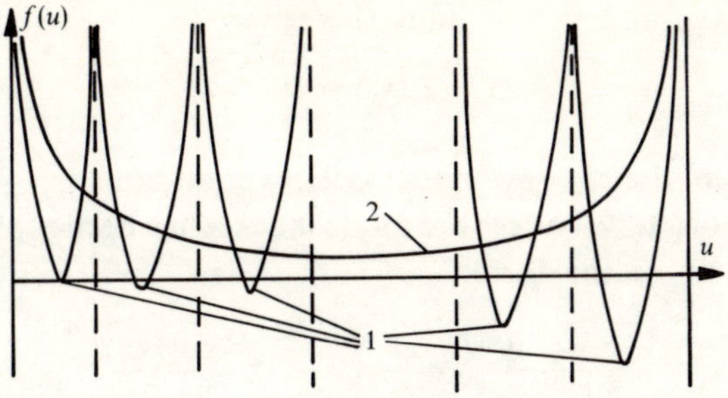

Figure 57

Solution of the waveguide equation for $n \neq 0$: curves 1 correspond to the functions η_1, curves 2 to η_2.

The fundamental waveguide equation also can be treated graphically in the limiting case of wavelengths close to the critical value ($s_2 \to 0$). The solution, however, can be derived by pure analytical methods in this case [149].

To simplify the notation, we write

$$q_1 = \frac{J_{n-1}(s_1)}{s_1 J_n(s_1)}, \quad q_2 = \frac{K_{n-1}(s_2)}{s_2 K_n(s_2)}. \tag{247}$$

Using the properties of cylindrical functions, we rewrite Eq. (232) in the form

$$q_1^2 - q_1\left[\frac{k_1^2 + k_2^2}{k_1^2} q_2 + n\left(\frac{2}{s_1^2} + \frac{k_1^2 + k_2^2}{k_1^2}\frac{1}{s_2^2}\right)\right] + \\ + \left[\frac{k_2^2}{k_1^2} q_2^2 + q_2 n\left(\frac{k_1^2 + k_2^2}{k_1^2}\frac{1}{s_1^2} + 2\frac{k_2^2}{k_1^2}\frac{1}{s_2^2}\right)\right] = 0. \tag{248}$$

Here q_1 is a rapidly varying function of s_1. Nevertheless, wherever s_1 enters the equation explicitly, q_1 may be treated as a constant.

Using the asymptotic behavior of q_1 for $s_2 \to 0$ (see Appendix 1), we obtain for $n \geqq 2$ two solutions of Eq. (248),

$$q_1^{(1)} = \frac{k_2^2}{(n-1)(k_1^2 + k_2^2)} \tag{249}$$

and

$$q_1^{(2)} = \frac{nk_1^2}{s_2^2(k_1^2 + k_2^2)}, \tag{250}$$

which specify the allowed values of the parameter s_1 corresponding to two hybrid modes. Since we are dealing with the case $s_2 \to 0$, the second solution, (250), goes to infinity, $q_1^{(2)} \to \infty$, and in the limit this gives

$$J_n(s_1) = 0. \tag{251}$$

The limit values of the root $q_1^{(1)}$ under critical conditions ($s_2 \to 0$) were derived in [139]. Making use of (A-22) (where the mode index n has been replaced with $(n-1)$), we rewrite Eq. (249) in the form

$$\frac{s_1 J_{n-2}(s_1)}{J_{n-1}(s_1)} = -(n-1)\frac{\varepsilon_1 - \varepsilon_2}{\varepsilon_2}. \tag{252}$$

If $\varepsilon_1 \simeq \varepsilon_2$, Eq. (252) reduces to

$$J_{n-2}(s_1) = 0. \tag{253}$$

Note that for $n = 2$ this equation coincides with the limiting equation for TE and TH waves [149]. The resulting similarities in the behavior of the TE_{0m}, TH_{0m}, and HE_{2m} modes in waveguides will be treated later on.

For $n = 1$, using properties of Hankel's function $K_n(s_2)$ for $s_2 \to 0$, we obtain for the roots of Eq. (248)

$$q_1^{(1)} = \frac{k_1^2 + k_2^2}{k_1^2 s_2^2} \to \infty, \tag{254}$$

$$q_1^{(2)} = \frac{2k_2^2}{k_2^2 + k_2^2} \ln(2/\gamma s_2) \to \infty. \tag{255}$$

These solutions define a parameter s_1 which coincides with the roots of first-order Bessel's function

$$J_1(s_1) = 0. \tag{256}$$

Considering the transverse electric and the transverse magnetic waves ($n = 0$), we conclude that the corresponding critical parameters are determined by the roots of the zero-order Bessel's function

$$J_0(s_1) = 0. \tag{257}$$

The solutions specifying the parameter s_1 and hence the principal characteristic of the mode—the propagation constant γ—for nearly critical wavelengths are summarized in Table 1. This table also lists the corresponding values of the parameter $P = H_z/E_z$. The data for various modes far from the critical conditions (axial modes) are given in Table 2. In the following, these results will be used when analyzing the field configuration corresponding to different modes.

Table 1

SOLUTION OF THE WAVEGUIDE EQUATION FOR NEARLY CRITICAL WAVELENGTHS

n	First solution: form of solution	First solution: P	First solution: mode	Second solution: form of solution	Second solution: P	Second solution: mode
$n = 0$	$J_0(s_1) = 0$	0	$TH_{0m}\ m = 1, 2, \ldots$	$J_0(s_1) = 0$	[illegible]	$TH_{0m}\ m = 1, 2, \ldots$
$n = 1$	$J_1(s_1) = 0$	-1	$HE_{1m}\ m = 1, 2, \ldots$	$J_1(s_1) = 0$	K_1^2/K_2^2	$EH_{1m}\ m = 1, 2, \ldots$
$n \geqq 2$	$\dfrac{s_1 J_{n-2}(s_1)}{J_{n-1}(s_1)} = -(n-1)\dfrac{\varepsilon_1 - \varepsilon_2}{\varepsilon_2}$	-1	$HE_{nm}\ m = 1, 2, \ldots$	$J_n(s_1) = 0$	K_1^2/K_2^2	$EH_{nm}\ m = 1, 2, \ldots$

Table 2

SOLUTION OF THE WAVEGUIDE EQUATION FAR FROM THE CRITICAL WAVELENGTH

n	First solution: form of solution	First solution: P	First solution: mode	Second solution: form of solution	Second solution: P	Second solution: mode
$n = 0$	$J_1(s_1) = 0$	0	$TH_{0m}\ m = 1, 2, \ldots$	$J_1(s_1) = 0$	∞	$TE_{0m}\ m = 1, 2, \ldots$
$n \geqq 1$	$J_{n-1}(s_1) = 0$	-1	$HE_{nm}\ m = 1, 2, \ldots$	$J_{n+1}(s_1) = 0$	$+1$	$EH_{nm}\ m = 1, 2, \ldots$

5. Energy flux along the waveguide. The optical diameter

The energy flux through unit surface area in unit time is expressed by the real part of the complex Poynting vector $\boldsymbol{S} = (c/4\pi)(\boldsymbol{E} \times \boldsymbol{H})$ perpendicular to the surface element. To find the energy propagating along the fiber axis, we should take the z-component of $\boldsymbol{S}$, i.e., use only the transverse components of $\boldsymbol{E}$ and $\boldsymbol{H}$,

$$S_z = \frac{c}{4\pi}\left[E_r H_\phi - E_\phi H_r\right]. \tag{258}$$

The transverse field components can be found from (237), but in order to obtain a clear picture of the radial dependence of S_z, the original expressions should be slightly modified. Using the properties of Bessel's functions, we find

$$E_r = i\frac{\gamma}{h_1}\left[\frac{1-P}{2}J_{n-1}(h_1 r) - \frac{1+P}{2}J_{n+1}(h_1 r)\right]F_{nc}; \tag{259}$$

$$E_\phi = i\frac{\gamma}{h_1}\left[- \frac{1-P}{2} J_{n-1}(h_1 r) - \frac{1+P}{2} J_{n+1}(h_1 r) \right] F_{ns}; \tag{259}$$

$$E_z = J_n(h_1 r)\, F_{nc};$$

$$H_z = - \frac{\gamma}{\mu\omega} P J_n(h_1 r)\, F_{ns};$$

$$H_\phi = i\,\frac{k_1^2 c}{\mu\omega h_1}\left[- \frac{(1 - P\gamma^2)/k_1^2}{2} J_{n-1}(h_1 r) - \frac{(1 + P\gamma^2)/k_1^2}{2} J_{n+1}(h_1 r) \right] F_{nc};$$

$$H_r = - i\,\frac{k_1^2 c}{\mu\omega h_1}\left[- \frac{(1 - P\gamma^2)/k_1^2}{2} J_{n-1}(h_1 r) - \frac{(1 + P\gamma^2)/k_1^2}{2} J_{n+1}(h_1 r) \right] F_{ns}.$$

After simple manipulations this gives

$$S_z = \frac{c^2}{4\pi}\,\frac{\gamma}{\mu\omega}\,\frac{k_1^2}{h_1^2}\,A_n^2\left[\frac{(1-P)(1 - P\gamma^2)/k_1^2}{4} J_{n-1}(h_1 r) + \right.$$

$$+ \frac{(1 - P^2\gamma^2)/k_1^2}{2} J_{n-1}(h_1 r)\, J_{n+1}(h_1 r) \cos 2\,(n\phi + \xi_n) +$$

$$\left. + \frac{(1+P)(1 + P\gamma^2)/k_1^2}{4} J_{n+1}(h_1 r) \right]. \tag{260}$$

Analysis shows [95, 149] that for close values of the refractive indices of the core and the casing $k_1^2/k_2^2 \cong 1$, $\gamma^2/k_1^2 \cong 1$, and $P = \pm 1$. The value $P = +1$ corresponds to the *EH* mode and $P = -1$ to the *HE* mode. In this case, the second term in (260), which describes the angular dependence of S_z, is zero, i.e., the energy flux is independent of ϕ. The radial distribution is written in the form

$$S_z \sim J_{n\pm 1}^2(h_1 r). \tag{261}$$

The minus in the subscript corresponds to the *EH* mode, and the plus to the *EH* mode.

The penetration of energy into the outer envelope of the fiber is described by a modified Hankel function $K_n(h_2 r)$, indicating an approximately exponential decrease of the field amplitude outside the fiber. Its analog is thus a negative exponential in the geometrical optics of plane surfaces.

Figure 58 shows two plane waves incident at a certain angle α_{nm} onto the side surface of a cylindrical fiber. The two waves experience total internal reflection. The wave normals are marked by arrows, whose length and direction determine the Poynting vector.

Superposition of the incident and the reflected waves produces a resultant wave

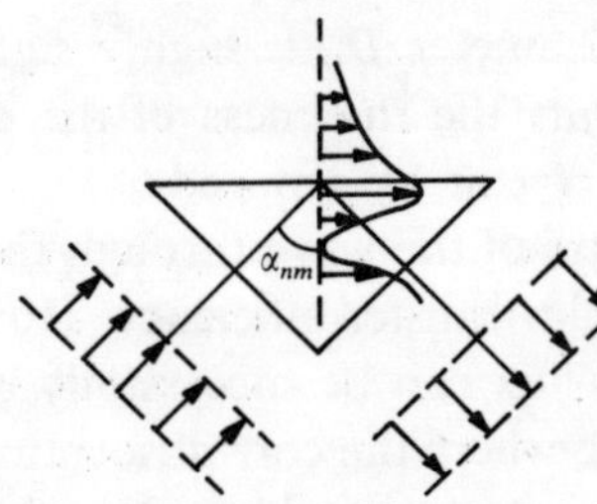

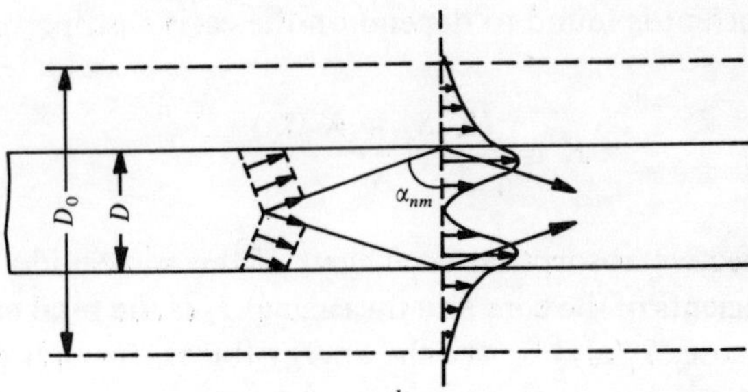

Figure 58

Penetration of light into the outer medium in case of total internal reflection from an interface between two dielectrics:

a plane interface, *b* cylindrical fiber, D_0 is the optical diameter.

whose constant-phase surfaces are normal to the interface and the Poynting vector is parallel to the waveguide axis; the amplitude of the Poynting vector is a sinusoidal function. In the outer medium (the casing), the wave front is also perpendicular to the interface, but the wave intensity falls off exponentially in the radial direction.

Thus as in the case of a plane interface (Figure 58*a*), part of the energy in a cylindrical waveguide (Figure 58*b*) is transferred along the interface in the outer medium. It is significant that almost the entire leakage flux propagates very close to the waveguide surface.

Because of this property of dielectric waveguides, it is advisable to introduce a new parameter which characterizes the depth of field penetration into the outer medium. The *optical diameter* D_0 is generally used as such a parameter [95]. It is defined as the diameter of the circular cross section perpendicular to the waveguide axis through which 99 percent of the total energy of a given mode is transferred. Mathematically, $r_0 = D_0/2$ is defined as follows:

$$\frac{\int_0^{2\pi}\int_0^{r_0} S_z(r,\phi)\, r\, dr\, d\phi}{\int_0^{2\pi}\int_0^{\infty} S_z(r,\phi)\, r\, dr\, d\phi} = 0.99. \tag{262}$$

Knowledge of the optical diameter D_0 is highly significant for the analysis of multifiber waveguides: G_0 limits the thickness of the casing for which the inter-coupling of the adjacent fibers can be ignored.

In most cases, the penetration of the wave through the core–casing interface has a detrimental effect on the light-transfer efficiency. However, when working with thin fibers, the outside energy flux can be successfully utilized for light transfer in the ultraviolet and the infrared, where the core absorption is uncomfortably high.

As the fiber diameter is reduced, the outside energy flux becomes substantial and the absorption coefficient is found to depend on the casing properties. For $n_2 \simeq n_1$, we have [96]

$$K = \frac{(K_1 S_1 + K_2 S_2)}{S_0}, \tag{263}$$

where K is some effective absorption coefficient of the waveguide, K_1 and K_2 are the absorption coefficients of the core and the casing, T_0 is the total energy flux transferred by the waveguide, S_1 and S_2 are the energy fluxes through the core and the casing, respectively:

$$S_1 = \int_0^{2\pi} \int_0^{D/2} S_z r \, dr \, d\phi$$

and

$$S_2 = \int_0^{2\pi} \int_{D/2}^{\infty} S_2 r \, dr \, d\phi.$$

The relationship between S_1 and S_2 depends on the waveguide parameters and the particular mode propagated. For angles α much greater than the critical angle for the particular mode, most of the transferred energy is confined to the core. As α approaches the critical angle, the outside flux may prevail. For $\alpha_{nm} < \alpha_{nm\,\mathrm{cr}}$, the particular mode does not propagate inside the fiber.

As we have noted above, the number of modes propagating in the waveguide depends on the parameter $R = (\pi D/\lambda)(n_1^2 - n_2^2)^{1/2}$. For $R < 2.405$ only the HE_{11} mode is propagated, and the effective absorption coefficient K is calculated without difficulty [96]. In other cases, the calculation of S_2 involves certain difficulties.

Figure 59 plots the ratio of the energy S_1 transferred through the fiber core to the total transferred energy T_0 [96]. The calculations have been carried out for HE_{1m} modes ($m = 1, 2, 3$). All the curves correspond to fibers with $n_2/n_1 = 0.8$.

We see from the figure that as the parameter R decreases, the contribution of the particular mode to the outside flux increases. For the critical value of R, the ratio S_1/S_0 for the particular mode is zero, which means that the entire energy flux associated with the mode has left the fiber core.

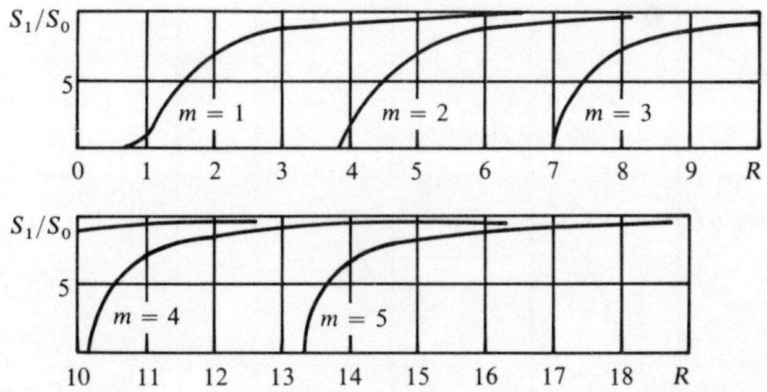

Figure 59

The ratio S_1/S_0 vs. the parameter R for the HE_{1m} mode.

Figure 60 plots the S_1/S_0 vs. R curves for two modes, HE_{11} and TE_{01}. The least critical value of R for TE_{01} is 2.405. Both curves correspond to the case $\delta = 1 - n_2^2/n_1^2 = 0.1$. Curve 2 for $R = 1.15$ shows an inflection point, and there is no cutoff frequency. The curve for the HE mode describes with adequate accuracy the variation of S_1/S_0 for smaller differences in the refractive indices of the core and the casing also; the fit is better the higher the value of R. Thus, for $R = 1$ and $\delta = 0.01$, about 17 percent of the total energy flux of the HE_{11} mode is transferred through the core. The corresponding value of S_1/S_0 read off the graph is 0.13. For $R = 2.4$, $S_1/S_0 = 0.827$, which is close enough to the value obtained from Figure 60 (0.824).

The above results can be used to calculate the K/K_1 curves, which characterize the gain in light transfer achieved by using the energy flux propagating through a casing whose absorption coefficient K_2 is less than the absorption coefficient K_1 of the core. For a fixed R, the light-transfer efficiency increases as the ratio K_2/K_1 diminishes.

These properties of thin dielectric waveguides are shown in Figure 61, which plots the curves of K/K_1 vs. R for various values of K_2/K_1.

These results were checked experimentally by Kapany et al. [96] using absorbing glass fibers with core diameters of 50, 5, 4, 3, and 2 μ. The casing material of all the fibers had a lower absorption coefficient. The fiber length varied from 25 mm to 110 mm. Similar experiments were also carried out with bundles of the same fibers. The dependence of emergent light intensity on fiber diameter and on the ratio of the core absorption coefficient to that of the casing revealed a good fit with the curves of Figure 61.

Theoretical and experimental investigations thus show that thin fibers can be successfully employed for light transfer in spectral regions where the core absorption is significant. The main difficulty in attaining a useful "outside" energy flux is technical: it is not easy to manufacture a high-quality core–casing interface. The surface precision requirements in this case are much more exacting than with thick fibers.

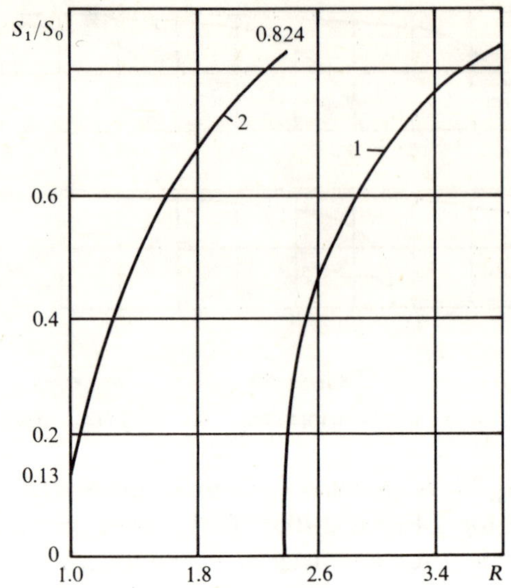

Figure 60

The ratio S_1/S_0 vs. the parameter R:
1) TE_{02} wave, 2) HE_{11} wave.

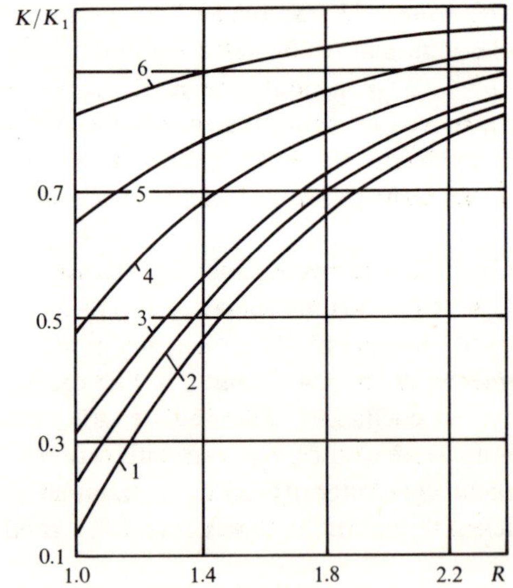

Figure 61

The ratio K/K_1 vs. the parameter R for various combinations of the core and the casing materials:

1) $K_2/K_1 = 0.05$; 2) $K_2/K_1 = 0.1$; 3) $K_2/K_1 = 0.2$; 4) $K_2/K_1 = 0.4$; 5) $K_2/K_1 = 0.6$; 6) $K_2/K_1 = 0.8$.

6. Field configuration in dielectric waveguides

As in metallic waveguides, the field configuration in dielectric fibers is generally represented by field lines. The line direction at any point characterizes the direction of the field, and the line density is proportional to the field strength. In what follows, we will only consider some particular modes in the fiber core. The field distribution in the fiber casing can be reconstructed using the boundary conditions for **E** and **H**. Since the field components in the casing are determined by the modified Hankel functions, which decrease monotonically with increasing r, the field line density in the casing will fall off rapidly in the radial direction.

The field configuration of the TE_{02} mode in the core of a dielectric waveguide for wavelengths far from the critical value [148] is shown in Figure 62. Since the magnetic field lines close in the direction of the z axis, they do not form closed curves in any of the cross-sectional planes. Dots and crosses in the longitudinal plane mark points of entrance and emergence of the electric field lines. This field pattern clearly shows that the electric field is perpendicular to the longitudinal plane.

In accordance with Table 2, the longitudinal magnetic component varies according to the zero-order Bessel function $J_0(h_1 r)$. The value of the parameter s_1 corresponds to the second root of the equation $J_1(s_1) = 0$, i.e., $s_1 = 7.02$.

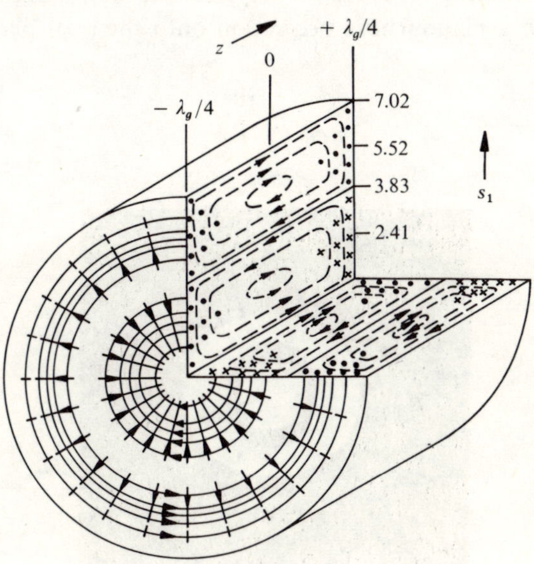

Figure 62

Field configuration of the TE_{02} mode in the core of a dielectric waveguide for $\lambda \ll \lambda_{cr}$: solid curves correspond to E; dashed curves with arrows, H.

The field configuration of this mode for a nearly critical wavelength can be similarly reconstructed. The parameter s_1 is determined in this case by the second root of the equation $J_0(s_1) = 0$, i.e., $s_1 = 5.52$ (Table 1). In other words, as λ approaches the critical wavelength, the field distribution is altered as if the waveguide interface has shifted from $s_1 = 7.02$ to $s_1 = 5.52$.

The field configuration corresponding to the TH_{02} mode is analogous to that shown in Figure 62, with the electric and the magnetic vectors interchanged.

Since the energy flux is determined only by the transverse components we see from Figure 62 that the intensity distribution on the exit end of the waveguide forms two concentric rings (Figure 63).

For mixed-type modes ($n \geqq 1$), the field distribution is greatly simplified when the difference in the refractive indices of the core and the casing is small ($P = \pm 1$) [148, 149]. From Eqs. (259), we write for the longitudinal components

$$E_z, H_z \sim J_n(h_1 r) \cos n\phi, \tag{264}$$

and for the transverse components

$$\begin{aligned} E_r &\sim (\pm 1)\, H_{n\pm 1}(h_1 r) \cos n\phi; \\ E_\phi &\sim J_{n\pm 1}(h_1 r) \sin n\phi. \end{aligned} \tag{265}$$

The plus sign corresponds to EH_{nm} modes, and the minus sign to HE_{nm} modes. The field line lies in a plane whose projection onto the (r,ϕ) plane is the solution of the equation

$$\frac{dr}{r\, d\phi} = \frac{E_r}{E_\phi}. \tag{266}$$

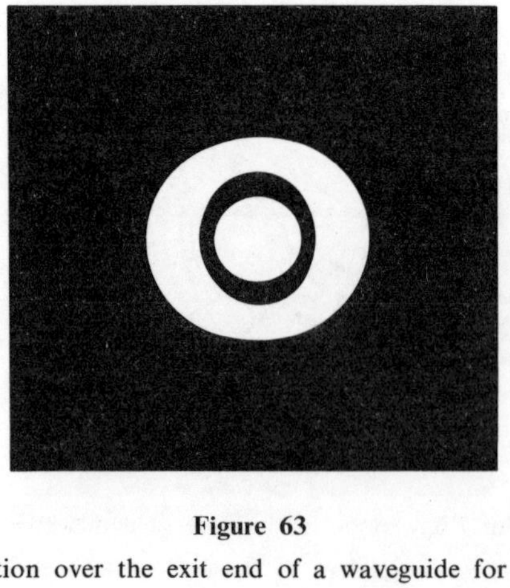

Figure 63

Intensity distribution over the exit end of a waveguide for the TE_{02} mode.

Thus, using (265), we find

$$\frac{dr}{r\,d\phi} = \pm \frac{\cos n\phi}{\sin n\phi},$$

and integration gives

$$r = C\,(\sin n\phi)^{\pm 1/n} \tag{267}$$

where C is an integration constant.

As before, the plus sign corresponds to the EH_{nm} mode, and the minus sign to the HE_{nm} mode. Varying the integration constant in (267), we obtain a family of curves which represent the electric field lines in the transverse plane. The magnetic field pattern can be similarly reconstructed. The direction of the magnetic lines makes an angle of $\pi/2n$ with the E lines.

Let us consider a few examples.

Eq. (267) for an HE_{1m} mode can be written in the form

$$r \sin n\phi = \text{const.} \tag{268}$$

This equation describes a family of straight lines parallel to the Ox axis. The field pattern for the HE_{12} mode is shown in Figure 64. Since the waveguide interface corresponds to $s_1 = 5.52$ (the second root of the equation $J_0(s_1) = 0$), this distribution is clearly applicable to $\lambda \ll \lambda_{cr}$. For critical wavelengths, s_1 is the solution of the equation $J_1(s_1) = 0$, i.e., $s_1 = 3.83$.

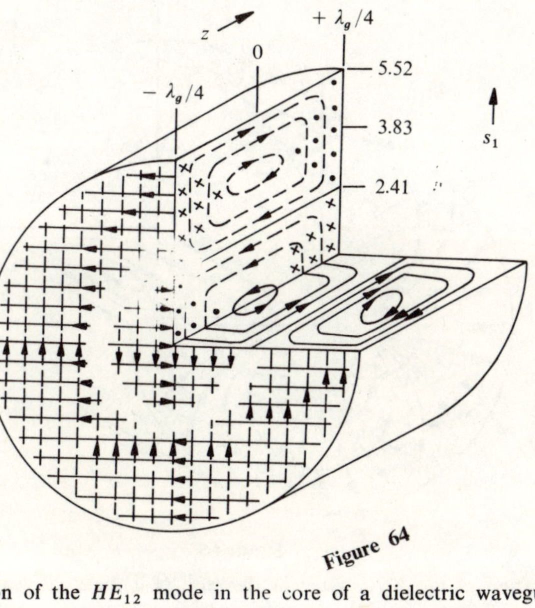

Figure 64

Field configuration of the HE_{12} mode in the core of a dielectric waveguide for $\lambda \ll \lambda_{cr}$: crossed solid lines with arrows correspond to E; dashed curves with arrows, H.

For the HE_{11} mode, for which no λ_{cr} exists, an increase in wavelength shifts the waveguide interface from $s_1 = 2.41$ to $s_1 = 0$.

A fundamentally different pattern of field lines is observed for the EH_{1m} modes. For the EH_{11} mode, Eq. (267) takes the form

$$r = C \sin \phi \tag{269}$$

or

$$x^2 + \left(y - \frac{C}{2}\right)^2 = \frac{C^2}{4}. \tag{270}$$

Eq. (270) describes a family of circles centered on the y axis and tangent at the origin to the x axis (i.e., these circles lie in the cross-sectional planes of the waveguide and are tangent to its axis). The field configuration of the EH_{11} mode at wavelengths far from the critical value is shown in Figure 65. The waveguide interface is defined by the first root of the equation $J_2(s_1) = 0$, i.e., $s_1 = 5.14$. Under critical conditions, $s_1 = 3.83$ at the interface (the first root of the equation $J_1(s_1) = 0$).

For the HE_{21} mode,

$$r = C\,(\sin 2\phi)^{-1/2}, \tag{271}$$

so that

$$xy = \text{const.} \tag{272}$$

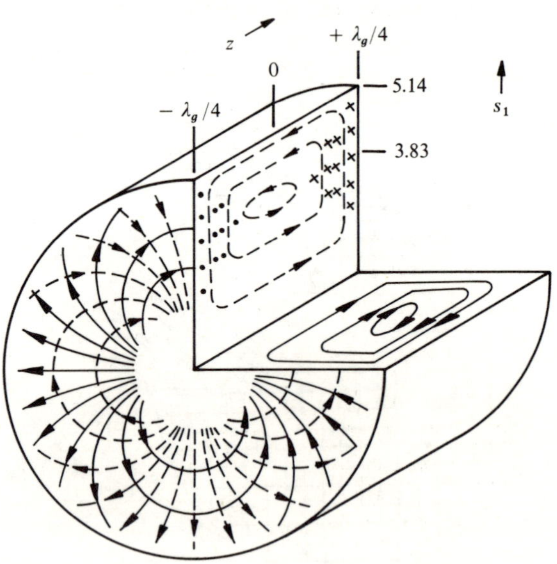

Figure 65

Field configuration of the EH_{11} wave in the core of a dielectric waveguide for $\lambda \ll \lambda$ solid curves with arrows, E; dashed curves with arrows, H.

Eq. (272) describes a family of hyperboles (Figure 66). As we have seen above, the critical value of the parameter s_1 for $n \geqq 2$ is determined by Eq. (252). When the difference between the refractive indices of the core and the casing is small, the right-hand side of Eq. (252) is small and we may approximately use the equation $J_{n-2}(s_1) = 0$. Hence, the critical value of the parameter for the HE_{2m} mode is approximately equal to the first root of the equation $J_0(s_1) = 0$, and thus coincides with the critical condition for the TE and TH waves. Exact equality of the critical parameters is achieved only when the difference $n_1^2 - n_2^2$ is infinitesimal; otherwise, the $s_{1\,\mathrm{cr}}$ of the HE_{2m} mode is somewhat higher than the corresponding value for the TE_{0m} and TH_{0m} modes. Nevertheless, the difference is quite insignificant. Thus, for $n_1 = 1.56$ and $n_2 = 1.52$, the first root of Eq. (252) is 2.427, whereas the corresponding root of the function $J_0(s_1)$ is 2.405.

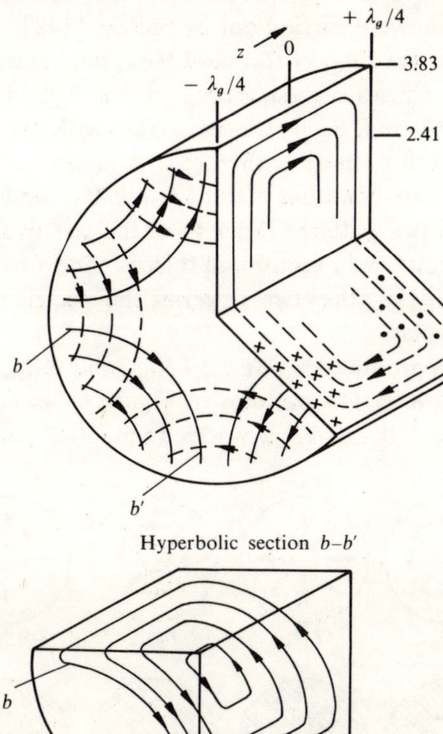

Hyperbolic section b–b'

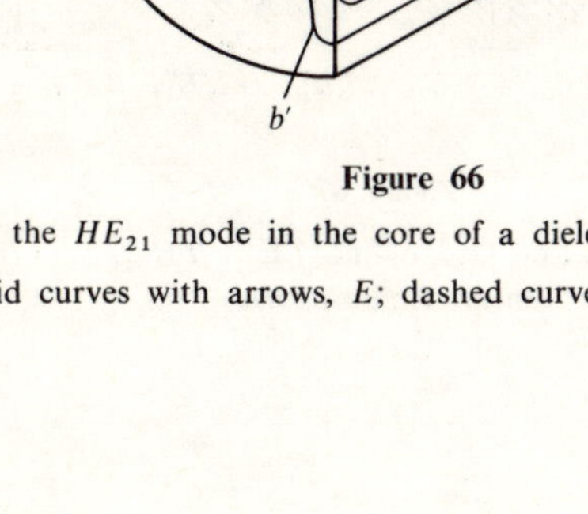

Figure 66

The field of the HE_{21} mode in the core of a dielectric waveguide for $\lambda \ll \lambda_{cr}$: solid curves with arrows, E; dashed curves with arrows, H.

However, the analogy between these modes is not restricted to a mere similarity in their critical parameters. Indeed, the radial dependence of the transverse components of the electric field for the TE_{0m} and TH_{0m} waves is expressed in the form

$$E_r = 0; \quad E_\phi \sim J_1(h_1 r) \quad (TE_{0m} \text{ wave}),$$

$$E_\phi = 0; \quad E_r \sim J_1(h_1 r) \quad (TH_{0m} \text{ wave}).$$

According to (265), a similar dependence is observed for the HE_{2m} modes. Since the Poynting vector is a function of the transverse field components only, these modes in general produce the same illumination distribution on the exit end of the waveguide. Since the TE, TH, and HE_{2m} waves have roughly equal phase velocities, a linear combination of the three modes may form in the waveguide. Careful experimental and theoretical analyses of the resulting illumination patterns on the exit end of the waveguide were carried out by Snitzer [148].

The electric vectors for the TE_{01}, TH_{01}, and HE_{21} modes are schematically shown in Figure 67*a*. The two figures corresponding to the "hybride" mode clearly show the double degeneracy of the HE_{21} mode associated with the cylindrical symmetry. This is characteristic of any mode with $n \geqq 1$. A schematic configuration of the electric field of the linear combinations of the TE_{01}, TH_{01}, and HE_{21} modes is shown in Figure 67*b*. The first two patterns differ from the last two in the relative alignment of the resultant electric field vector and the zero line. For the first two patterns, they are parallel, and for the other two patterns, the electric field and the zero line are mutually perpendicular.

Experimental patterns produced by TE_{0m}, TH_{0m}, and HE_{2m} modes for $m = 1, 2, 3$ are shown in Figure 68, *a–c*. The patterns in Figure 68, *d–f* correspond to superposition of modes with $n = 0$ and HE_{2m} modes. In practice, pure modes are observed

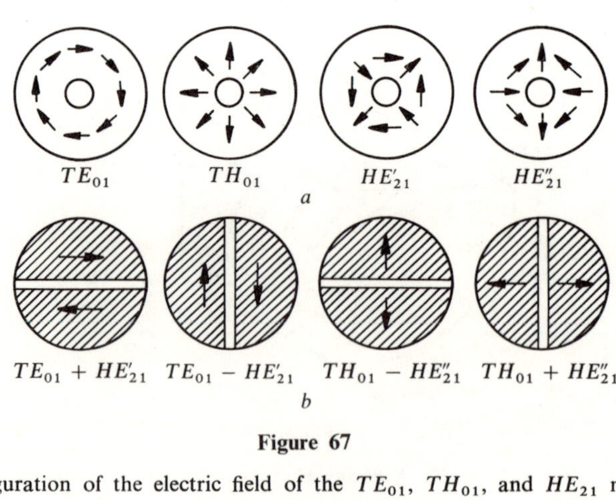

Figure 67

Schematic configuration of the electric field of the TE_{01}, TH_{01}, and HE_{21} modes and their linear combinations.

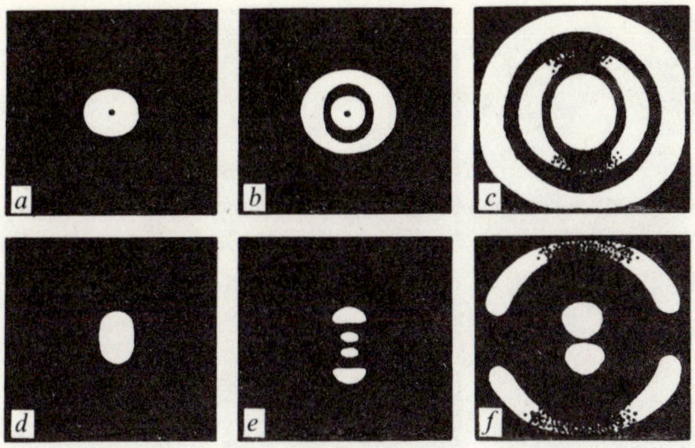

Figure 68

Configuration of TE_{0m}, TH_{0m}, and HE_{2m} modes for various m:

a $m = 1$; *b* $m = 2$; *c* $m = 3$; *d–f* illumination patterns produced by a combination of TE_{0m} and TH_{0m} modes with HE_{2m} mode.

only when the waveguide is illuminated by a strictly axial beam. Generally, configurations of the shape shown in Figure 68, *d–f* are observed.

More complex patterns of low-order modes and their combinations are shown in Figure 69 [148]. Some of these patterns were obtained by illuminating the waveguide with polarized light and examining the exit end through an analyzer.

The illumination patterns of elliptical-section waveguides are shown in Figure 70. The zero lines in this case are confocal ellipses and hyperbolas, rather than concentric circles and radial lines.

7. A bundle of thin fibers

Cross-coupling between adjacent fibers constitutes a serious problem for light transfer through fiber bundles. This interference is particularly significant in thin fibers, when the wave-optics phenomena prevail.

The analysis of a large number of closely packed waveguides is not a simple problem. No detailed calculations have been carried out so far, and only some fragmentary results have been published. Snitzer [148] considered the modes in a system of two close dielectric waveguides. The waveguides exchange energy as a result of field leakage into the fiber casing. This is a resonant coupling, and each mode is transmitted to the adjacent waveguide regardless of the presence of other modes.

The various modes of a composite waveguide can be represented as a combination of the modes of the individual fibers. If the parameters of the modes centered in each fiber are known, the resultant mode of the entire bundle can be derived using the

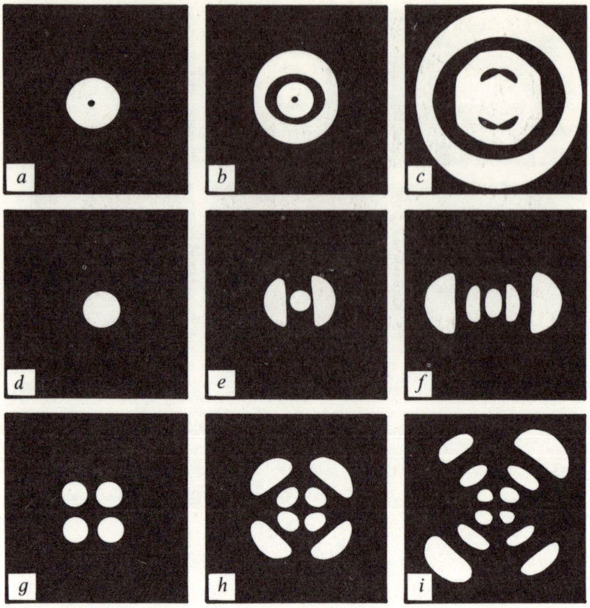

Figure 69

Configurations of some low-order modes in a circular waveguide:

a–c patterns corresponding to TE_{0m}, TH_{0m}, or HE_{2m} modes with $m = 1, 2, 3$; *d* HE_{11} mode; *e* EH_{11} or $HE_{12} + HE_{31}$; *f* EH_{12} or $HE_{13} + HE_{32}$; *g–i* EH_{1m} or HE_{3m} for $m = 1, 2, 3$.

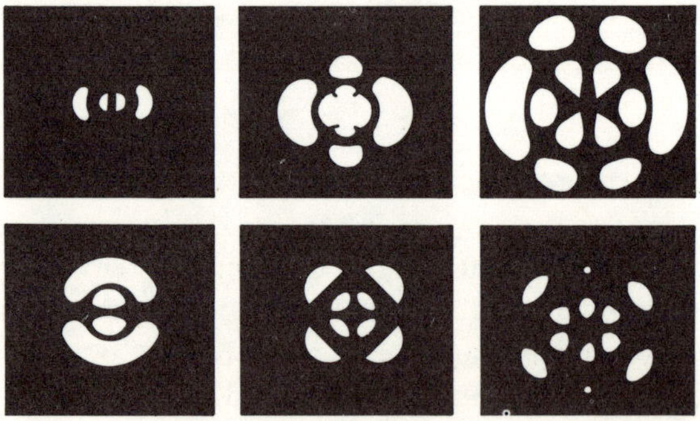

Figure 70

Specimen photographs of typical modes in an elliptical cross section waveguide.

perturbation theory by a method similar to the derivation of the molecular orbitals as linear combinations of the atomic orbitals.

The coupling between the fibers in a bundle is enhanced as the outside flux of each fiber increases. We should naturally consider the dependence of this process on the parameters of the individual fibers and the light wavelength.

The optical diameter and the contribution from the energy flux propagating along the fiber surface are described by the parameter

$$R = \frac{\pi D}{\lambda}(n_1^2 - n_2^2)^{1/2} = \frac{\pi D}{\lambda} A, \tag{273}$$

where A is the numerical aperture of the fiber

Let us consider two cases: $D \gg \lambda$ and $D \simeq \lambda$ ($D \ll \lambda$ is difficult to achieve for technical reasons).

1. $D \gg \lambda$. Since the fiber apertures are of the order of unity or somewhat less than 1, the parameter R is large and the optical diameter is virtually equal to the geometrical diameter. The entire light flux propagates inside the core, and only a very minor fraction leaks into the casing. This points to a weak coupling between the fibers, and the effect is confined to the closest neighbors only.

2. $D \approx \lambda$. The parameter R is small. The optical diameter may exceed the geometrical diameter of the fiber core, and the outside energy flux gains in importance. In closely packed bundles, the outside flux may penetrate into the adjacent fibers, and the cross-coupling will become acute.

For simplicity, let us consider the case when only the end face of a single fiber in a bundle is illuminated.

D and A entering the expression for R are kept constant. Then R depends on the wavelength λ only. As λ increases, R decreases, and a higher proportion of energy leaks into the nearby fibers, i.e., the red part of the spectrum is more susceptible to cross coupling. Experimentally, this effect is reflected in the appearance of characteristic color on the exit end of the bundle when white illumination is used [95]. The exit ends of the fibers adjacent to the illuminated fiber acquire a blue color, and the color gradually changes to red for fibers lying farther toward the periphery.

For a fixed R, the main contribution to the outside leakage flux is from the highest order modes. Characteristic mode selection is therefore observed in the radial direction over the bundle cross section: near the illuminated fiber, the same modes are propagated as in its core, whereas farther away from that fiber, the low-order modes disappear and at the periphery only the highest order modes remain. This pattern has been observed experimentally [95].

Let us now consider the case when the wavelength remains fixed. This does not involve any experimental difficulties: the constant wavelength can be achieved using a monochromator and a micro-objective focusing part of the exit slit onto the end face of one of the fibers. The ratio between the energy flux in the illuminated fiber and in the adjacent fibers at a fixed λ is determined by the diameter D and the numerical aperture A of the fibers. Eq. (273) shows that of two bundles made up of fibers with identical numerical aperture, the bundle with thicker fibers ensures a lower cross-coupling (and thus a higher contrast at the exit end).

The above considerations are supported by the experimental curves of Figure 71, borrowed from [95]. The horizontal axis gives the distance in microns, reckoned on the exit end from the axis of the illuminated fiber. The integrated light flux corresponding to different end areas is laid off the vertical axis. The first experimental point on each curve corresponds to the flux emerging from the illuminated fiber, the second point also incorporates the flux through the exit ends of the six nearest neighbors, etc. Curves 2 and 3 correspond to bundles made up of individual fibers of constant numerical aperture $A = 1.13$ but different diameter ($D_3 = 4\mu$, $D_2 = 2\mu$). The fiber parameters of the first bundle are $A = 0.49$ and $D_1 = 3.5\,\mu$.

Figure 71 provides a clear illustration of the significant scattering of luminous energy into the adjacent fibers with the decrease in D and A. This factor is of funda-

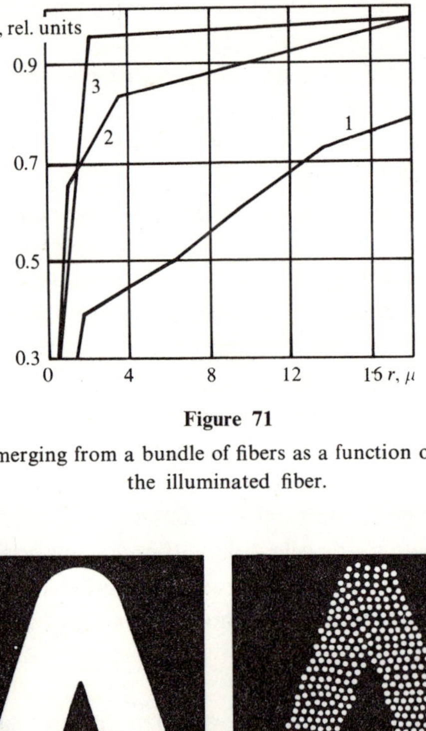

Figure 71

Integrated luminous flux emerging from a bundle of fibers as a function of the distance from the axis of the illuminated fiber.

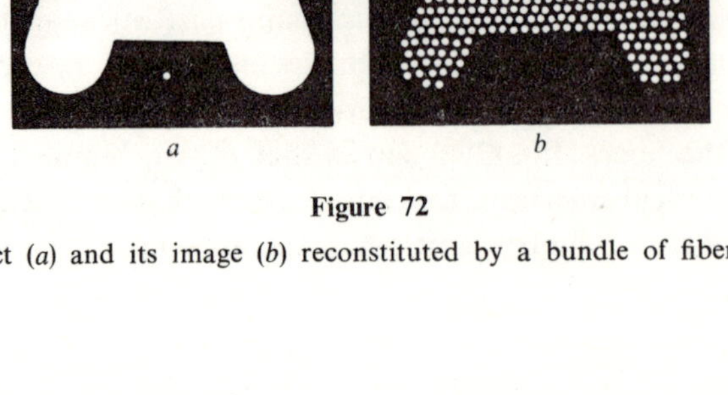

a *b*

Figure 72

Luminous object (*a*) and its image (*b*) reconstituted by a bundle of fibers with $D = 1\,\mu$.

mental importance and it must be taken into consideration in the manufacture of bundles of micron fibers. However, to ensure high resolution, the decrease in D should be accompanied by an appropriate increase in the aperture of the individual fibers. If the aperture is greater than unity, an adequate contrast of the transmitted image can be obtained even with bundles made up of very thin fibers. This is borne out by Figure 72, which shows a photograph of the luminous object and its image transmitted by a bundle of micron fibers with numerical aperture $A = 1.15$.

4 The Principal Characteristics of Fiber Bundles

1. General background

A single fiber only transmits information regarding the magnitude of the luminous flux incident on its entrance end, since the illumination distribution in the exit plane is independent of the entrance illumination distribution. Therefore, a single fiber transforms at most one image element. To transmit the entire image, bundles of uniformly packed fibers should be used. The image is then made up of a large number of sufficiently small dots, not unlike the image on the television screen or a half-tone photograph in a newspaper.

The primary image should lie in the plane of the entrance end of the bundle or near it. This is accomplished by using suitable optical systems or by bringing the object, e.g., a drawing, in direct contact with the entrance end of the bundle.

A bundle in which the relative arrangement of the fibers is the same on both ends is called *coherent*.* Inside the bundle, however, the fibers may be arranged in an arbitrary configuration.

Images can be transmitted only by means of coherent bundles, whereas noncoherent bundles are suitable only for transmission of illumination.

To improve the flexibility of the bundle, it is advisable to bond only the ends of the fibers, leaving the whole fiber length free or properly lubricated [128]. Bundles of this kind must be enclosed in a special flexible casing which protects the fibers against mechanical damage and prevents the loss of lubricant.

The principal optical characteristics of a coherent bundle are the light-transfer efficiency, the numerical aperture, the resolving power, and the image contrast.

The light-transfer efficiency is determined by the absorbing properties of the fibers and their quality. The main losses of light intensity are attributed to absorption in the fiber core. For example, the screen of a cathode-ray tube manufactured from fused fibers in the shape of a disk 12.7 mm thick transmits about 90% of the

* The term *coherent bundle* is not particularly suitable, but it is accepted in the literature.

incident radiation, whereas a light pipe made of the same fibers 2.134 m long transmits only half of the incident energy [128].

Besides absorption, there are other energy losses which are characteristic of the individual fibers and of the bundle as a whole. Together with absorption, they determine the total light-transfer efficiency of the fiber device.

The numerical aperture of a coherent bundle is determined by the numerical apertures of the individual fibers. It depends on the ratio of the refractive indices of the core and the casing.

The resolving power is a function of the fiber diameter, the casing thickness, and the packing of the fibers in the bundle. For an ideally parallel packing, the fiber bundle will resolve two lines separated by a distance slightly greater than the diameter of a single fiber. If the resolving power is expressed in lines per mm, it is approximately equal to half the number of fibers packed linearly in 1 mm [3]. It is therefore not enough to reduce the fiber diameter to the permissible minimum: the thickness of the interlayer between the fibers should also be made as small as possible. This, in its turn, improves the system luminosity, since the image transfer efficiency is inversely proportional to the fraction of the cross sectional area occupied by the fiber casings and the unfilled gaps. Depending on the fiber packing, the relative loss of the effective bundle area with a decrease in fiber diameter fluctuates between 16 and 20%. We see from Figure 73 that, keeping the total cross-sectional area of the bundle constant, the effective area markedly decreases with the decrease in D [115].

The effective area is increased by using hexagonal or square cross-section bundles. The manufacture of such fibers, however, involves considerable technical difficulties.

The contrast of the transmitted image depends on the quality of the individual fibers and their packing. If the fibers are packed at a distance equal to or less than 2 λ, there is considerable light leakage between the fibers, so that the background intensity increases and the image contrast decreases [92].

For closely packed thin fibers, light leakage constitutes a serious problem. Therefore, in every individual case, the requirements of image contrast are used as a basis for the choice of fiber diameter and casing thickness. Other characteristics of the fiber bundles are of course also taken into consideration, e.g., the resolving power.

The background illumination is caused not only by leakage but also by the scattering of light which hits the entrance plane of the bundle at supercritical incidence angles (this applies mainly to the outer part of the light cone). Readily penetrating from one fiber into another, these rays create an additional background in the exit plane, markedly lowering the image contrast [89].

The scattered light can be successfully suppressed by coating the fiber casings with an absorbing film. We see from Figure 74 [146] that this measure is quite effective. The photograph shows a circular fiber disk about 4 cm in diameter and 1 cm thick, fused from single- and double-coat fibers of the same numerical aperture. The double-coat fibers are packed so that they form the letters SM. When the disk is illuminated

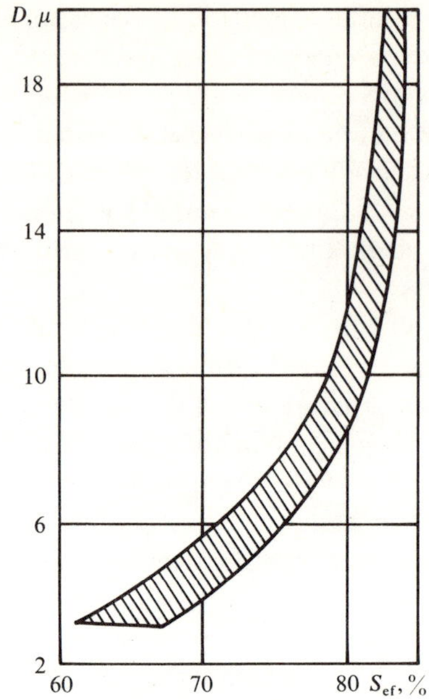

Figure 73

Effective bundle area S_{ef} vs. fiber diameter D for constant total bundle cross section. The right edge corresponds to hexagonal packing of fibers, and the left edge to square packing.

with diffuse light and the exit end is viewed inside the image-carrying cone, the letters SM remain invisible against the general bright background (Figure 74*a*). This indicates that the transfer efficiency of both types of fibers is almost the same. If, however, the end face is viewed from outside the image cone, the letters SM become clearly visible against the bright background (Figure 74*b*).

A bundle of regularly packed fibers can be heated to its softening point, pressed, and sintered into a monolithic vacuum-dense slab. The cross section of the individual fibers in this slab depends on the thermal properties of the core and the casing. Cylindrical fibers acquire a polyhedral cross section when fused (Figure 75), and the casing fills the empty spaces, forming a honeycomb [3]. While hot, the bonded bundles can be stretched into conical slabs [138]. The image projected onto the large base of the cone is transmitted in reduced form (Figure 76), and the illumination in the plane of the small base markedly increases. In the opposite direction, the cone functions as a magnifying system.

As the angle of inclination of the rays in a tapered conical fiber changes, some light penetrates through the lateral surface and the image contrast deteriorates [89]. For diverging beams, this effect is considerably less pronounced. It may be completely eliminated if every individual fiber is coated with an additional opaque envelope.

Figure 74

The effect of scattered light on image contrast:

a the transferred light beam is observed inside the cone of image-carrying rays; *b* the transferred beam is observed from outside the image cone.

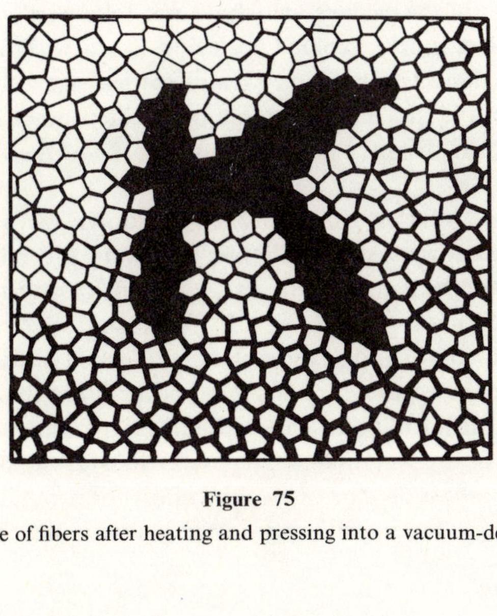

Figure 75

The shape of fibers after heating and pressing into a vacuum-dense slab.

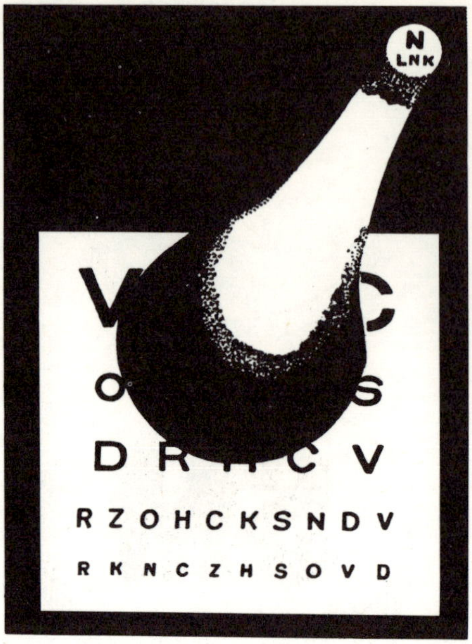

Figure 76
Image transmission by a conical fiber bundle.

2. Leakage effects

Part of the information transmitted by fiber light pipes is inevitably lost because of the finite thickness of the casings, which do not take part in light transfer. These losses could have been partly avoided by using fibers without any insulating envelopes. In this case, however, we are risking light leakage between adjacent fibers. Light leakage is possible not only in points of direct contact between the adjacent fibers, but whenever the distance between the surfaces of the two fibers is less than 2λ.

Leakage at a plane interface following a breakdown of the conditions of total internal reflection was discussed in Chapter 1. Using the relations of Chapter 1, we can calculate the dependence of the leakage intensity on the width of an air gap between two identical glass plates ($n = 1.5$). The results of the calculations carried out for natural (solid curve) and plane-polarized (dashed curves) radiation incident at 80° [94] are shown in Figure 77. We see from the figure that for natural light the leakage effect is significant only for gap widths $t \leqq \lambda/4$, whereas for polarized light the effect is stronger for the p-component (curve 2) than for the s-component (curve 1).

It is not easy to analyze the corresponding effect for a bundle of fibers, since the intensity of the leakage flux depends on numerous variables. They include the dia-

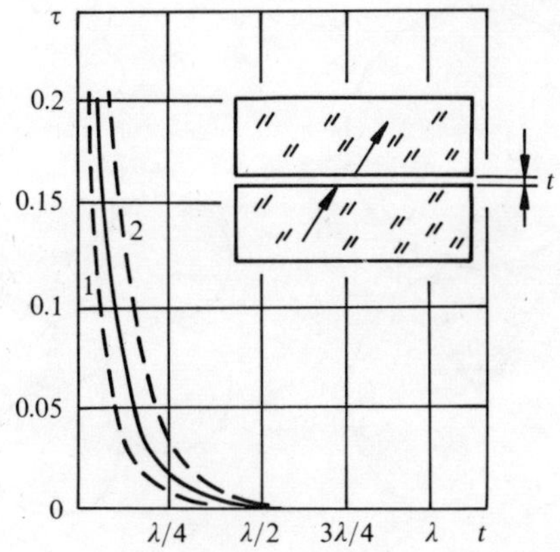

Figure 77

Leakage coefficient τ as a function of the air gap width t.

meter, the length, and the refractive index of the fibers; the wavelength of the transmitted radiation, its polarization and incidence angle at the interface; the contact between the adjacent fibers and the refractive index of the medium in the gaps (this may be air, in particular), etc. Diffraction effects complicate the problem even further.

Using various simplifying assumptions, we will now estimate the upper bound value of the leakage flux for the case of maximum contact between adjacent fibers, when $t \leqq \lambda/4$. We assume total leakge in the contact region, irrespective of the angle of incidence on the cylindrical surface.

If the incident luminous flux is uniformly distributed over the cylindrical surface, the calculations reduce to the determination of the angular width of the leakage sectors in the fiber cross section. To determine the width of the leakage sector, we have to establish a relationship between the angle β, the radius r of the fiber, and the wavelength of the incident light. We see from Figure 78a that, for $r \gg \lambda$,

$$AC \simeq r\frac{\beta'}{2}, \quad BC \simeq AC\frac{\beta}{4}. \tag{274}$$

Since the angles β and β' are nearly equal, we may write

$$\frac{\lambda}{4} \approx r\left(\frac{\beta}{2}\right)^2,$$

whence

$$\beta \simeq \left(\frac{\lambda}{r}\right)^{1/2}. \tag{275}$$

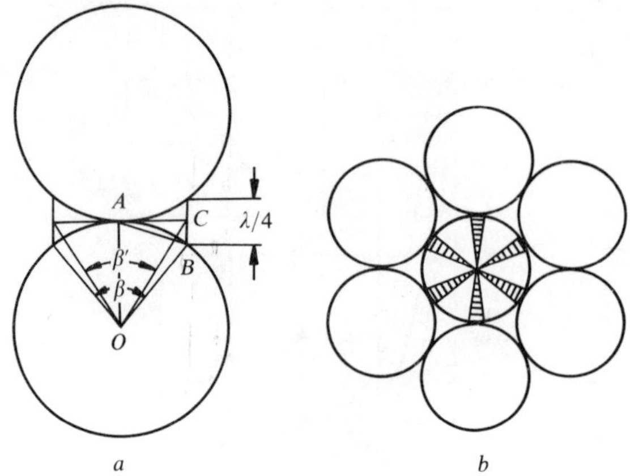

Figure 78

Angular width of the leakage sector:

a leakage angle; *b* a cross section through a bundle of hexagonally packed fibers (the six leakage sectors are hatched).

In case of hexagonal packing of the fibers, there are six leakage sectors in each fiber (Figure 78*b*), and the ratio of the leakage flux to the total flux in the central fiber is given by

$$\delta = \frac{6\beta}{2\pi} = \frac{3}{\pi}\left(\frac{\lambda}{r}\right)^{1/2}. \tag{276}$$

For $\lambda = 6000$ Å and $r = 25\,\mu$, $\delta \approx 15\%$; for the same wavelength and $r = 5\,\mu$, $\delta \approx 33.5\%$, i.e., δ increases as the radius decreases.

This analysis gives a rough estimate of the leakage effect, especially since we considered only the meridional rays, neglecting the oblique rays. Moreover, it was assumed that reverse leakage through the contact sectors and reflection losses at the cylindrical surface are second-order effects, which do not change the final result.

A more exact analysis of the losses associated with leakage to the adjacent fibers can be carried out if we calculate what fraction of the luminous flux incident on the central fiber emerges through the six surrounding fibers [92]. Remembering that leakage does not vanish only for gap widths $t = 2\,\lambda$, we use Figure 78*a* to find the angular leakage sector

$$\beta = \cos^{-1}\frac{r - \lambda}{r}. \tag{277}$$

As expected, this result differs from that obtained for $t \leqq \lambda/4$.

A schematic plot of leakage as a function of t is shown in Figure 79*a*. The graph ignores the light leakage into the next 12 fibers, since the effect is insignificant for diameters much greater than the diffraction limit.

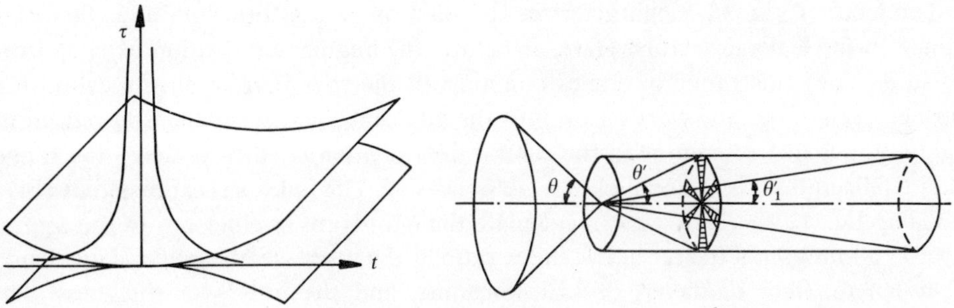

Figure 79

Curves of leakage vs. gap width between fibers (*a*) and cones of light rays propagating in the fiber (*b*).

The light flux entering the central fiber can be split into three components. One of these are the rays confined inside a cone with an opening angle $2\theta'_1$, which are not reflected from the lateral surface (Figure 79*b*). If the light flux is uniformly distributed inside a cone of angle $2\theta'$, Eq. (158) gives for this component

$$T_1 = A_1 \int_0^{\theta'_1} (1 - R)^2 \exp(-Kl \sec\theta') \sin\theta' d\theta', \qquad (278)$$

where $A_1 = (1 - \cos\theta'_1)^{-1}$ is a normalization constant determined from the condition $T_1 = 1$ for $R = K = 0$; R and K are the reflection coefficient for the fiber end surface and the absorption coefficient of the fiber material; l is the fiber length.

Another component comprises the rays in the six leakage-free sectors. The corresponding cross-sectional sectors of the cylinder are the plain sectors in Figure 79*b*. Eliminating the first component, we obtain in this case

$$T_2 = C_1 A_2 \int_{\theta'_1}^{\theta'} (1 - R)^2 \exp(-Kl \sec\theta') \sin\theta' \, d\theta', \qquad (279)$$

where $A_2 = (\cos\theta'_1 - \cos\theta')^{-1}$ is a normalization constant and

$$C_1 = 1 - 3\beta/\pi.$$

The coefficient C_1 determines the fraction of the total luminous flux accommodated by the integration region.

The third and the last component comprises the rays in the six leakage sectors. The corresponding sectors are cross-hatched in Figure 79*b*. To find the luminous flux corresponding to these sectors, angular integration has to be supplemented with integration over the thickness t of the interlayer. The flux is found to be

$$T_3 = \frac{C_2 A_2}{2\lambda} \int_{\theta'_1}^{\theta'} \int_0^{2\lambda} (1 - R)^2 (1 - \tau)^n \exp(-Kl \sec\theta') \sin\theta' \, d\theta \, dt, \qquad (280)$$

where $\tau = \frac{1}{2}(\tau_{\parallel} + \tau_{\perp})$ is the leakage coefficient for unpolarized light.

The factor $C_2 = 3\beta/\pi$ characterizes the fraction of the total luminous flux contained in the leakage sectors. Here, as before, the angular integration extends from θ'_1 to θ', since this range of angles contains all the rays that reach the cylindrical surface of the fiber and partly leak into the adjacent fiber. After the first reflection, the luminous flux remaining in the central fiber is attenuated by a factor $1 - \tau$, and after η reflections it is attenuated by a factor $(1 - \tau)^{\eta}$. The index η is expressed by (147).

Using Eqs. (278)–(280), we can calculate the light-transfer efficiency of the central fiber as a function of the refractive index ratio at the interface, the angle of incidence, fiber length, fiber diameter, optical constants, and the interlayer thickness. The resulting expressions also apply to the case of polarized incident radiation.

We have so far assumed that the individual fibers and the bundle as a whole are not bent or twisted. If this is not so, the solution of the problem is greatly complicated, since for a flexible fiber we cannot establish a fixed line or sector of contact, and the numerical values of the coefficients C_1 and C_2 are difficult to find. In our calculation of T_2 and T_3, we ignored scattering losses attributable to cylindrical surface inhomogeneities. If these losses are significant, a factor $\rho^{\eta} = [1 - K'(\theta')]^{\eta}$, where $K'(\theta')$ is the scattering coefficient, should be introduced in the integrands in (279) and (280).

Using Eqs. (278)–(280), we write for the light transfer of a fiber surrounded by six identical fibers

$$T = A_3 \left(\frac{T_1}{A_1} + \frac{T_2}{A_2} + \frac{T_3}{A_3} \right), \tag{281}$$

where $A_3 = (1 - \cos\theta')^{-1}$ is a normalization coefficient.

In case of nonuniform illumination of the entrance end, the integrand in (278)–(280) should contain a function which describes the illumination distribution.

To establish the effect of leakage on the contrast of a point image formed by a linear fiber bundle, we have to determine the fraction of the flux which is transferred as a result of leakage through the six fibers surrounding the original image-carrying fiber. We see from Figure 78*b* that the successive lines of contact of the six adjacent fibers with the central fiber occur at angular distances of 60°. As a result of multiple reflections, the ray alternately enters two diametrically opposite leakage sectors (Figure 80), losing part of its energy in each reflection. Taking the leakage coefficient to τ, the fraction of the energy remaining in the central fiber after η reflections is $(1 - \tau)^{\eta}$. The part of the luminous flux penetrating from the central fiber into each pair of diametrically opposite fibers is expressed by the series

$$\tau + \tau(1 - \tau) + \tau(1 - \tau)^2 + \ldots + \tau(1 - \tau)^{\eta - 1}.$$

The fraction of the flux emerging from each pair is

$$T''(\theta') = \tau\{1 + (1 - \tau) + (1 - \tau)^2 + \ldots + (1 - \tau)^{\eta - 1}\} \times$$
$$\times \exp(-Kl\sec\theta') = [1 - (1 - \tau)^{\eta}] \exp(-Kl\sec\theta'). \tag{282}$$

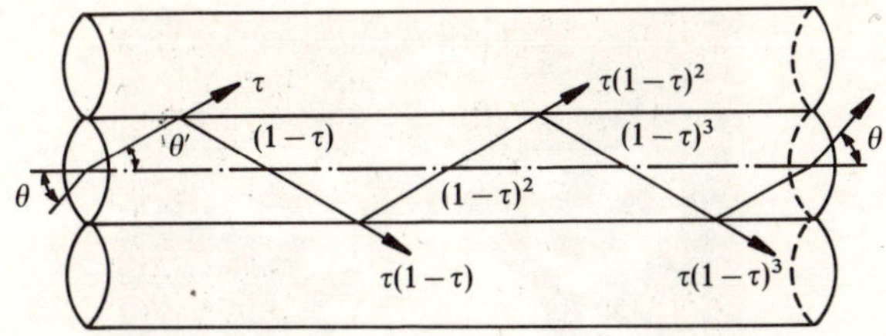

Figure 80

The path of meridional rays in fibers with leakage.

This is by no means the true result, since it was derived ignoring the interpenetration of light in all the optically coupled fibers of the bundle.

To determine the fraction of the luminous flux emerging from each of the adjacent fibers, Eq. (282) should be integrated over the entire light cone and over the effective thickness of the gap. This integration gives

$$T' = \frac{A_3 C_3}{2\lambda} \int_0^{\theta'} \int_0^{2\lambda} [1 - (1 - \tau)^n] \exp(-\, Kl \sec \theta') \sin \theta' \, d\theta' \, dt, \qquad (283)$$

$$\text{where } C_3 = \frac{\beta}{2\pi}.$$

Thus given τ and T', we can find the intensity of the luminous fluxes emerging from the central fiber and the six adjacent fibers. According to Kapany [92], the intensity ratio of these fluxes is 3300 for fibers 50 μ in diameter and 2.5 mm long. To verify experimentally the results of these calculations, a point source was projected onto the entrance end of one of the fibers and the exit end was viewed through a low-power microscope [92]. The energy distribution curve of the "point" image depends on the state of contact between the fibers. It is therefore very important to choose a suitable bundle segment, one with especially regular packing and high-quality optical contact between the fibers. If the experiment is properly conducted along these lines, cross-coupling between the fibers in the bundle is avoided and Eq. (283) provides an adequate approximation to the true flux.

The results of [92] are shown in Figure 81 in the form of a photomicrograph of the corresponding part of the bundle and a diagram giving the distribution of illumination in the point image. For some unknown reasons, the diagram clearly does not fit the experimental intensity ratio (3500:1), although an analysis of photomicrographs reveals an adequate qualitative agreement between the theory and the experiment.

For short fibers thicker than 25 μ in diameter, light leakage due to the breakdown of the total internal reflection is thus insignificant, and optical insulation of the fibers is not absolutely essential.

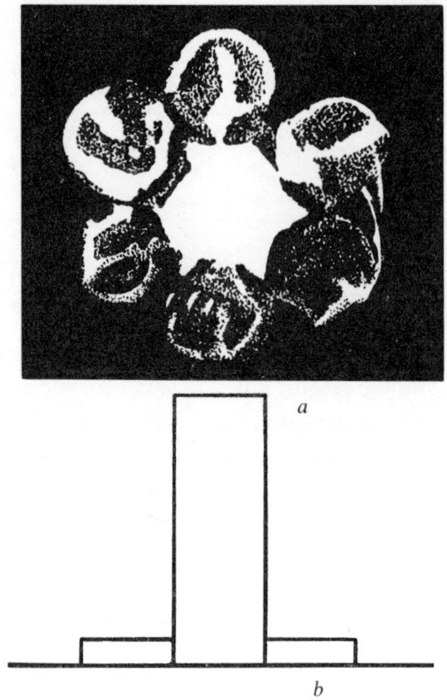

Figure 81

Distribution of illumination in the image of a point source:

a photomicrograph of the point source image; *b* illumination diagram in the end plane of a central fiber and two adjacent fibers, with their centers lying on the same line.

3. Resolving power

As a result of multiple reflections inside the fiber, uniform illumination is observed on the exit end, which is proportional to the total light intensity incident on the entrance end. The fine features of the transmitted image are therefore lost irretrievably within the limits of each fiber. Details corresponding to the blind parts of the bundle —gaps and fiber casings—are also lost. The resolving power of a fiber element is therefore dependent on the packing arrangement and also on the orientation of the imaged object.

Let us determine the resolving power of bundles with hexagonal and square packing of fibers. In the case of hexagonal packing (Figure 82*a*), the centers of the adjacent fibers form the vertices of a regular hexagon, and in square packing, they form the vertices of a square (Figure 82*b*).

In accordance with the theory of information, the function $F(x)$ describing the spatial distribution of information in the direction x, which does not contain fre-

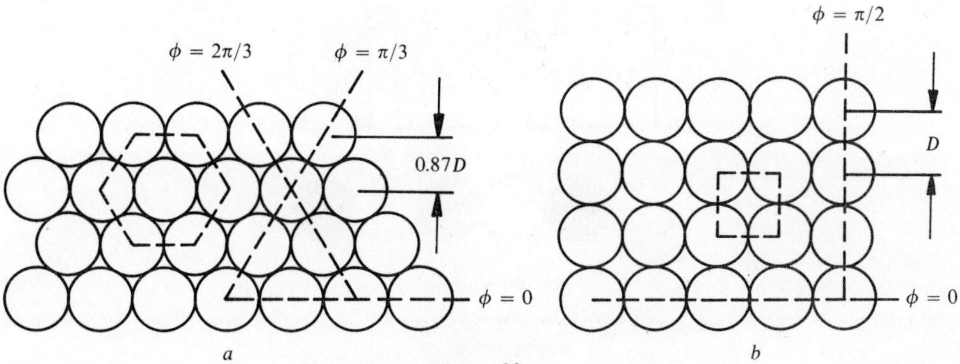

Figure 82

Packing of fibers into bundles:

a hexagonal, *b* square.

quencies higher than N lines per mm, is completely specified by its values at a series of discrete points spaced at intervals of $N/2$ mm [155, 142]. Indeed, let the information distribution in the direction x be such that there are N lines per mm (Figure 83). If the information is read at intervals of $1/N$, all the transmitted points are detected, although we know nothing about the possible information content between the successive read points. Therefore, in order to resolve two successive sendings and to fully recover the distribution in the direction x, the read operation must be performed at least twice as frequently.

Keeping in mind the above remarks, let us consider the transmission of the image of a grid by a bundle of adjacent fibers of diameter D (Figure 84). Since each fiber transmits at most a single image element, the minimum information read period is D. Therefore, the minimum spacing of the grid whose image can be transmitted by the fiber bundle ensuring a complete resolution of the individual lines will be $D/2$. For smaller grid spacings, the same uniform illumination is obtained in the plane of the exit ends of all the fibers, and no image is formed.

The maximum resolving power of a fiber system which transmits a certain information distribution is thus determined by the real read period in the corresponding

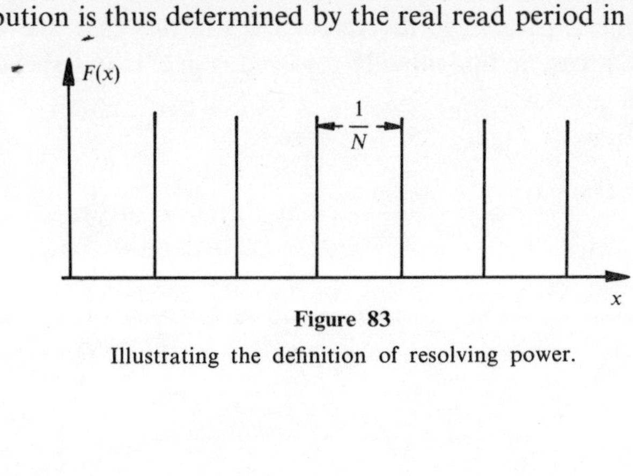

Figure 83

Illustrating the definition of resolving power.

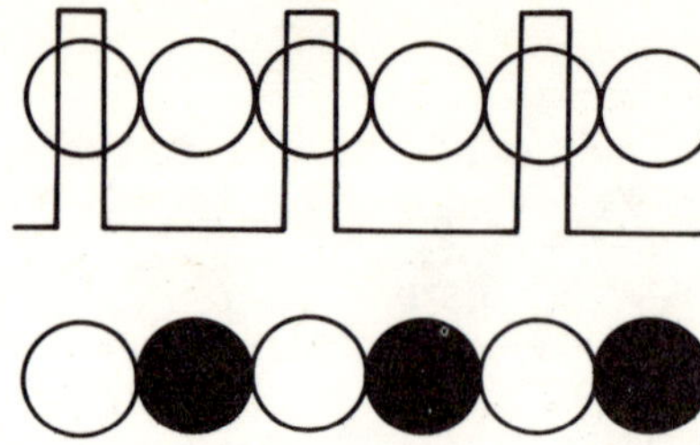

Figure 84

Transmission of the image of a grid where the line spacing is double the distance between the centers of adjacent fibers.

direction. This period, for any direction in the plane of the exit end of the fiber bundle, can be expressed in general form in terms of the fiber diameter D [66].

We see from Figure 85 that, in the case of hexagonal packing, rotation of the bundle about any of the fiber axes through an angle $\pi/3$ gives again the original configuration. Regardless of the particular symmetry of the transmitted image, the resolving power of the system should not change as a result of these rotations. Periodicity and symmetry properties of fiber bundles should therefore be taken into consideration inside the angle delimited by the vectors OT_1 and OT_2 which join the center of the given fiber with the centers of two adjoining fibers (Figure 85). Then, for any direction OO', we have

$$OO' = mOT_1 + nOT_2, \tag{284}$$

where m and n are integers.

If T_1T_1' and T_2T_2' are vectors parallel to OO', there are $n - 1$ fiber centers between T_1T_1' and OO', and $m - 1$ centers between T_2T_2' and $OO' - m - 1$. Straight lines through these centers parallel to OO' partition the sector between the vectors T_1T_1' and T_2T_2' into $m + n$ equal segments of length p. This p is in fact the period of the fiber structure in the particular direction.

Suppose that the grid lines to be imaged alternate in the direction perpendicular to OO', i.e., each line is parallel to the vector OO'. In this case, the minimum grid spacing for lines that can be individually resolved is $p/2$. If this spacing is reduced, all the fine details are lost.

Using the notation of Figure 85, we write

$$\frac{D \sin \alpha_1}{n} = -\frac{D \sin \alpha_2}{m} = p, \quad \alpha_2 = \alpha_1 - \tfrac{1}{3}\pi.$$

Therefore

$$\frac{\sin \alpha_1}{n} = \frac{\sin \alpha_1 - \sin \alpha_2}{n + m} = \frac{p}{D}.$$

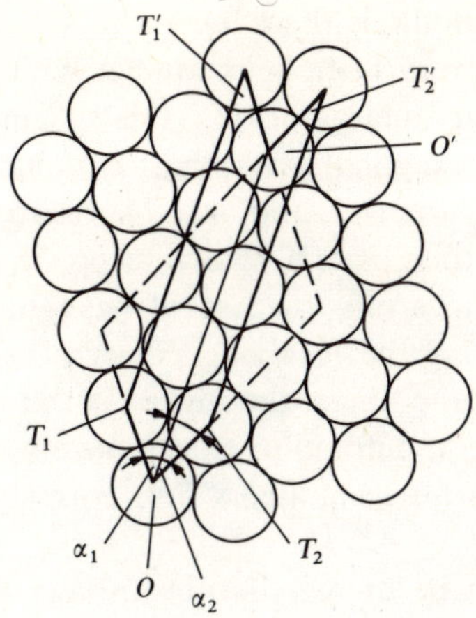

Figure 85
The period of a hexagonally packed bundle.

Since

$$(n + m) \sin \alpha_1 = \tfrac{1}{2} n \sin \alpha_1 + \frac{n\sqrt{3}}{2} \cos \alpha_2.$$

we have

$$\tan \alpha_1 = \frac{n\sqrt{3}}{n + 2m}$$

Hence

$$p = \frac{D \sin \alpha_1}{n} = D \frac{\sqrt{3}/(n + 2m)}{\{1 + [3n^2/(n + 2m)^2]\}^{1/2}} = \frac{D\sqrt{3}}{2\,(m^2 + n^2 + mn)^{1/2}}. \tag{285}$$

For $m = 0$ and $n = 1$, $p = D\sqrt{3}/2 \approx 0.87\,D$, which corresponds to the directions $\phi = 0, \frac{1}{3}\pi, \frac{2}{3}\pi$ (Figure 82*a*). For $m = 1$ and $n = 1$, $p = 0.5\,D$, which corresponds to $\phi = \frac{1}{6}\pi, \frac{1}{2}\pi, \frac{5}{6}\pi$. If, however, $m = 1$ and $n = 2$, we have $p = D\sqrt{3}/2\sqrt{7} \approx 0.327\,D$, etc. We can thus always choose a certain direction in the bundle cross section for which p is as small as desired compared to D. However, as p is decreased below the value $p = D$, the image quality deteriorates, since the illumination distribution in the plane of the given element (e.g., a line of the grid) of the final image does not correspond to the illumination distribution in the plane of that element in the transmitted object. Indeed, an increase of the resolving power in one direction entails a certain loss of resolving power in the perpendicular direction, since the distance between the centers of the fibers lying on the same line parallel to the given direction increases.

Since the transmitted image is a two-dimensional structure, the fiber bundle should have an adequate resolving power in two mutually perpendicular directions.

The resolving power of a bundle is therefore expressed in terms of the spacing between the layers in which every fiber is in contact with two adjacent fibers in the same layer. For example, for square packing, this spacing is equal to the fiber diameter D (Figure 82*b*), for hexagonal packing, it is equal to 0.87 D, and the corresponding resolving powers are $D/2$ and $D/1.74$, respectively. A bundle of fibers each 50 μ in diameter will thus have a resolving power of 10 lines/mm in case of square packing and 11.5 lines/mm in case of hexagonal packing.

The basic factor influencing the resolving power of a fiber system is clearly the fiber diameter. The thinner the fibers, the higher is the resolving power. However, in bundles of very thin fibers, diffraction effects become significant, causing cross-illumination of the ends of adjacent fibers and correspondingly lowering the resolving power.

The resolving power can be increased substantially without reducing the fiber diameter. The two ends of the bundle (and for short bundles, the entire bundle) are made to oscillate parallel to the image planes [99]. To stress that the information in this case is transferred by a moving fiber, this method is called *dynamical scanning.*

The reason for the increase of resolving power by dynamical scanning becomes clear if we remember the existence of blind gaps in the bundle cross section. By making the bundle scan the image, we drive the active parts of the fibers across all the details of the transmitted image.

A similar result can be accomplished by moving the image. Synchronously vibrating glass plates mounted before the entrance and the exit ends of the fiber element are used. A ray hitting a plane-parallel plate at a nonzero angle is displaced, while retaining its original direction of propagation. The amount of displacement depends on the thickness of the plate, its refractive index, and the angle of incidence. If such a plate is mounted in front of the entrance end of the bundle and is then made to oscillate parallel to the fiber end, the transmitted image is displaced in the plane of the bundle cross section. The glass plate can be replaced with a glass wedge, which displaces the light rays like a prism.

Dynamical scanning is fairly easy to achieve for fused fiber optics elements. However, with flexible bundles, the oscillatory motion of the two ends is difficult to synchronize, and dynamical scanning is not a trivial undertaking.

Using the results of [100], we will now estimate the gain in resolving power due to dynamical scanning. Let the illuminance in the plane of the exit end be described by a two-variable periodic function $G(x,y)$. If $\boldsymbol{d}_1$ and $\boldsymbol{d}_2$ are the corresponding vectors of periodicity in the directions x and y, the condition of periodicity of the function $G(x, y)$ can be written as

$$G(x, y) = G[(\boldsymbol{x} + a_1\boldsymbol{d}_1) + (\boldsymbol{y} + a_2\boldsymbol{d}_2)],$$

where a_1 and a_2 are integers.

This function can be expanded in a double Fourier series

$$G(x, y) = \sum_{h_1 h_2} C_{h_1 h_2} e^{2\pi i(h_1 x/d_1 + h_2 y/d_2)}, \tag{287}$$

where

$$C_{h_1 h_2} = \frac{1}{d_1 d_2} \int_0^{d_1} \int_0^{d_2} G(x, y)\, e^{-2\pi i(h_1 x/d_1 + h_2 y/d_2)} dx\, dy. \tag{288}$$

The elementary lattice described by the vectors $\boldsymbol{d}_1$ and $\boldsymbol{d}_2$ is conveniently replaced by the reciprocal lattice vector

$$\boldsymbol{W} = w_1 \boldsymbol{i} + w_2 \boldsymbol{j}, \tag{289}$$

where $w_1 = 1/d_1$, $w_2 = 1/d_2$, and $\boldsymbol{i}$ and $\boldsymbol{j}$ are the unit vectors along the axes x and y, respectively. Then

$$\boldsymbol{W}_{h_1 h_2} = h_1 w_1 \boldsymbol{i} + d_2 w_2 \boldsymbol{j} \tag{290}$$

and

$$\boldsymbol{W}_{h_1 h_2} \boldsymbol{r} = h_1 w_1 x + h_2 w_2 y. \tag{291}$$

We can thus rewrite (287) and (288) in the form

$$G(x, y) = \sum_{h_1 h_2} C_{h_1 h_2} \exp\left[2\pi i (\boldsymbol{W}_{h_1 h_2} \boldsymbol{r})\right]; \tag{292}$$

$$C_{h_1 h_2} = w_1 w_2 \int_0^{1/w_1} \int_0^{1/w_2} G(x, y) \exp\left[-2\pi i (\boldsymbol{W}_{h_1 h_2} \boldsymbol{r})\right] dx\, dy. \tag{293}$$

As we have noted above, the minimum grid spacing which ensures complete resolution of the lines by the bundle is limited by the fiber diameter. The fiber bundle thus acts as a filter. The frequency spectrum of the image on the exit end is obtained by multiplying the frequency spectrum of the transmitted image and the frequency response of the fiber, which may be written in the form [66, 137]

$$Y(W) = \left[\frac{2J_1(\pi W D)}{\pi W D}\right]^2, \tag{294}$$

where W is the spatial frequency of the fiber bundle, D is the fiber diameter, J_1 is a first-order Bessel's function of the first kind.

Using (294), we write the distribution of illumination in the plane of the entrance end in the form

$$G'(x, y) = \sum_{h_1 h_2} C_{h_1 h_2} Y(\boldsymbol{W}_{h_1 h_2}) \exp\left[2\pi i (\boldsymbol{W}_{h_1 h_2} \boldsymbol{r})\right], \tag{295}$$

where

$$Y(\boldsymbol{W}_{h_1 h_2}) = \left[\frac{2J_1\{\pi D[(h_1 W_1)^2 + (h_2 W_2)^2]^{1/2}\}}{\pi D[(h_1 W_1)^2 + (h_2 W_2)^2]^{1/2}}\right]^2. \tag{296}$$

Let us now apply the above relations to the case of dynamical scanning of some oriented periodic function [100] shown in Figure 86 by the solid curve. We have from the figure

$$w_1 = \frac{\sin\theta}{N}, \quad w_2 = \frac{\cos\theta}{N}.$$

$$\boldsymbol{W}_{h_1h_2}\boldsymbol{r} = \frac{h_1 x \sin\theta}{N} + \frac{h_2 y \cos\theta}{N}$$

and

$$|\boldsymbol{W}_{h_1h_2}| = \frac{1}{N}(h_1^2 \sin^2\theta + h_2^2 \cos^2\theta)^{1/2}. \tag{297}$$

Suppose that in the coordinates x', y' the function G is expressed in the form

$$G(x', y') = A \sin^2 \frac{\pi x'}{N} = \frac{A}{2}\left(1 - \cos\frac{2\pi x'}{N}\right). \tag{298}$$

Rotating the axes through an angle θ, we obtain

$$G(x, y) = \frac{A}{2}\left[1 - \cos\left(\frac{2\pi \sin\theta}{N} x\right)\cos\left(\frac{2\pi\cos\theta}{N} y\right) - \right.$$
$$\left. - \sin\left(\frac{2\pi\sin\theta}{N} x\right)\sin\left(\frac{2\pi\cos\theta}{N} y\right)\right]. \tag{299}$$

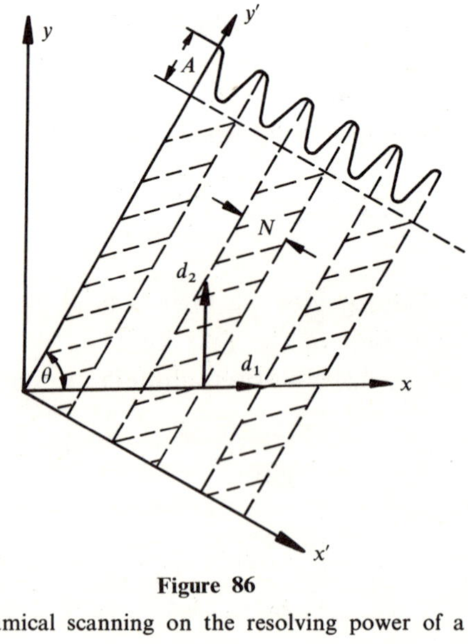

Figure 86

The effect of dynamical scanning on the resolving power of a fiber system.

Choosing fixed θ and N, we can find the function $G'(x, y)$. For $\theta = 90°$,

$$G(x, y) = \frac{A}{2}\left(1 - \cos\frac{2\pi x}{N}\right). \tag{299a}$$

Taking $C_{h_1 h_2}$ from (293) and $Y(W_{h_1 h_2})$ from (296), we obtain*

$$G'(x) = \frac{A}{2}\left[1 - \left(\frac{2J_1(\pi D/N)}{\pi D/N}\right)^2 \cos\frac{2\pi x}{N}\right]. \tag{300}$$

Analyzing this expression we conclude that the function $G'(x)$ produces the same illumination distribution along the axis x as (299a), but with a different depth of modulation, which is determined by the coefficient before $\cos(2\pi x/N)$.

Bessel's function entering this expression is an oscillating function which vanishes for certain values of the argument. We see from (300) that for $J_1 = 0$ the exit end of the fiber bundle is uniformly illuminated. The first zero of the function $J_1(z)$ obtains for $z = 3.83$. Therefore, the condition $z = \pi D/N = 3.83$ gives the minimum spacing of the image lines which can be transmitted by dynamical scanning,

$$W_{max} = \frac{1}{N_{max}} = \frac{3.83}{\pi D} = \frac{1.22}{D}.$$

The function $J_1(z)$ also vanishes for higher values of the argument. In these cases, however, we get $N \ll D$, i.e., two or more lines fall inside a single fiber. These lines naturally cannot be resolved, and we should therefore consider the behavior of $J_1(z)$ for its first root only.

We see from the above results that for square packing

$$\frac{W_{max\,dyn}}{W_{max\,stat}} = \left(\frac{1.22}{D}\right) 2D = 2.44,$$

and for hexagonal packing

$$\frac{W_{max\,dyn}}{W_{max\,stat}} = \left(\frac{1.22}{D}\right) 1.74D = 2.12.$$

Thus the application of dynamical scanning with square packing increases the resolving power of a fiber bundle by a factor of 2.44, and with hexagonal packing by a factor of 2.12. This result was checked using the experimental setup shown in Figure 87.

The image of a test pattern—a periodic structure with elements of various illuminations—is projected onto the entrance end of a bundle with a $50 \times 25\ mm^2$ rectangular cross section, made up of fibers 1 mm in diameter and 30 mm long.

* Eq. (300) differs from the corresponding expression in [100] in the power of the coefficient before $\cos(2\pi x/N)$.

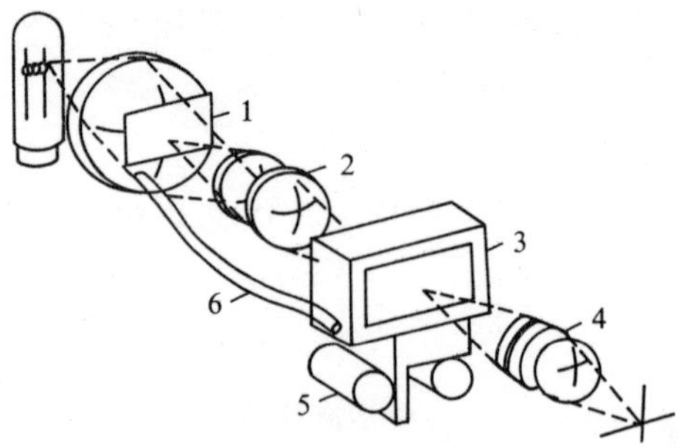

Figure 87

An experimental setup for checking the effect of dynamical scanning on the resolving power of a fiber element:

1) test object, 2) condenser, 3) fiber bundle, 4) objective, 5) vibrator, 6) a single fiber.

A vibrator moved the bundle in a plane perpendicular to its axis. A single fiber, placed in contact with the bundle, served as a light pipe recording the trajectory of the scanning bundle. The image formed on the exit end was photographed. The maximum vibration frequency reached 5 Hz, and during the exposure time the bundle would complete about 80 cycles.

Photographs of the test pattern and its image obtained with a static bundle and with dynamical scanning reveal a quite satisfactory transmission for a scanning amplitude of $4D$. If the scanning is done in a circular sense, the transfer quality is poorer, since not all the points of the test pattern are scanned to the same extent. The image formed by a static fiber shows excessive graininess, adequately resolving only the large features. Similar results, although obtained with test patterns of a different kind, are shown in Figure 88. Comparison of the static and the dynamical-scanning images shows that the latter technique has marked advantages.

The image of a grid of equidistant lines of various widths and intensities, obtained by dynamical scanning with a bundle of 1 mm dia. fibers was analyzed densitometrically to establish a quantitative relationship between the fiber diameter and the image line width. We see from Figure 89 that the width of all the lines is virtually the same, being close to the fiber diameter. For comparison, the dashed lines indicate the true width of the grid lines. This result clearly illustrates the dependence of the resolving power on diameter.

In practice, the image is often transformed by a succession of two or more numerous elements. This combined arrangement, however, has a lower resolving power. For example, for two successive elements of equal resolving power, the resulting static resolution drops to 0.707 of the original value. For a succession of three identical elements, it falls to 0.578, etc. [115]. The decrease in the resolving power of a fiber system with the increase in the number of components is attributable to the inevitable misalignment of the various elements one relative to another. As a result,

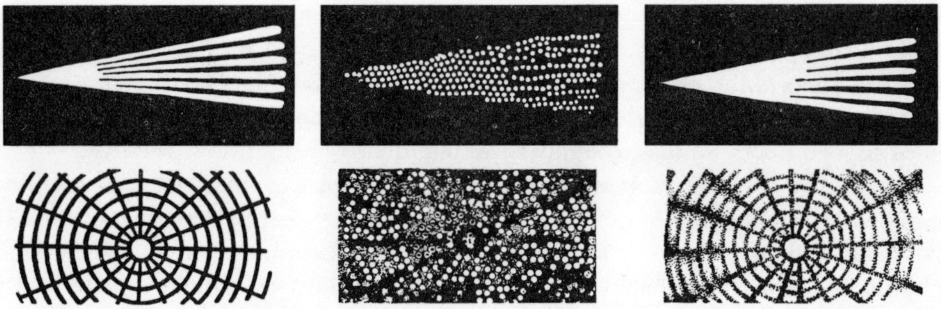

a b c

Figure 88

The increase of resolving power due to dynamical scanning:

a test object, *b* image of test object with static bundle, *c* image of test object obtained by scanning bundle; the maximum distance between the lines of the top test object is 1.4 *D*, the distance between the circles of the bottom test object is 0.75 *D*.

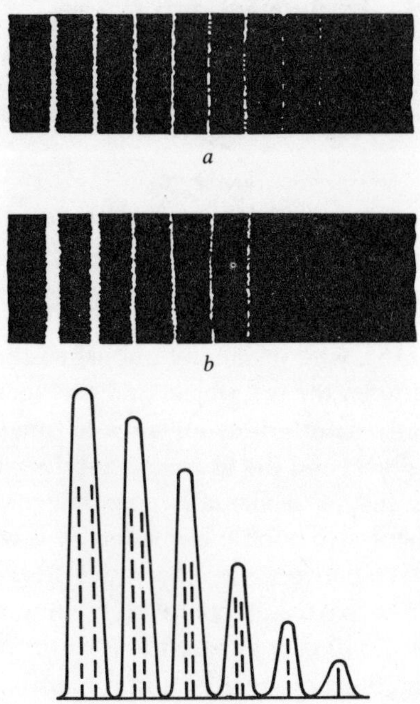

a

b

c

Figure 89

The transmitted image of lines of various widths (left to right: 0.5 *D*, 0.25 *D*, 0.13 *D*, 0.063 *D*, 0.031 *D*, 0.07 *D*):

a object, *b* image, *c* densitometric curve of intensity distribution in the final image.

the light emerging from a certain fiber of the first element may hit, at least partially, a blind spot on the entrance end of the next element or it may split between two adjacent fibers of this element. Therefore, while the primary fiber carried information regarding a single image element, the same function in the second bundle is fulfilled by two fibers, so that, roughly speaking, the resolving power is halved.

The decrease of the resolving power observed when two fiber elements are used is shown in Figure 90. Figure 90*a* is a photograph of a test pattern containing 114 lines/mm, taken with a 70 × microscope with diffuse illumination. The middle photograph (Figure 90*b*) shows the test image transmitted through a plate of 5-μ fibers of 1.01 numerical aperture, placed in direct contact with the test object. Figure 90*c* was obtained with two such plates, arbitrarily superimposed. The resolution of the fine details at the center, where the feature density is the highest, is clearly poorer.

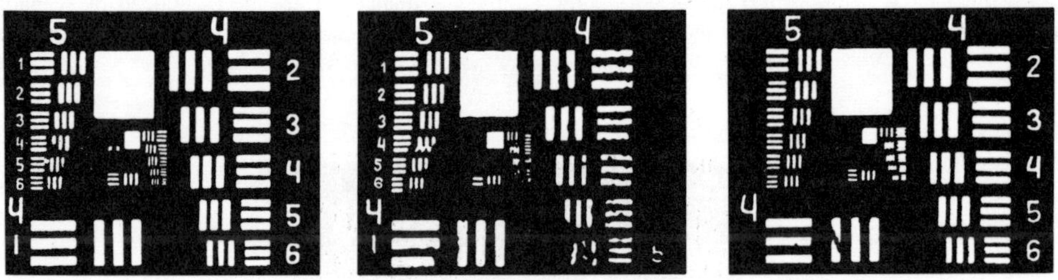

a *b* *c*

Figure 90

Variation of resolving power with the increase in the number of fiber elements: *a* photograph of test object, *b* image transmitted through a single fiber plate, *c* image transmitted through two fiber plates. Magnification, 70 ×.

Krakows and Baker [115], who originally published these photographs, indicate that imperfect contact between the test object and the fiber plate has a drastic effect on image quality. It greatly deteriorates for gaps as small as a few hundredths of a millimeter between the object and the fiber element. Dynamical scanning markedly improves the image, and ensures a virtually constant resolving power irrespective of whether one or two successive bundle elements are used.

Imperfect optical contact between the successive fiber-optics elements leads to Fresnel losses. They can be partly or completely eliminated by using immersion liquids or optical coating. Coating is apparently a better policy, since it reduces the reflection coefficient to fractions of a per cent in a relatively wide spectral region.

The density of the transmitted information and the resolving power of the bundle are thus mainly determined by the fiber diameter. The reduction of the fiber diameter is limited by diffraction, which causes intensive leakage of radiation through the fiber surface [144]. Diffraction is particularly harmful for thin fibers with diameters small compared to the wavelength of transferred light. Experimental studies show that bundles of 5-μ fibers still possess a sufficiently high resolving power, but further reduction of the fiber diameter leads to a marked deterioration of the transmitted image. In the visible spectrum, fiber diameters of 1 μ are inadvisable.

4. Image contrast

Image contrast is defined as the ratio

$$C = \left| \frac{I_{max} - I_{min}}{I_{max} + I_{min}} \right|, \tag{301}$$

where I_{max} is the intensity corresponding to the maximum illuminance of a certain element of the image; I_{min} is the intensity corresponding to the minimum illuminance of the adjacent element.

For example, if the image is made up of a succession of dark and light stripes, I_{max} corresponds to the center of a light stripe, and I_{min} to the center of the adjacent dark stripe. For $I_{min} = 0$, the image contrast reaches its maximum value of 1.

The contrast of the transmitted image is intimately linked with the resolving power of the system. It should therefore depend on the orientation of the image elements relative to the principal directions of the bundle. The image orientation is particularly significant if the width of the image elements is close to the fiber diameter.

Consider a light beam of periodic spatial structure incident on an elementary layer of closely packed fibers. The intensity distribution in the beam is shown in Figure 91. The illumination peaks and valleys are of the same width, both equal to the fiber diameter. Assuming that the peaks are centered on the respective fibers and that there is no leakage in the bundle, we obtain an image with $C = 1$. If the grid is displaced relative to the fiber element, the contrast diminishes, since each bright element is split between two adjacent fibers. The contrast is zero if the grid is displaced so that the peak midpoints coincide with the points of contact of the fibers.

C as a function of the grid displacement relative to the fiber centers can be estimated as follows. Let X be the overlap between the image element and the fiber end in the direction of the axis x (Figure 92); the illumination distribution in this

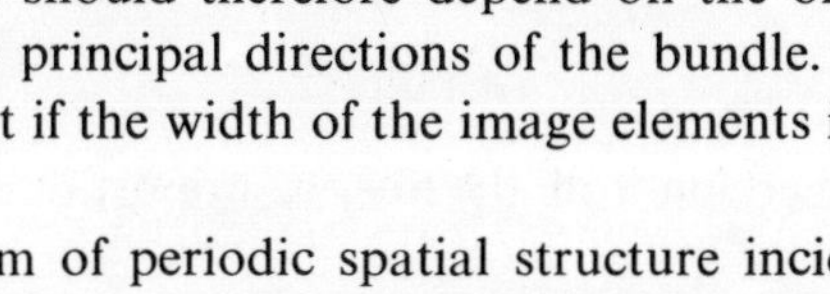

Figure 91

The contrast of a grid image transmitted by an elementary fiber bundle.

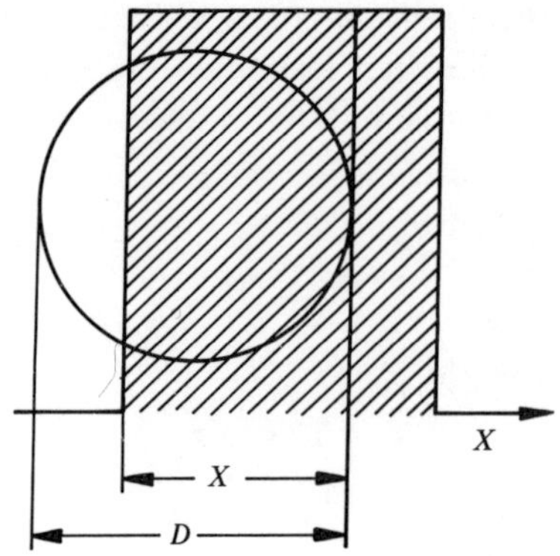

Figure 92

Distribution of illumination in the plane of the entrance end of a fiber produced by a grid image.

overlap is described by the function $G(X)$. Now let the shape of the entrance end be described by the function $F(x)$. The origin coincides with the fiber center. The luminous flux reaching an element dx of the entrance end is thus

$$G(X)F(x)dx.$$

The luminous flux emerging from the fiber is proportional to the incident flux, i.e.,

$$\Phi(X) \sim \int_0^D G(X)\,F(x)\,dx.$$

In what follows, we ignore reflection and scattering losses, i.e., the two fluxes are equal.

Since the fibers are circular cylinders, the function $F(x)$ may be expressed in the form

$$F(x) = 2\,[x\,(D-x)\,]^{1/2}, \quad 0 \leqq x \leqq D. \tag{302}$$

In accordance with figure 91, the function $G(X)$ is taken in the form of a square pulse of unit amplitude. Since X is the overlap interval between the image element and the fiber end (Figure 92), the luminous flux incident on the entrance end is

$$\Phi(X) = K\,2\int_{D-X}^{D} [x\,(D-x)\,]^{1/2}\,dx, \tag{303}$$

where K is a normalization constant, defined by the equality

$$K\,2\int_0^D [x\,(D-x)\,]^{1/2}\,dx = 1,$$

which implies that the luminous flux reaching the fiber reaching the fiber cross section is unity.

From (303) we have

$$\Phi(X) = K\,2 \int_{D-X}^{D} \sqrt{\left(\frac{D}{2}\right)^2 - \left(x - \frac{D}{2}\right)^2}\, dx = K\,2 \int_{a-X}^{a} \sqrt{a^2 - t^2}\, dt, \tag{303a}$$

where

$$a = \frac{D}{2}, \quad t = x - \frac{D}{2}.$$

Finally this gives

$$\Phi(X) = \frac{1}{2} + \frac{2}{\pi D^2}(2X - D)\left[X(D - X)\right]^{1/2} + \\ + \frac{1}{\pi}\sin^{-1}\left(\frac{2X}{D} - 1\right). \tag{304}$$

From (304) we see that for $X = 0$, $\Phi(X) = 0$, and for $X = D$, $\Phi(X) = 1$, i.e., as X increases from 0 to D, the emergent light flux increases from 0 to 1. A plot of the function $\Phi(X)$ is shown in Figure 93. The illumination of a row of fibers (like the luminous flux shown in Figure 91) as a function of X is thus described by a saw-tooth curve.

To establish the variation of the image contrast, let us consider the variation of the illumination in the end plane of two adjacent fibers.

Let the original image correspond to that shown in Figure 94*a*. In case of no leakage, the contrast should be 1. Shifting the test object in the direction marked by the arrow, we reduce the illumination of the first fiber and increase that of the second. For $X = 0.5\,D$, both illuminations are equal and the contrast is zero. If X is further increased to $X = D$, the contrast again rises to unity, since the illumination of the second fiber is maximum and that of the first zero. The variation of contrast and illumination of both fibers as a function of the displacement X is shown in Figure 94*b*.

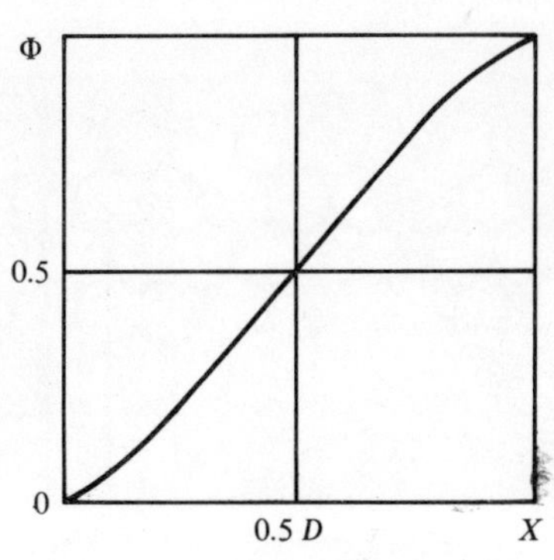

Figure 93

The light flux emerging from a fiber vs. the displacement X of the grid line relative to the fiber center.

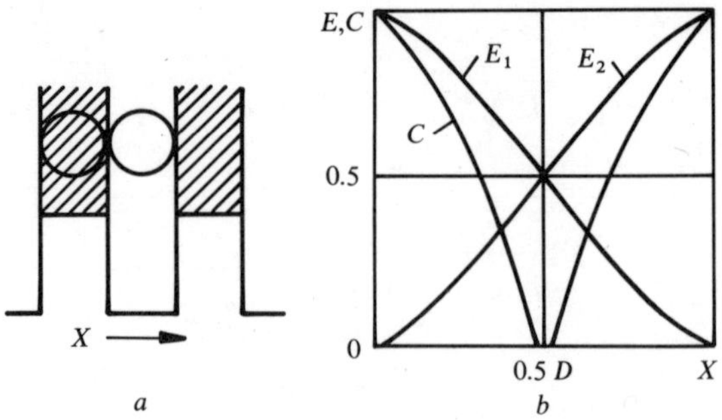

Figure 94

The relative position of grid lines and two fibers (a) and the variation of the illuminations E_1, E_2 of these fibers and the image contrast C as a function of the displacement X (b).

Let us now consider a more realistic case, when the image of a grid is transmitted by a bundle of hexagonally packed fibers. With this packing (Figure 95, right), the image contrast in the horizontal direction is described by the curve shown in Figure 94b. The contrast between adjacent rows of fibers shows a phase difference of π.

An entirely different picture is observed when the lines of the test grid are oriented in the azimuthal direction $\phi = \pi/3$ (Figure 95, left). The test pattern in this case cannot be displaced to produce maximum illumination in the plane of one fiber and zero illumination on the end of the adjacent fiber.

Let us examine the variation of contrast in this case. The original position is that shown in Figure 96a: the illumination of the first fiber is zero and that of the second fiber is close to unity, so that the contrast is maximum. As the test pattern is displaced, the contrast decreases. For $X = 0.43\ D$, the contrast drops to zero, since the illumination of both fibers is the same. Further displacement of the test object increases the illumination of the second fiber, while the illumination of the first fiber continues decreasing. The contrast nevertheless increases, reaching its maxi-

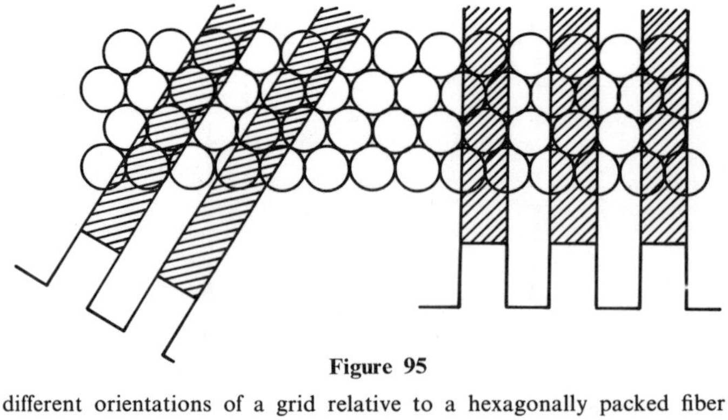

Figure 95

Two different orientations of a grid relative to a hexagonally packed fiber bundle.

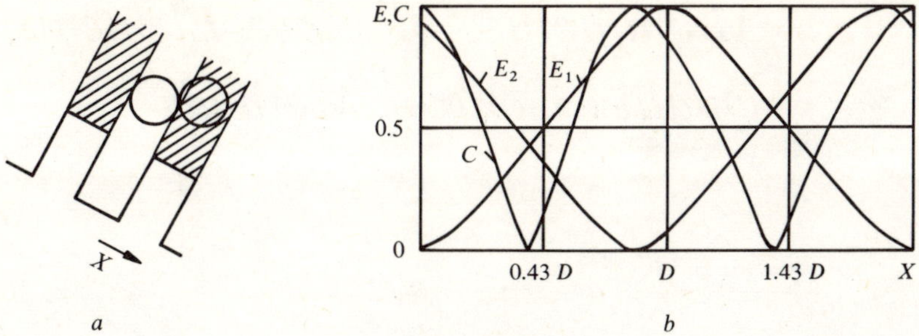

Figure. 96

Relative orientation of grid lines and two fibers in case of maximum contrast (*a*) and curves plotting the illuminations E_1, E_2 of two adjacent fibers and the contrast C as a function of the displacement X (*b*).

mum for $X = 0.87\,D$. For $0 \leqq X \leqq 2D$, the functions $E_1(X)$, $E_2(X)$, and $C(X)$ are plotted in Figure 96*b*.

Zero leakage is assumed, although the fibers do not have optical insulation. Since leakage is inevitable in such fibers, the actual contrast will be somewhat less than shown by the curves. Theoretical and experimental studies [97] carried out with test objects of sinusoidal illumination distribution [101, 102] show that if the intensity of the luminous flux emerging from the six surrounding fibers is 5–10% of the flux intensity emerging from the central fiber, the image quality remains quite satisfactory.

5 Manufacture of Fiber Elements and Quality Control Methods

1. Manufacture of glass fibers and bundles

To manufacture a single fiber, a glass rod with a high refractive index is fitted into a glass tube of low refractive index, and the assembly is softened in a cylindrical furnace. The glass, softened to a sufficient degree, is then drawn into a fiber of the required diameter.

The fiber drawing setup is shown in Figure 97. Above the vertical cylindrical furnace 1, there is a feeding contraption 2 into which the glass rod 3 is loaded. The drawn fiber 4 is wound onto a traction drum 6. The gap between the walls of the glass tube—the casing—and the core rod need not be particularly small [97].

The optimum furnace temperature is chosen by trial and error, depending on the grade of the glass used.

The same figure shows a small furnace 5 of special design, which contains molten aluminum. If necessary, the fiber is drawn through the metal bath acquiring a thin protective metal film in the process.

Drawing assemblies of this design will produce long fibers with diameters ranging from a few millimeters to several microns. The fiber diameter depends on the dimensions of the tube and the rod, the rate of feed of the glass into the furnace, the furnace temperature, and the drawing speed.

Bundles of fibers can be manufactured by close packing of the individual threads (Figure 98). A long glass fiber is densely wound onto a circular template with a rectangular cross-section groove along the edge. The groove dimensions are determined by the required bundle diameter. The resulting toroidal bundle of fibers is fastened in several places by special brackets; the template is then disassembled, and the bundle is cut up into sections of desired length. The cuts are generally heated to the softening point and pressed until bonded. The fastening brackets then can be removed. The fiber casings fuse, retaining the coherent packing of the fibers in the bundle. The packing coherence was checked in [88] for a square cross-section bundle made up of fibers 0.06 mm in diameter. A knife-edge was slowly moved across the entrance

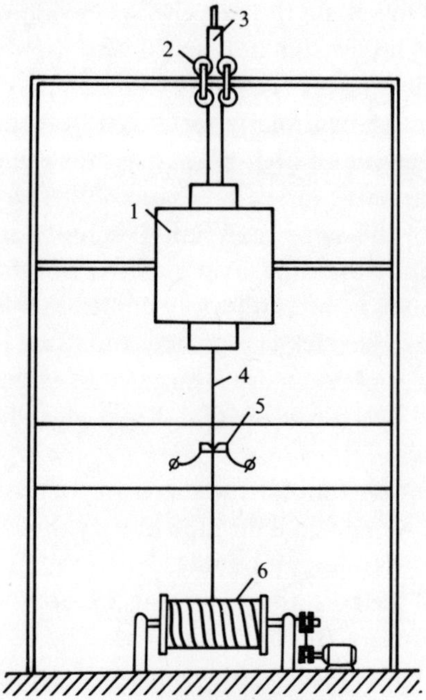

Figure 97

A machine for drawing of single fibers.

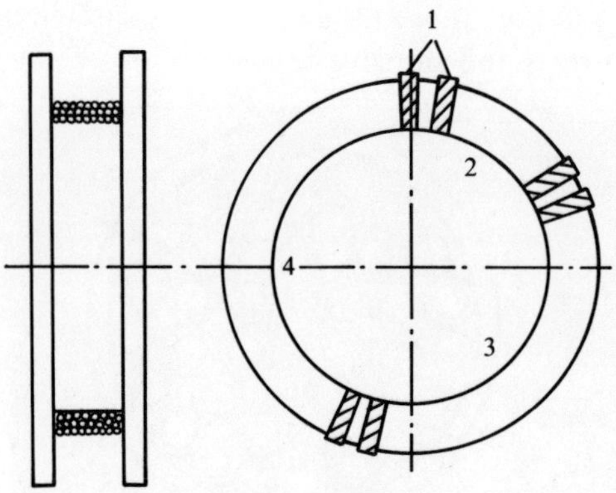

Figure 98

Manufacture of fiber bundles by close packing of fibers into a grooved template:

1) clamps, 2, 3, 4) fibers.

end of the bundle, and its image on the exit end was examined through a low-power microscope. The scanned image did not reveal any discontinuities, which proved the coherent packing of fibers over the entire bundle cross section.

Multithread fibers also can be directly produced in furnaces of the above design [6, 98]. We start with a bundle of high-refractive-index glass rods, each encased in a glass tube of a lower refractive index n_2 (Figure 99). The entire bundle is then inserted into a large tube of the same material. The individual tubes fusing between themselves and with the encasing tube form a matrix of refractive index n_2, in which light pipes of refractive index n_1 are uniformly embedded. Multithread fibers manufactured in this way can then be packed regularly and fused into a single thick bundle. The same method can be applied to draw bundles of other than circular cross section, e.g., multithread ribbons made up of colored glass fibers (these may be used in color television [22]).

The last method suggested for the manufacture of thick bundles is preferable to the previous method if we need a light pipe of very thin fibers, which are virtually impossible to pack into a coherent bundle.

The refractive indices of the core and the casing are not the only factors to be taken into consideration in choosing the fiber material. The temperature expansion coefficients, the softening temperatures, the chemical compatibility of the various glasses, and other factors must be taken into consideration.

A highly important characteristic is the transparency, which together with the numerical aperture primarily determines the choice of the fiber material [7, 146]. Since the increase in the refractive index of the core raises the numerical aperture and lowers the transparency (especially in the blue part of the spectrum, see Figure 100), to ensure faithful color transmission we often have to partly sacrifice the numerical aperture and prefer high transparency glasses of moderate refractive index. For short fiber elements, e.g., those used to face the surface of electron-beam tubes, the numerical aperture is the decisive factor.

Figure 99

An assembly of glass rods and tubes for the manufacture of a multithread fiber.

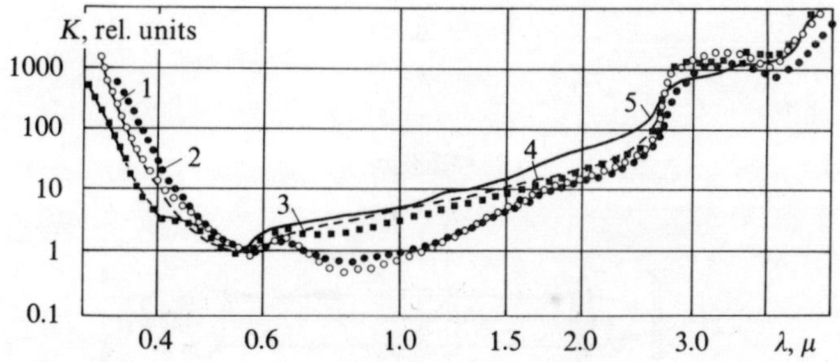

Figure 100

Spectral curves of the relative absorption coefficient of optical glasses:
1) flints, 2) heavy flints, 3) crowns, 4) heavy crowns, 5) baryte crowns.

Short bundles of high quality glass can be applied in a wide spectral interval of wavelengths, from 0.35 to 1.8 μ. Fiber elements based on arsenic trisulfide are being developed for the longer wavelengths of the infrared spectrum.

2. Quality control of fibers

Quality control of fiber surface and cylindrical geometry

The light transfer of a fiber is largely limited by losses associated with total internal reflections. Surface defects can be regarded as a departure from the ideal cylindrical shape of the fiber, which are detected and quantitatively assessed by usual interferometric methods. A diagram of the interferometric setup used in [93] is shown in Figure 101. The controlled fiber 5 is mounted in one of the interferometer arms. A standard surface 4 (either plane or cylindrical) is mounted in the second arm. Equal-thickness fringes localized on the surface of the test fiber image are viewed through the eyepiece. A narrow band filter is used to ensure monochromatic illumination.

Various interference patterns are observed depending on the geometry of the standard surface and the relative position of its image and the image of the fiber. Some of these patterns are shown in Figure 102. Comparison of the interference patterns shows that a high-quality cylindrical surface is most suitable as a comparison standard, e.g., the surface of a perfect glass fiber. Indeed, we see from Figure 102, *a* and *c*, that although the two patterns are identical, the equal-thickness interference fringes obtained with two cylindrical surfaces are more widely spaced, so that a larger area can be controlled at any one time. The density of the interference fringes in this case is determined by the ratio of the diameters of the fiber and the standard cylinder, whereas with a plane standard surface, the fringe density depends on the fiber diameter only.

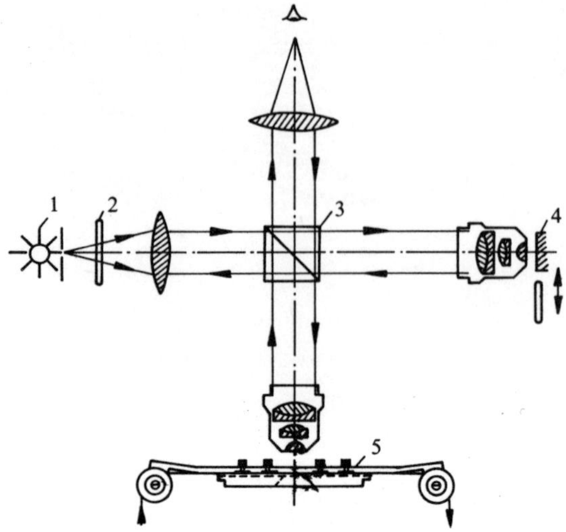

Figure 101

Interferometric apparatus for controlling the surface quality and the cylindrical geometry of fibers: 1) light source, 2) filter, 3) light-splitting cube, 4) standard surface, 5) test fiber.

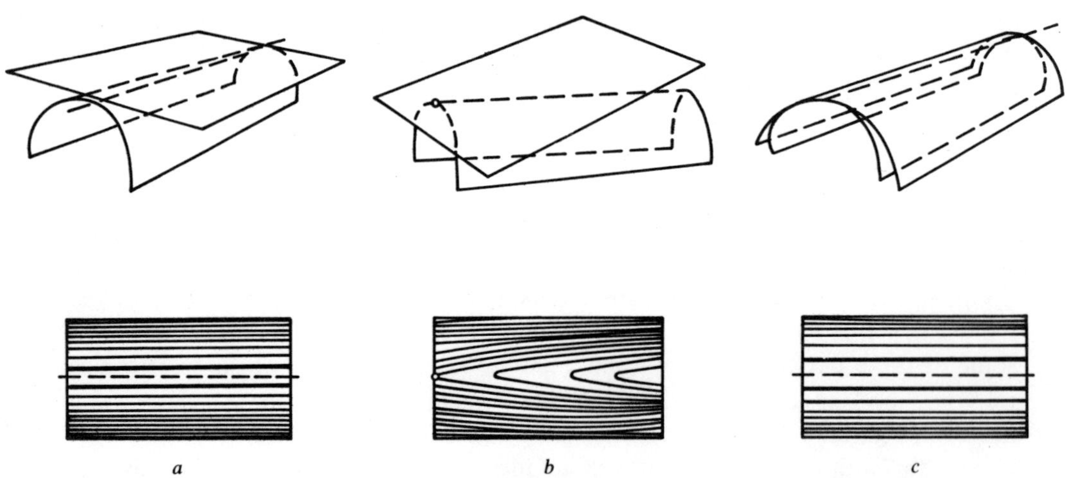

Figure 102

Interference patterns for standard surfaces of various geometries in various positions relative to the image of the test fiber:

a plane surface parallel to the fiber axis; *b* plane standard set at an angle to the fiber axis; *c* cylindrical standard.

This method will control both the surface quality and the cylindricity of the fibers. If the fiber has a perfect surface but shows a slight conical tapering, the interference pattern will change: the equal-thickness fringes will not be strictly parallel.

The interference pattern observed with the interferometer shown in Figure 101 characterizes the quality of a small part of the surface, limited by the instrumental field of view. To analyze the fiber surface along its entire length, the fiber should be moved uniformly in front of the objective, without the slightest vibrations. This naturally constitutes a serious obstacle to regular interferometric measurements.

Interferometric analysis of glass fibers 25 μ in diameter carried out by this method did not reveal any significant deviations from cylindrical geometry or any large local surface imperfections.

Some additional information on fiber defects was obtained in [93] by electron diffraction techniques. Almost all the specimens were found to be pitted with small depressions a few hundred Å in diameter and about 150 Å deep. The surface density of the pits is variable from one specimen to another, but on the average it is the same for freshly drawn and air-aged fibers. According to the authors, the density and the size of these surface defects do not affect the light transfer efficiency of the fiber.

Magnitude and homogeneity of the refractive index

Depending on the grade of glass and the fiber manufacture technology, certain deviations from the rated refractive index of the source material may be noted. There are also volume inhomogeneities in the refractive index distribution which affect the light-transfer efficiency: they cause additional light scattering. Homogeneity of the refractive index is highly significant for long fibers.

The refractive index of thick and short fibers can be monitored by interferometric, refractometric, and polarimetric methods, without damaging the fibers. Fibers examined between crossed polaroids [93] mostly reveal radially symmetric internal stresses.

Fairly long fibers cannot be tested by this method without breaking. One or several fibers are therefore generally extracted from each batch and cut up into short test specimens.

As an example, Figure 103 shows a microphotograph taken with a Twyman interferometer. It provides indication of the distribution of the refractive index and the homogeneity of the interface in a specimen cut from a 50-μ glass fiber [97]. The curvature of the interference fringes is apparently associated with internal stresses, which arise because of differences in the thermal properties of the core and the casing material or because of excessive fiber drawing rates. Analysis of such interferograms will reveal imperfections in the core–casing interface which lead to a marked increase in reflection losses.

The refractive index gradient along the axis can be readily determined by two-

Figure 103

Interference pattern of an encased fiber (Twyman interferometer).

beam interferometry. Specimens of equal length are cut from different parts of the fiber; one of these segments is placed in one of the interferometer arms, and the other segments are inserted one after another in the second arm. The interferograms give the difference in the mean refractive indices of the test segments.

The Lebedev polarization interferometer is highly suitable for these measurements (Figure 104). The interfering beams are split by two quartz plates Q_1 and Q_2 of equal thickness, whose principal planes are aligned parallel to each other. The plates are mounted between crossed nicol prisms N_1 and N_2, their principal planes making an angle of 45° with the plane of the plates.

A plane-polarized beam passing through Q_1 is split into two rays: ordinary and extraordinary. The half-wave plate P_1 rotates the plane of polarization of each ray through 90°. The paths of the rays in Q_2 are therefore symmetric to those corresponding to plate Q_1. The two rays thus emerge Q_2 with zero phase difference and total extinction is observed in the field of view.

If a transparent object with volume inhomogeneities is placed between Q_1 and Q_2, the light emerging from the optical system will be elliptically polarized. A similar effect is observed if two, and not one, inhomogeneous objects are placed in the interferometer arms.

The analyzer comprises a fixed quarter-wave plate and a rotating nicol N_2 with a graduated limb showing the rotation angles. The plate converts the light into linearly polarized, with the plane of polarization rotated through a certain angle relative to the initial direction; this angle is a function of the path difference. Having measured the angle of rotation of N_2 to total extinction, we can calculate the corresponding path difference and hence the difference in the refractive indices.

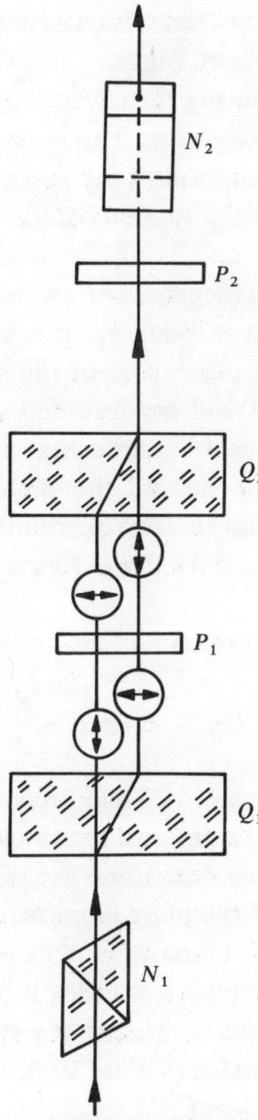

Figure 104
Optical diagram of the Lebedev polarization interferometer.

The Lebedev interferometer is a unique instrument due to the exacting quality standards met by the quartz plates Q_1 and Q_2. Because of the small fiber diameter, the beam splitting introduced by the quartz plates may reach as little as a few hundredths of a millimeter. This distance is determined by the plate thickness, and the size of the plates may therefore be small. In this case, a regular polarization microscope equippped with some extra components can be used as an interferometer [15].

The path difference in the Lebedev interferometer is measured to within $\lambda/100$, i.e., for sufficiently short specimens, the refractive index fluctuations can be controlled to three and even four significant digits.

A convenient method of measuring the refractive index of fibers is based on Fresnel diffraction at the interface of two media. The main advantage of this method is that it does not require special preparation of the specimen, and fibers of any shape may be used. We will consider one of the versions of this method, known as the Obreimov method [14, 29].

Consider an optically homogeneous plate of thickness d and refractive index n_2 immersed in a liquid of refractive index n_1. If a sufficiently wide parallel beam of wavelength λ is incident on the plate, part of the rays will pass through the liquid. Interference of diffracted waves will be observed along the geometrical shadow of the plate. If the path difference of the interfering rays is equal to a whole number of wavelengths λ, the field of view behind the plate will be uniformly illuminated. When the path difference is equal to a whole number of half-wavelengths, an interference minimum in the form of a dark fringe forms along the edge of the geometrical shadow.

The condition of illumination maxima and minima is written in the form

$$(n_2 - n_1)\,d = m\frac{\lambda}{2}, \tag{305}$$

where even m correspond to maxima, and odd m to minima.

If the dispersion curve of the liquid and the thickness of the test plate are known, edge diffraction can be applied to determine the dispersion curve of the plate. Conversely, if the refractive index of the plate is known, we can find the refractive index of the liquid. In this method, a plane-parallel plate is used as a standard. Its refractive index n_{st} should be close to the refractive index n of the test object immersed in the same liquid. The cell with the liquid is placed in a monochromatic light beam behind the exit slit S_p of the monochromator (Figure 105). To ensure high accuracy ($\sim 10^{-5}$) the liquid temperature is stabilized.

By changing the wavelength of the incident light, we can find the wavelengths λ_1 and λ_2 at which the refractive index of the liquid n_{liq} is equal to the refractive index of the standard and the specimen, respectively. In the former case, the contour of the standard vanishes from the field of view, and in the latter case the specimen vanishes. If the specimen is of nonuniform thickness, the entire outline will vanish only at the one wavelength for which $n_{liq} = n$. The standard, on the other hand, vanishes each time the path difference of the interfering waves is equal to a whole number of half-wavelengths (305). To identify the zero-order fringe $m = 0$, i.e., the case $n_{liq} = n_{st}$, the top corner of the standard plate 2 is ground flat, as shown in Figure 105. At this point, the standard vanishes only for $m = 0$.

The measurements are carried out as follows. The wavelengths $\ldots, \lambda_{-3}, \lambda_{-2}, \lambda_{-1}, \lambda_0, \lambda_1, \lambda_2, \ldots$, at which the shadow rim of the standard vanishes are marked

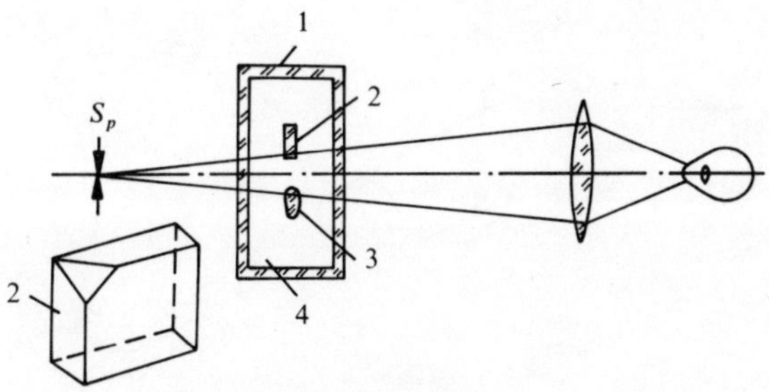

Figure 105

Schematic diagram of the refractive index measurement by Obreimov's method:

1) cell, 2) standard, 3) specimen, 4) liquid.

on the monochromator drum (λ_0 corresponds to the zero-order fringe). Given the thickness of the standard and its dispersion, we can calculate the difference in the refractive indices of the standard and the liquid at the corresponding wavelengths and thus plot the dispersion curve of the liquid. The refractive index of the test specimen at the particular wavelength coincides with the refractive index of the liquid if the specimen rim vanishes completely.

To find the dispersion curve of a specimen of an arbitrary shape, a selection of liquids should be used. However, the fiber ends can be always made plane parallel, so that one liquid with a known dispersion curve is sufficient for fiber quality control.

This method will rapidly establish the axial uniformity of the refractive index in fibers. In cases of high optical homogenity of the fiber, the shadow fringe of the specimens cut from different parts of the fiber should vanish at the same wavelength.

Some authors measure the homogeneity of the refractive index of fibers by a combination method which also incorporates certain features of multibeam interferometry. The experimental setup in this case is more complex, but the results are characterized by higher accuracy.

Consider the setup described in [93], where fringes of equal chromatic order were observed following multiple reflections in a Fabry–Perot interferometer. The test specimen was placed in an immersion liquid whose dispersion curve crosses the dispersion curve of the fiber in the middle part of the visible spectrum. The cell containing the liquid and the specimen is placed between the plates of a Fabry–Perot interferometer (Figure 106), which is illuminated by a parallel beam of white light. The image of the object is projected by a microobjective onto the entrance slit of the spectrograph. Fringes of equal chromatic order are thus observed simultaneously for the specimen and the immersion liquid. For wavelengths at which the refractive indices of the liquid and the specimen do not coincide, the interference fringes show a kink at the specimen–liquid interface. If the specimen is homogeneous, the magnitude of the kink for various λ can be used to calculate the dispersion curve

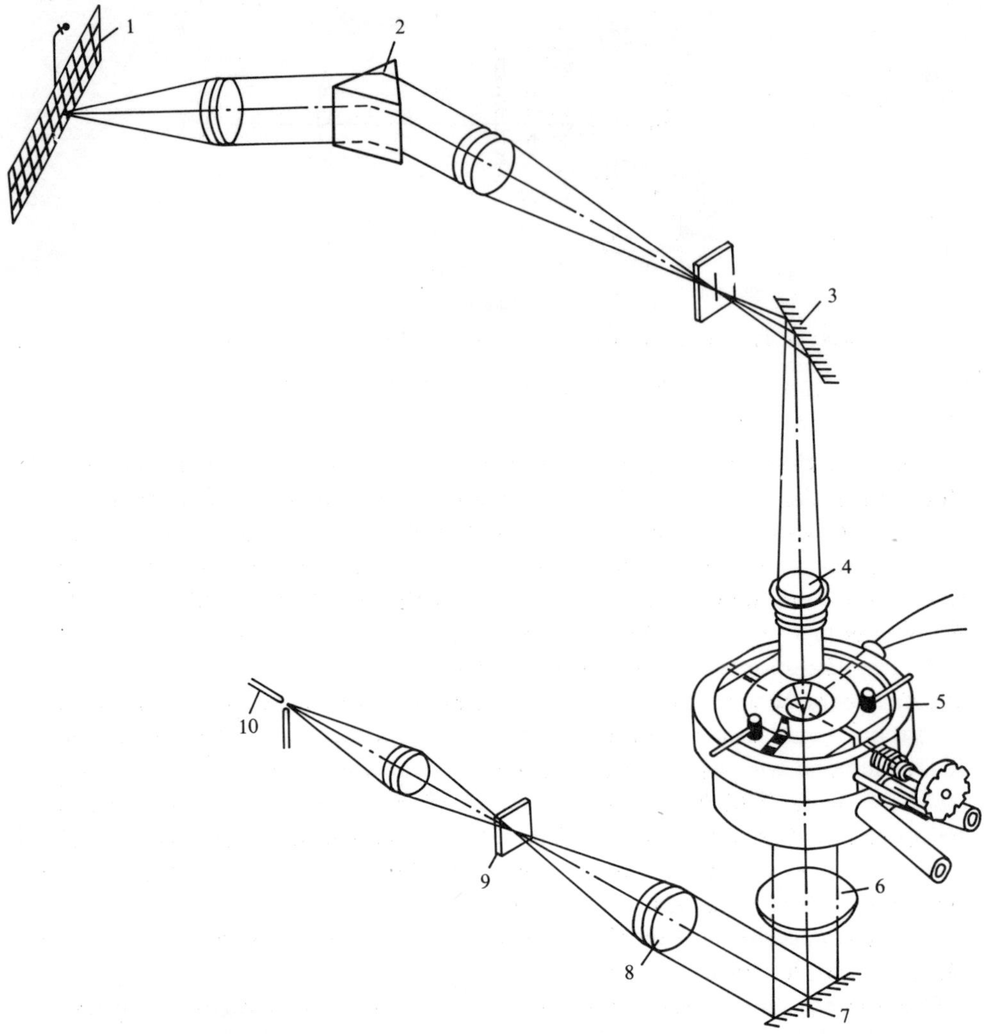

Figure 106

Optical diagram of the instrument for dispersion analysis of microscopic objects:

1) interference pattern, 2) spectrograph, 3) mirror, 4) microobjective, 5) table with specimen, 6) polarizer, 7) mirror, 8) collimator, 9) aperture stop, 10) light source.

of the specimen material. The calculations require knowledge of the exact thickness of the test specimen. In [93], the following method was used.

The temperature dependence of the liquid dispersion curve was first determined. Then the wavelength λ_0 corresponding to the "zero" point, i.e., the point when the refractive index of the specimen is equal to that of the liquid, was determined experimentally at various temperatures. The lines of equal chromatic order observed at this wavelength do not show any kink at the interface. If the specimen possesses high optical homogeneity, the interference fringes will be straight. This method gave refractive indices of transparent microobjects with an accuracy of not less than ± 0.0001.

If the specimen contains inhomogeneities, the interference fringes will never be parallel. Local variations of the refractive index can be determined by comparing different sections of the specimen with the surrounding liquid.

Birefringence

Fibers become birefringent because of inhomogeneities in the refractive index. The magnitude of birefringence is determined by the fiber material and the manufacture technology.

Studies of glass fibers of various diameters carried out in [93] with a polarization microscope show that on the whole the birefringence is negligibly small. It is much more pronounced, however, in fibers manufactured from various synthetic materials. As an example, Figure 107 shows photomicrographs illustrating the effect of birefringence for fibers of various diameters. The same figure shows the radial distribu-

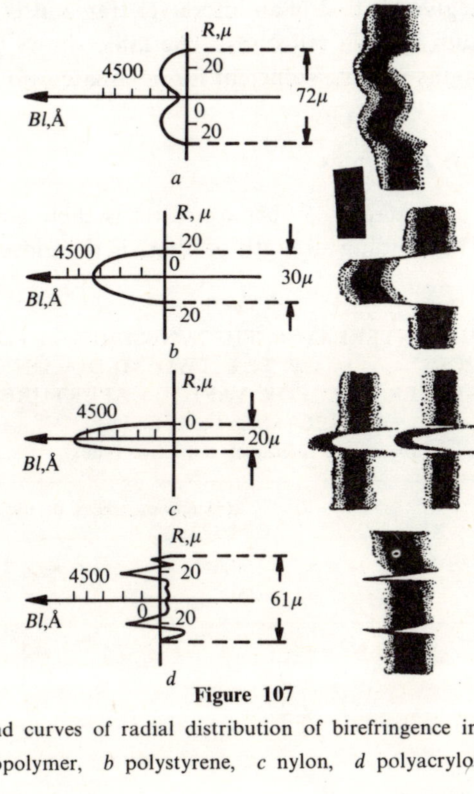

Figure 107

Photomicrographs and curves of radial distribution of birefringence in some synthetic fibers: *a* copolymer, *b* polystyrene, *c* nylon, *d* polyacrylonitrile.

tion curve of the birefringence, obtained from a quantitative analysis of microphotometer tracings. The horizontal axis gives the values of Bl (B is a birefringence characteristic, l is the specimen length) and the vertical axis indicates the fiber radius. Figure 107 shows that the birefringence effect is significantly different in fibers made of different materials: in nylon fibers, for example, the birefringence is several orders of magnitude higher than in polystyrene fibers. Moreover, various fibers differ not only in the magnitude of birefringence, but also in its radial distribution.

Casing thickness

To ensure maximum image contrast, the fiber casing should be as thin as possible. The lower limit is set by cross-coupling effects between adjacent fibers, i.e., by the depth of penetration of electromagnetic waves into the second medium following total internal reflection.

The minimum casing thickness was determined in [97] proceeding from the breakdown of the conditions of total internal reflection for plane-parallel interfaces. The corresponding results for various relative objective apertures (i.e., various incidence angles) and three values of the core-to-casing refractive index ratio are given in Table 3. The thickness given is that of an interlayer transmitting 0.001 % of the light into the outside medium in each reflection. The table shows that casing thickness of one or two wavelengths is quite sufficient for adequate optical insulation.

Mechanical strength of fibers

One of the principal advantages of fiber elements is their ability to transfer light along curved paths. The bending strength of fibers is therefore one of the most sig-

Table 3

THE MINIMUM INTERLAYER THICKNESS (IN λ) FOR VARIOUS REFRACTIVE INDICES $n_1 = n_3$ OF THE TWO MEDIA ON EITHER SIDE OF THE INTERLAYER AND FOR VARIOUS APERTURES OF THE PROJECTING OBJECTIVE

(n_2 is the interlayer refractive index)

Relative objective aperture	Refractive indices of the layers		
	$n_1 = n_3 = 1.60$; $n_2 = 1.00$	$n_1 = n_2 = 1.80$; $n_2 = 1.00$	$n_1 = n_2 = 1.80$; $n_2 = 1.50$
1:1	λ	$\lambda/2$	λ
1:2	$\lambda/2$	$\lambda/2$	λ
1:5	$\lambda/2$	$\lambda/2$	$\lambda/2$
1:10	$\lambda/4$	$\lambda/4$	$\lambda/2$

nificant parameters in fiber optics. Since the light-transfer efficiency of a fiber is virtually unaffected up to radii of curvature exceeding the fiber radius by a factor of about 20, it is the brittleness of the fibers that limits their application to devices where the optical image is transferred through a complex-configuration highly curved channel.

The minimum radius of curvature for which the fiber breaks is determined by its diameter and by the core and casing materials. The corresponding data [6] are plotted in Figure 108. The sloping straight lines in this figure are the lines of equal stresses, calculated from the equality

$$\sigma = E\frac{D}{2R}, \tag{306}$$

where D is the fiber diameter, R is the radius of curvature of the fiber, E is the elasticity modulus, equal to 5528 kg/mm^2 for the Soviet-made TF5 glass and to 8130 kg/mm^3 for K17 glass.

The dashed lines correspond to K17 glass, and the solid lines to TF5 glass. We see that while thick fibers—rods of 0.7–2 mm in diameter—break for radii of curvature $R = (200\text{–}300)\,D$, thinner ($D = 20\text{–}100\ \mu$) fibers can be bent to much smaller radii $R = (20\text{–}140)\,D$.

For synthetic fibers, the problem of bending strength does not arise because of their exceptional flexibility.

3. Ulexite fiber elements

The optical properties of some natural minerals are not unlike the properties of artificial fiber bundles. One of these peculiar minerals is ulexite $NaCaB_5O_9 \cdot 8H_2O$ [162]. The fibrous structure of this mineral is generally fairly randomized; however, some large specimens show a strict parallel alignment of the fibers. If free from extraneous impurities, these specimens are sufficiently transparent, and will give a fair resolving power, reaching 10 lines/mm (Figure 109).

Ulexite has a triclinic lattice. Its refractive indices are $n_\alpha = 1.491$, $n_\beta = 1.501$ and $n_\gamma = 1.520$. The optical axis C sometimes does not quite coincide with the OY axis. In this case, the intermediate refractive index n_β is slightly different from the above figure. If the two axes strictly coincide, each fiber, even if of ideally circular cross section, will transmit light like an elliptical cross section fiber because of the intrinsic biaxiality of the crystal.

The remarkable point is that the problem of optical insulation has been fully solved in the natural ulexite fiber bundles. The parallel alignment of the individual fibers in the bundle only fixes the position of the OY axis (the refractive index n_β),

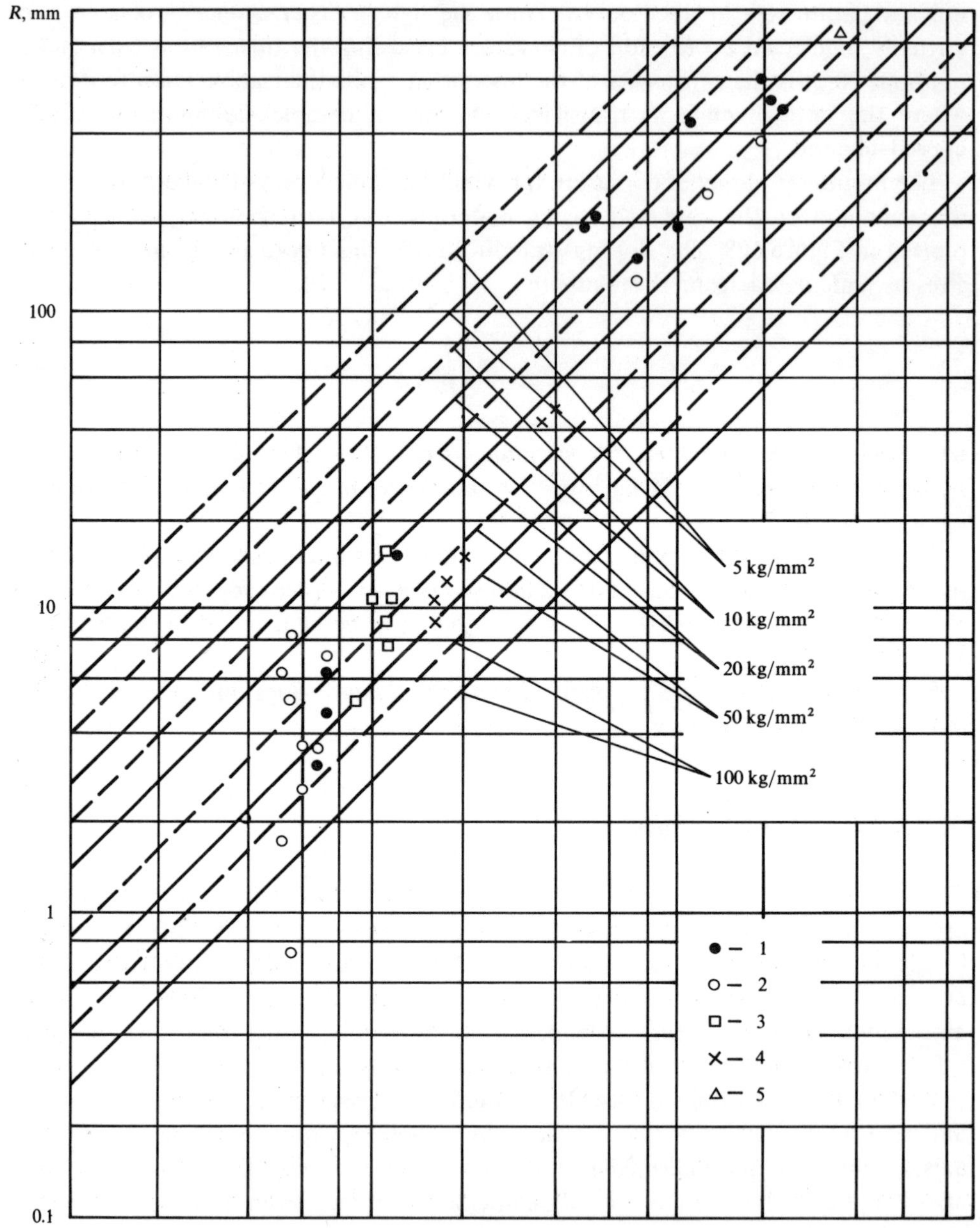

Figure 108

The ultimate radius of curvature of a light pipe vs. its diameter:

1) fiber and rods of TF5 glass; 2) encased fiber (TF5 glass core, K17 casing); 3) 25-thread bundle made of the same fibers; 4) VNIISV fiber; 5) fibers of K8 glass.

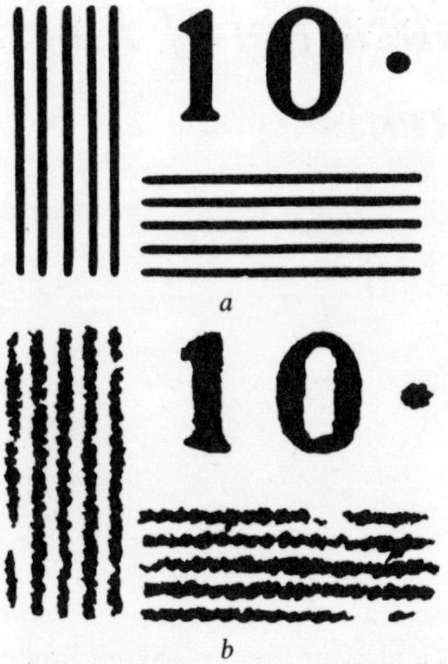

Figure 109

Image transmission by an ulexite bundle:

a object, *b* image.

and the other two directions are, in general, quite arbitrary. Therefore, the direction of the maximum refractive index n_γ in one fiber may coincide with some intermediate value of n in the adjacent fiber, in particular it may even be equal to n_α. As a result, an interface between media of different refractive indices is formed. The maximum difference in the refractive indices of the core and the casing may reach $n_\gamma - n_\alpha = 0.029$, which corresponds to a numerical aperture $A = 0.30$. In reality, however, A is somewhat lower, varying from fiber to fiber depending on their mutual orientation. Moreover, it should be remembered that of two adjacent fibers, only one is optically insulated in the particular direction, whereas for the other fiber the conditions of total internal reflection are not satisfied at all ($n_\alpha < n_\gamma$). Therefore, the numerical aperture of the entire bundle is equal to some statistical average of the numerical apertures of the constituent fibers.

6 *Application of Fibers in Optical Systems*

1. General remarks

The principal task of any optical system is to ensure faithful transmission of the image of various objects. In most cases, we are dealing with three-dimensional objects, and the imaging process involves projection (whether reduced or magnified) onto a plane perpendicular to the main optical axis. Because of aberrations, high-quality images of objects (not only three-dimensional, but even plane objects set perpendicularly to the optical axis) are very difficult to obtain. Aberrations are corrected by increasing the complexity of the optical system, and inevitably raising its price.

The formation of a plane image must meet the conditions of stigmatism, i.e., every point in the object space should be represented by a point, or a small disk, in the image space. However, even if the image is stigmatic, its quality may still be deficient. It is often required to reproduce faithfully the ratio of distances between different points of the object, to reproduce without distortion various angles in the object (e.g., a photograph of locality taken for purposes of cartography), etc. Distortion, however, cannot be corrected without markedly complicating the design of the optical system.

Some shortcomings of optical systems, including distortion, can be corrected with the aid of fiber elements. It is not always that fiber elements will replace correcting lenses, but they are quite useful for a number of particular problems considered below.

In Chapter 4 we have seen that the resolving power of a fiber element is determined by the diameter of the component fibers. The thinner the fibers, the finer the reproduction of details by the elements. Since the fiber diameter has a fixed lower-bound limit, combination of fiber elements with ordinary optical systems may often lead to an undesirable reduction of the overall resolving power. In these cases, fiber elements natually should be avoided.

2. Correction of distortion [103, 104] and field curvature

We distinguish between two different types of distortion correcting devices. One is intended for eliminating pincushion distortion, and the other for barrel distortion. These correcting devices are fairly difficult to manufacture, since tapered conical fibers are not enough: the fibers of the required shape should be packed in a certain way in bundles. So far, no industrial solution has been found to this problem. The situation is much better concerning fiber elements correcting field curvature aberrations, and we will discuss these elements in some detail.

In optical systems, especially photographic objectives, the Gaussian plane of the image is made to fit as close as possible the surface of the actually formed image. Unfortunately, this requirement is not always easy to meet.

In a number of systems, field curvature is such that the image is formed on a surface whose points lying off the optical axis are impermissibly far from the Gaussian plane. The Gaussian plane itself is tangent to the real image surface.

The field curvature of photographic objectives must be reduced to a minimum, since the photographic plate or film is always planar. In stigmatic lenses, field curvature will destroy the sharpness of the image off the principal optical axis and good focusing will be obtained only in a small patch of film around the optical axis. The image quality can be improved only by using a more complex and more expensive lens. However, by increasing the number of component lenses, we lower the speed of the objective, a highly undesirable side effect for systems used under low light conditions, e.g., in twilight.

Fiber optics has been successfully applied to correct field curvature of photographic objectives [44, 105, 122, 146]. The principle of a fiber correcting device is clear from Figure 110. This is a concave element made of constant cross-section fibers. The concave end has the shape of the Petzval surface on which the image is formed, and the plane end is the exit end of the system. The image transferred to the plane end can be contact-printed on a photographic plate or some other light-sensitive element.

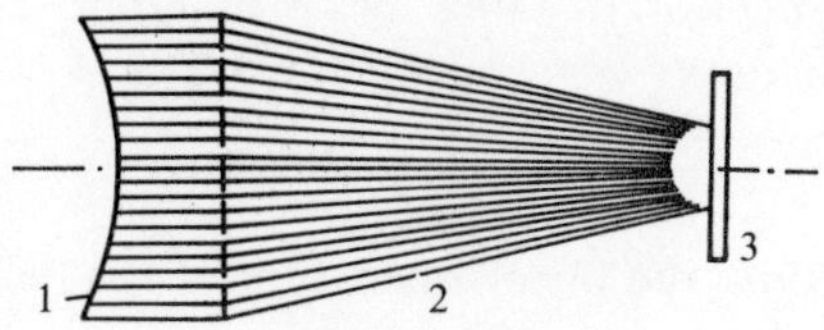

Figure 110

A fiber field curvature corrector:

1) correcting device, 2) conical fiber bundle, 3) photographic plate.

If the image should be magnified or reduced, a conical fiber bundle with plane ends is fitted behind the fiber corrector. Figure 110 shows a field curvature correcting device with a fitted conical element reducing the image scale. All the fibers in the conical bundle should have the same ratio of entrance to exit end radii, since otherwise additional distortion will be introduced in the transferred image.

The principal light losses in a field correcting device are the Fresnel losses at the two ends and absorption losses inside the fibers. Both categories of losses can be reduced to a minimum by optical coating of the fiber ends and an appropriate choice of the fiber glass.

Since even the best modern objectives transmit not more than 70–80% of the incident light, fiber elements can be advantageously used in some cases to correct the field curvature and to simultaneously increase the speed of the lens (by avoiding a multicomponent lens design).

The action of a fiber field corrector is described in [146]. The photographs taken by the authors with and without their fiber correcting device constitute the best proof of the efficacy of fiber elements in photography.

The imperfections in the image transmitted by fiber elements stem mainly from the mosaic structure of the bundles. The resolving power of optical systems with fiber correcting devices cannot exceed a certain value set by the fiber diameter. The best results are therefore attainable with very thin fibers. However, the lower-bound value of the fiber diameter is fixed by diffraction phenomena which also lower the resolving power. It would seem that a fiber correcting device should not introduce additional errors into the optical system, besides lowering the resolving power. Nevertheless, the field curvature corrector does introduce a certain additional distortion in the system. The inevitability of this additional distortion follows from Figure 111 which shows the exit pupil of the lens system and an axial section through the fiber field corrector. A ray emerging from the center of the pupil at an angle u to the optical axis crosses the Petzval surface (this is the corrector surface) and the Gauss plane at the points A and B, respectively. When the corrector is introduced, the ray meets the Gauss plane at the point C (and not B), whose position in the plane is fixed to within the fiber diameter. Thus, when the ray is intercepted by a fiber on the Petzval surface, additional distortion inevitably arises, since for different fibers the interception distance BC is different. This distance is related to the angle u by the equality

$$BC = AC \tan u.$$

Without fiber correction, the linear distortion of the lens systems is given by the difference

$$\Delta l = l' - l_0,$$

where l' is the true size of the image in the Gauss plane, and l_0 corresponds to the

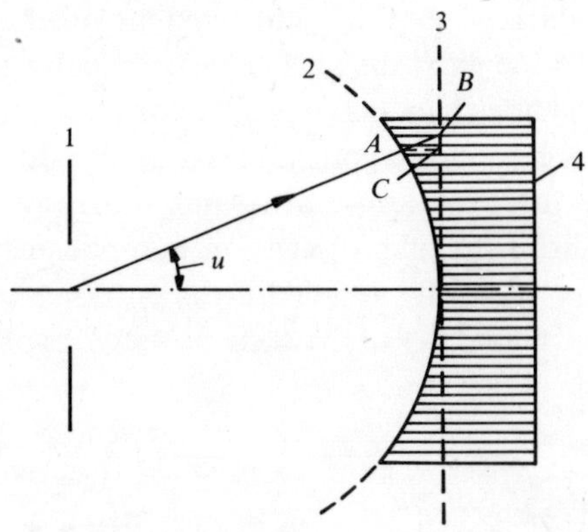

Figure 111

Distortion introduced by the fiber field flattener:

1) exit pupil of the lens optical system, 2) Petzval surface, 3) Gaussian image plane, 4) fiber field flattener.

size of the ideal image derived from the equations of Gaussian optics, without distortion. If the system magnification is V, and the object size is l, we have $l_0 = lV$.

With a fiber field corrector, the total image distortion is

$$\Delta l_p = \Delta l - AC \tan u,$$

and the relative distortion is given by

$$\frac{\Delta l_p}{l'} = \frac{\Delta l}{l'} - \frac{AC \tan u}{l'}. \tag{307}$$

In most cases, the distortion introduced by the field corrector is very small. For a lens with $f = 25$ mm and a Petzval surface with a radius of 37 mm, the values of $\Delta l_p/l'$ for $u = 20°$ does not exceed 3% [105]. Additional distortion, in excess of the permissible level, can be eliminated by suitable lens correction. This policy is adopted when the optical system without a fiber correcting device is on the whole ineffective or uneconomical. The same result can be achieved by introducing one more fiber element into the system, a distortion corrector. This approach is sometimes more profitable than lens correction.

The field corrector may lower the speed of the optical system. This shortcoming is readily avoided by using fibers whose numerical aperture is greater than the numerical aperture of the optical system.

The correcting device described above is intended for field curvature correction of photographic systems. Another device, not unlike the Fresnel lens, can be designed to correct the field curvature of any optical system. Let us consider the action of this device and its possible practical design.

Consider a sufficiently narrow cone of light incident on the entrance end of the fiber at a certain angle θ to the axis; the axis is perpendicular to the two fiber ends. The rays emerging from the fiber may make positive or negative angles with the axis, depending on whether they have undergone an even or an odd number of reflections inside the fiber (for the axis ray, these angles are θ and $-\theta$, respectively) (Figure 112*a*). This uncertainty in the direction of propagation complicates the design and calculation of optical systems. It can be avoided if the entrance end of the fiber is cut at such an angle that the principal ray of the cone propagates parallel to the cylindrical surface (Figure 112*b*).

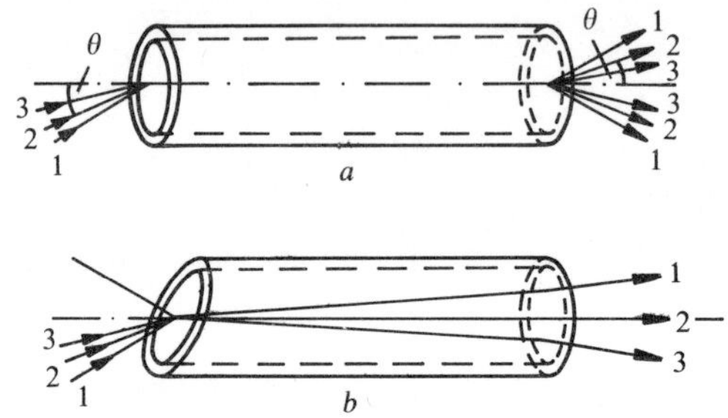

Figure 112

Emergence of rays from a fiber with an entrance end cut at different angles to the axis: *a* perpendicular to the axis, *b* at an angle to the axis.

Since the fiber diameters are small, the solid angle of the incident light cone will also be small. All the rays refracted at the entrance end will then propagate within a narrow cone whose surface does not touch the cylindrical surface (this is possible only if the fiber is relatively short).

To retain the correspondence between the entrance and the exit pupils of the optical system on the whole, i.e., together with the correcting device, field lenses L_1 and L_2 are used in front and behind the system (Figure 113*a*). The image formed by the first lens lies on a spherical surface of curvature R_{p1}, and the image formed by the second lens lies on an analogous surface of curvature R_{p2}. The two surfaces are identified in Figure 113*a* by their radii or curvature.

The front and the rear ends of the field corrector should have such a configuration that the elementary light beams hitting the individual fibers propagate along the fiber axis, the entire light beam emerging through the exit pupil of the optical system.

If R_{p2} were a plane surface, the image formed by the lens L_2 would have a nonzero curvature and no field correction would be achieved in the final image. Therefore, the second end of the correcting fiber plate should have a suitable curvature such that the final image is plane. Since every ray passing through the exit pupil of the lens L_1 should enter upon emerging from the fiber the entrance pupil of the

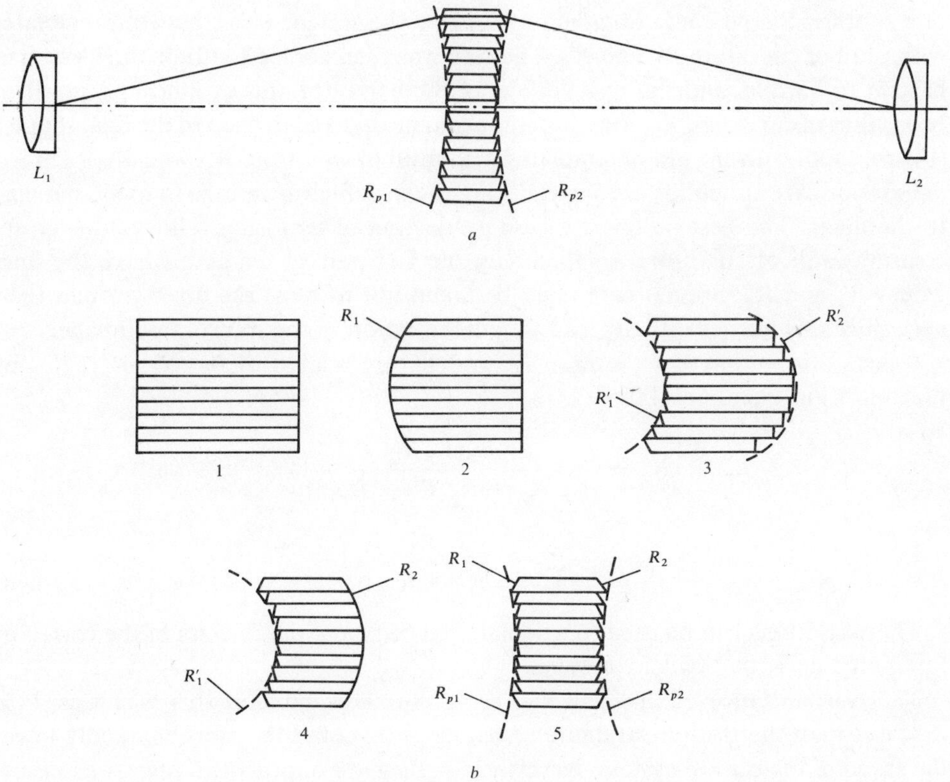

Figure 113

Fresnel-lens fiber field flattener:

a functional diagram, *b* stages of the manufacturing process.

lens L_2, the fiber ends should be designed in the form of spherical surfaces with radii of curvature approximately equal to $F_1(n - 1)$ and $F_2(n - 1)$ for the entrance and the exit ends, respectively (F_1 and F_2 are the focal distances of the lenses L_1 and L_2, n is the refractive index of the fiber material).

The first field correcting device was built by the authors of [105]. It was made of fibers 200 μ in diameter and 12 mm long each; the total cross-section area was 6.25 cm^2. The fibers had been drawn from glass with $n = 1.61$ and bonded into a monolith slab with Canadian balsam; the condition of total internal reflection was thus ensured.

The manufacturing process is illustrated by Figure 113*b*. One of the surfaces of the slab 1 was ground and polished to a spherical segment of radius R_1. As a result, the slab took the form designated as 2 in Figure 113*b*. The slab was then heated to the softening point of the Canadian balsam to allow the fibers to glide freely along one another. Shape 3 was obtained by locking the softened slab between templates with radii of curvature R_1' and R_2'. The slab was then cooled to allow the Canadian balsam to harden, and the rear surface of the corrector was ground and polished to a radius of curvature R_2. The slab was again heated and formed to the final shape 5 (Figure 113*b*) with the aid of templates with radii of curvature R_{p1} and R_{p2}.

Extreme care should be exercised during the forming operations to avoid damage to the fibers. The best policy is to use a selection of templates with gradually decreasing radii of curvature, so that only the last pair of templates have the final values R_1' and R_2'. Special care must be taken not to twist the fibers around their axis, since this may eventually lead to a deformation of the transmitted image.

Calculations show that the radii R_1' and R_2' are related to R_1, R_2, R_{p1}, R_{p2} by the following equalities [105]:

$$R_1' = \frac{R_{p1}R_{p2}R_2}{R_{p2}R_2 - R_{p1}R_2 + R_{p1}R_{p2}};$$
$$R_2 = \frac{R_1R_1'}{R_1 - R_1'}. \qquad (308)$$

These relations can be used to calculate the principal parameters of the corrector for matching with the lens system.

The transmittance of the field curvature corrector built in this way was 75% (5% less than the theoretical figure). Devices of this kind therefore markedly lower the speed of the optical system. Nevertheless, they are of practical interest as means of improving the image quality.

In conclusion, one further point should be stressed. It is clear that the image quality improves as the fibers of the correcting device are made thinner. The curve plotting the number of fibers vs. the aperture angle of the optical system, assuming that each fiber carries one unit of information, is shown in Figure 114. We see that a corrector designed for a system with an aperture angle of about 30° should have about 3.5×10^6 fibers. This is a very large number, but devices of this kind definitely can be built.

3. Optical fibers in a stellar spectrograph

The main difficulty of stellar spectroscopy is primarily the low intensity of the objects of study. Even under excellent atmospheric seeing conditions with a high-speed telescopic system, the quantity of light energy hitting the photographic plate of a stellar spectrograph is not always sufficient for forming a good spectrogram. Since the entrance slit of the spectral instrument cuts a narrow slit in the image of the planetary disk, even prolonged exposures do not lead to the desirable results: under

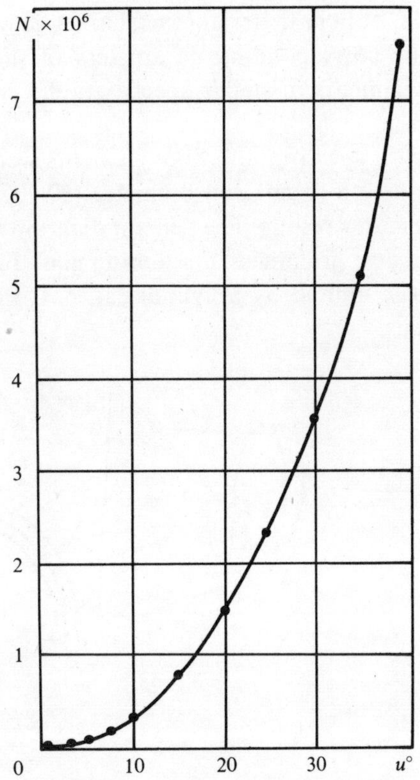

Figure 114

Number of corrector fibers vs. the aperture angle of the optical system.

optimum observation conditions, only 80–85% of the luminous flux forming the image in the plane of the spectrograph entrance slit are transmitted to the plate. Fiber optics was found to ensure much better results in this case [94, 106].

The problem to be solved calls for the conversion of a circular image into a linear image. The fibers at the entrance end of the bundle are correspondingly bonded in a circle, whereas the fibers at the exit end (the one coinciding with the entrance slit of the spectrograph) are arranged in a straight line of length equal to the length of the slit. Since the total fiber area is the same for the two ends, the planetary light collected by the telescope is effectively utilized to a much greater extent. Absorption losses in the fibers and the losses associated with the blind spots of the bundle cross section are relatively small. Fresnel losses at the ends can be reduced to a minimum by optical coating.

If the fiber aperture exceeds the aperture of the collimator objective and is completely filled with incoming light, additional losses associated with beam divergence arise. Appropriate selection of fiber material (see Chapter 2) will give bundles whose aperture angles match the corresponding parameters of the spectral instrument. The application of fiber elements in stellar spectrographs clearly opens wide and promising new vistas.

The first experiments with spectrograph fiber elements were undertaken in [94]. Despite the considerable length (6 cm) and diameter (50, 100, 200 μ) of the fibers, the author obtained satisfactory results. The general diagram of the system is shown in Figure 115. To avoid image graininess, a scanning glass plate was used, and the height of the spectrum was limited by a cylindrical lens. Photoelectric recording was done without the lens.

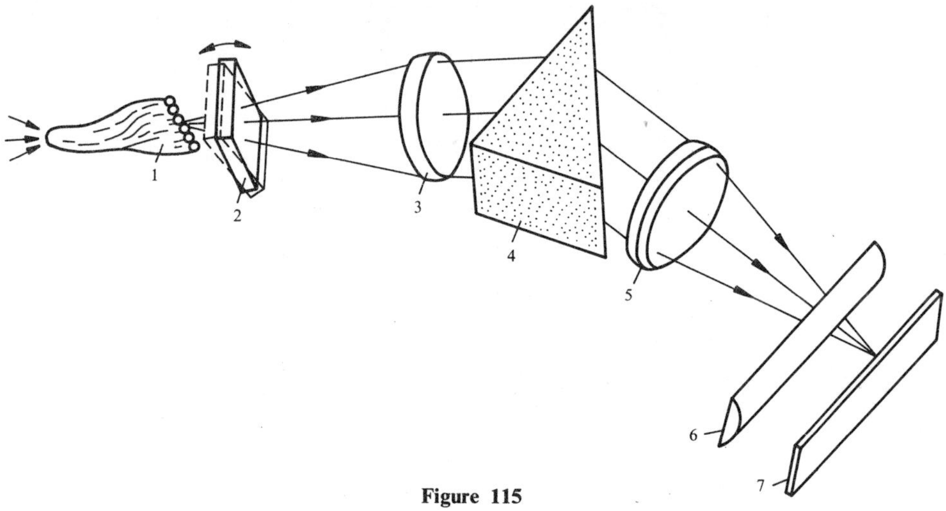

Figure 115

Optical system of a stellar spectrograph:

1) fiber funnel, 2) glass plate, 3) collimator lens, 4) prism, 5) camera objective, 6) cylindrical lens, 7) photographic plate.

4. Point-target autocollimator

Autocollimating methods of monitoring the parallel alignment of two or more reflecting surfaces, measuring the wedge angle, the radii of curvature, and the focal distances of optical elements and systems are widespread in laboratory work and in industry.

The accuracy of these methods depends on autocollimator design. Sometimes, the accuracy can be improved only by greatly complicating the instrument design and raising its cost. However, it has been recently shown [151] that a single conical fiber will markedly simplify the autocollimator design, without detrimentally affecting its performance.

An autocollimator using a fiber point target has no sighting scales or cross-hairs, no light-splitting systems or other optical elements, and the fiber can be manufactured in any laboratory.

To elucidate the advantages of autocollimators with a fiber element, let us compare the optical systems of a conventional autocollimator (Figure 116*a*) and the point-target autocollimator described in this section (Figure 116*b*). In the conventional autocollimator, a cross drawn on the dark background of a horizontal plate 5 is brightly illuminated by the source 6 and sends a light beam to a light-splitting cube 3. Part of the light is reflected toward the objective 2 and is then directed in a parallel beam to the mirror 1. This is feasible only if the cross is set at the focal distance from the objective. The light beam reflected by the mirror 1 will then pass through the objective 2 and the cube 3, forming an image of the cross in the plane of the vertical plate 5 mounted in the focal plane of the objective. If the mirror is strictly perpendicular to the autocollimator optical axis, the image of the horizontal cross coincides with the cross on the vertical plate.

The point-target autocollimator shown in Figure 116*b* operates on the same principle. The light source 5 illuminates a conical fiber 4, whose narrow end faces the objective 2. A divergent light cone leaving the fiber passes through the objective, and upon reflection from the mirror 1 returns to the light source. If the exit end of the fiber lies in the focal plane of the objective, and the mirror 1 is perpendicular

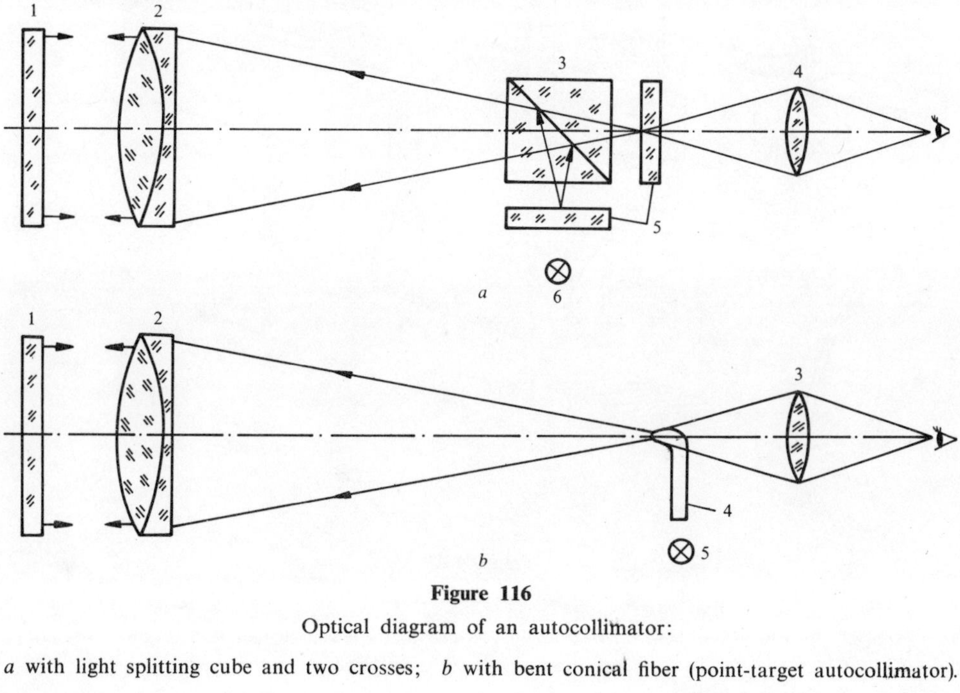

Figure 116

Optical diagram of an autocollimator:

a with light splitting cube and two crosses; *b* with bent conical fiber (point-target autocollimator).

to the optical axis of the system, the image of the end face should coincide with the end face itself. This coincidence is easily detected by watching through the eyepiece 3 for the disappearance of a bright spot.

In the point-target autocollimator (Figure 117), the conical fiber was manufactured from an ordinary cylindrical fiber 75 μ in diameter. The fiber was first heated to the softening point, and then drawn and bent at a right angle. The narrow end was formed by cutting with scissors. All the different operations were performed with a micromanipulator under a binocular microscope. In the final version, the diameter of the tapered end was approximately 10 μ, i.e., the area ratio of the wide to the narrow end was approximately 56. This ensured a considerable difference in the illumination of the exit end compared to the entrance end of the fiber, thus greatly simplifying the task of bringing the source and the image to coincide.

The parallel alignment of two reflecting surfaces is monitored by estimating the deviation from parallel alignment in fractions of the diameter of the bright spot. The spot is viewed through a low-power microscope, or, if higher accuracy is desired,

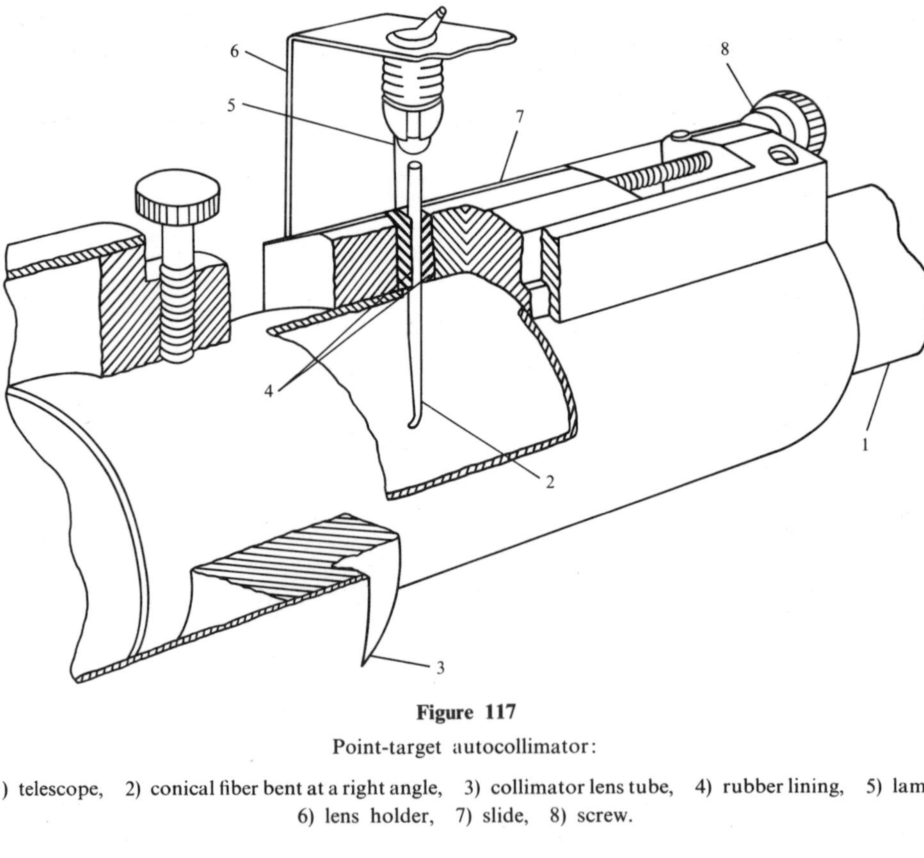

Figure 117

Point-target autocollimator:

1) telescope, 2) conical fiber bent at a right angle, 3) collimator lens tube, 4) rubber lining, 5) lamp, 6) lens holder, 7) slide, 8) screw.

through a 50 × magnification telescope, which gives the image of the tapered end of the fiber in the form of a sharply outlined disk.

Experiments show that parallel alignment can be determined with this autocollimator with an accuracy of about 3″ (converting the fractions of the spot diameter into angular units). It is assumed that the operator is capable of detecting a shift of half a diameter (or a slightly greater fraction). In some cases, this accuracy is insufficient for autocollimator measurements, but it can be markedly increased by using improved reading techniques. According to [151], the accuracy probably can be increased to 0.1″.

5. Dielectric cylinder refractometer

In the previous section we considered the application of a single fiber in an optical instrument. Now we will describe the design of a refractometer using a thick cylindrical rod of optical glass as its principal functional element. Although a thick rod is not quite a fiber-optics element, the theoretical tools and the experimental features of thick-rod optical waveguides are the same as for thin fibers.

A dielectric cylinder refractometer utilizes the dependence of the cylinder light-transfer efficiency on the refractive index of the surrounding medium. If the ambient refractive index is close to that of the cylinder, the light-transfer efficiency is low. The light cone propagating along the cylinder will lose much of its energy by penetrating into the outer medium.

If the end losses and the slight absorption inside the optical waveguide are ignored, the light-transfer efficiency will be determined entirely by the aperture angle θ_{cr} which is related to the refractive index n_1 of the cylinder and the refractive index n_2 of the outer medium by (133). Any ray hitting the entrance end of the cylinder at an angle $\theta > \theta_{cr}$ will already lose a large part of its energy after the first reflection from the cylindrical surface. For a fixed solid angle of the incident cone, the transmittance of the cylinder is a function of the refractive index of the outer medium, so that we can find the unknown refractive index of the medium by measuring the transmittance of the cylinder and using its known refractive index [107].

We see from Figure 118 that the incident light cone can be partitioned into three components. One of these three components propagates inside the cylinder without

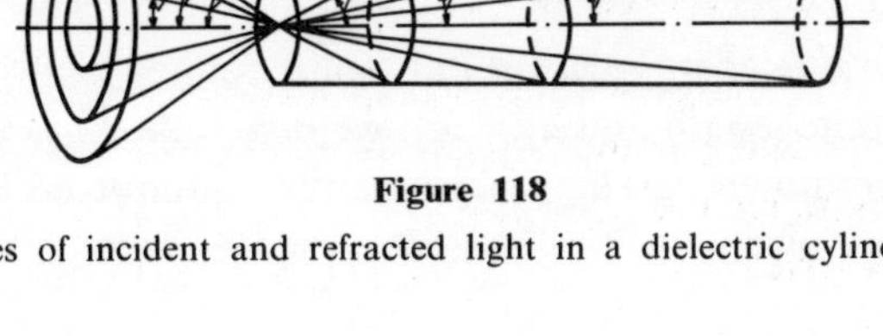

Figure 118

Cones of incident and refracted light in a dielectric cylinder.

any reflections, and therefore it does not carry any information regarding the object of study. The limiting incidence angle for these rays is

$$\theta_0 = \sin^{-1}\left\{\frac{n_1}{[1 + (2L/D)^2]^{1/2}}\right\}, \tag{309}$$

where L and D are the length and the diameter of the cylinder, respectively.

The second component includes rays with incidence angles lying in the interval $\theta_0 < \theta < \theta_{cr}$, which satisfy the conditions of total internal reflection. The angle θ_{cr} can be calculated from Eq. (133). It is these rays that transport the required information concerning the outer medium.

The incidence angles of the third component satisfy the inequalities $\theta_{cr} < \theta < \theta_1$. The corresponding rays penetrate into the outer medium, and do not add anything to the useful information transported down the cylinder by the second component.

Eq. (133) shows that the light transfer of the cylinder decreases as n_2 diminishes, since at the same time θ_{cr} decreases and a progressively larger part of the light flux escapes into the outer medium. In the limiting case $n_2 = n_1$, i.e., $\theta_{cr} = 0$, the light transfer is entirely determined by the first (unreflected) component. Thus, under conditions of minimum light transfer, the refractive indices n_1 and n_2 are equal. Further growth of n_2 should lead to an increase in transmittance. We thus see that the dependence of the cylinder transmittance on the refractive index of the outer medium can be applied to find n_2.

To estimate the accuracy of the method, we introduce the following assumptions [108]: the aperture of the incident light cone and its intensity are constant in time; losses due to absorption in the cylinder and reflection at its ends are negligible; the third component of the incident flux does not make an appreciable contribution to the light-transfer efficiency; the energy distribution in the incident cone is uniform.

If all these requirements are satisfied, the energy emerging from the cylinder is proportional to the solid angle Ω which corresponds to a cone with an opening half-angle θ_{cr}. Since

$$\Omega = \int_0^{2\pi} d\phi \int_0^{\theta_{cr}} \sin\theta\, d\theta = 2\pi(1 - \cos\theta_{cr}), \tag{310}$$

the energy leaving the cylinder is

$$E(n_1, n_2) = 2\pi E_0(1 - \cos\theta_{cr}) = 2\pi E_0[1 - (1 - n_1^2 + n_2^2)^{1/2}], \tag{311}$$

where E_0 is the luminous energy within unit solid angle.

For a cylinder of finite length and $n_1 = n_2$, we may take $\theta_{cr} = \theta_0$, i.e., the detector will pick up the minimum energy E_{min} which can be found from Eqs. (309) and (311) taking $L/D \gg 1$. This gives

$$E_{\min} = 2\pi E_0 (1 - \cos \theta_0) = 2\pi E_0 \left\{ 1 - \left(1 - \frac{n_1^2}{1 + (2L/D)^2} \right)^{1/2} \right\} =$$

$$= \frac{\pi E_0}{4} \left(\frac{D n_1}{L} \right)^2. \tag{312}$$

The sensitivity of the method is characterized by the derivative dE/dn_2, and we may write

$$S = -\frac{dE(n_1, n_2)}{dn_2} = \frac{2\pi E_0 n_2}{(1 - n_1^2 + n_2^2)^{1/2}}. \tag{313}$$

The minus sign corresponds to a decrease of light transfer with an increase in n_2. The relative sensitivity is

$$\frac{S}{E} = \frac{n_2}{A(1 - A)}, \tag{314}$$

where $A = (1 - n_1^2 + n_2^2)^{1/2}$.

Analysis of Eq. (314) shows that the relative sensitivity has finite values only for n_2 satisfying the inequalities

$$(n_1^2 - 1)^{1/2} \leqq n_2 \leqq n_1. \tag{315}$$

If n_2 takes on the extreme values, i.e., $(n_1^2 - 1)^{1/2}$ or n_1, we have $S/E \to \infty$ and the sensitivity of the method reaches its maximum value. The dependence of S/E on n_2 for various n_1 is shown in Figure 119.

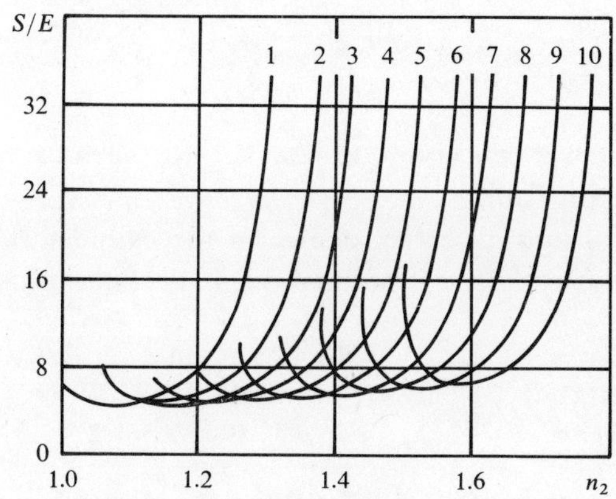

Figure 119

Relative sensitivity S/E of the refractometer vs. the refractive index n_2 of the outer medium for various cylinder refractive indices n_1:

1) $n_1 = 1.333$; 2) $n_1 = 1.400$; 3) $n_1 = 1.450$; 4) $n_1 = 1.50$; 5) $n_1 = 1.55$; 6) $n_1 = 1.60$; 7) $n_1 = 1.65$; 8) $n_1 = 1.70$; 9) $n_1 = 1.75$; 10) $n_1 = 1.80$.

Writing n_2 in the form

$$n_2 = n_1 - \delta \tag{316}$$

and taking $2\pi E_0 = 1$, we replace Eqs. (311), (312), and (314) with convergent series which are obtained when the right-hand sides are expanded in powers of the small parameter δ:

$$E = n_1\delta + \frac{n_1^2 - 1}{2}\delta + \ldots \simeq n_1\delta; \tag{317}$$

$$S = n_1 + (n_1^2 - 1)\,\delta + \ldots \simeq n_1; \tag{318}$$

$$\frac{S}{E} = \frac{1}{\delta}\left[1 + \frac{n_1^2 - 1}{2n_1}\delta + \ldots\right] \simeq \frac{1}{\delta}. \tag{319}$$

The condition $\theta_0 < \theta_{cr}$ in this case is replaced with the inequality

$$\frac{L}{D} > \left(\frac{n_1}{8\delta}\right)^{1/2}, \tag{320}$$

which is readily derived from Eqs. (309), (311), and (316).

Let the minimum relative change in energy detectable by the recording instrument be much less than $2\pi E_0$, i.e.,

$$\frac{\Delta E}{E} = K \ll 1. \tag{321}$$

Using Eq. (313), we obtain for the smallest change in n_2 to which the instrument responds

$$|\Delta n_2|_{\min} = \frac{\Delta E\, E}{SE} = \frac{K}{S/E} = K\delta. \tag{322}$$

Using the last equality, we can compute $|\Delta n_2|_{\min}$ for various values of K and δ. The results are listed in Table 4.

We see from Table 4 that a correct choice of the cylinder refractive index will markedly increase the sensitivity of the method. It becomes much higher than the

Table 4

MINIMUM DETERMINABLE VALUES OF THE REFRACTIVE INDEX n_2 AS A FUNCTION OF K AND δ

δ	$K = 0.01$	$K = 0.02$	$K = 0.05$	$K = 0.1$
0.0001	0.000001	0.000002	0.000005	0.00001
0.001	0.00001	0.00002	0.00005	0.0001
0.01	0.0001	0.0002	0.0005	0.001
0.1	0.001	0.002	0.005	0.01

sensitivity of ordinary refractometers. Moreover (see Eq. (320)), for small δ, the ratio L/D has a reasonable value. Thus, for $\delta = 0.001$, $L/D \simeq 100$, which is quite practicable.

The theory of the cylinder refractometer with allowance for absorption and Fresnel losses was developed by Kapany and Pike [108]. The dependence of S/E on n_2 for the dielectric cylinder with $n_1 = 1.50$ which follows from this theory is plotted in Figure 120 (curve 2). For comparison, the same figure gives curve 1, calculated without allowance for the above factors. For $n_2 \to (n_1^2 - 1)^{1/2}$ curve 1 goes to infinity, and curve 2 goes to zero. For $n_2 \to n_1$, both curves coincide. This indicates that the rapid growth of sensitivity for the case $n_2 \to (n_1^2 - 1)^{1/2}$, which emerges from the elementary theory, is not realistic.

Inequality (320) is intensified if losses on the cylindrical surface are taken into consideration,

$$\frac{L}{D} \gg \left(\frac{6n_1}{\delta}\right)^{1/2}. \tag{320a}$$

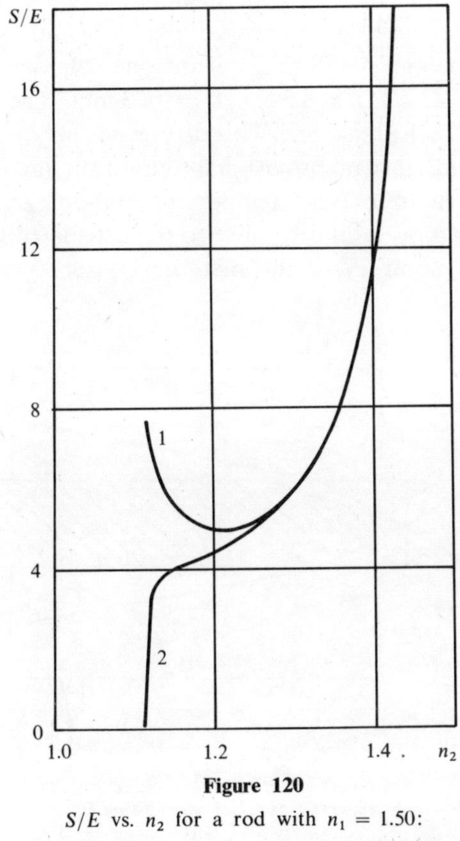

Figure 120

S/E vs. n_2 for a rod with $n_1 = 1.50$:
1) elementary theory, 2) rigorous theory.

The rigorous and the elementary theory, with the additional condition (320*a*), give identical results for $\delta \leqq 0.1$.

The experimental verification of the theory was carried out by the same authors using the setup shown in Figure 121. The light from an incandescent lamp was focused by a microobjective on the entrance end of the rod, making a spot 0.5 mm in diameter. The numerical aperture corresponded to $\theta = 31°$. A photomultiplier was used as a detector, and its amplified signal was recorded by a milliammeter.

The cylinder with a known refractive index was placed in a light-tight cell so that its ends slightly emerged beyond the cell walls. To avoid leakage of the liquid, the rod was packed with Teflon rings in the cell nests. The temperature of the liquid was varied to achieve small changes in the refractive index. Since the temperature dependence of the refractive index was known, objective data were obtained for checking the reliability of the method. The temperature dependence of the cylinder material was also needed. All the data required for the comparison of the theory with the experiment were available [108].

The results of calculations and the experimental curve are shown in Figure 122. The straight lines correspond to theoretical results for cylinders with various refracive indices. The experimental data are given for one cylinder only, with the parameters $n_D = 1.5235$ (at 23°C), $L = 8.5$ cm, $D = 1.73$ mm. The experimental points fall somewhat below the calculated line. This divergence may be due to the temperature gradient in the liquid, since no provision for liquid stirring was made. In general, no measures were taken to prevent temperature inhomogeneities. Nevertheless, the experimental values of n_2 fall fairly close to the calculated figures, this being an indication of adequate accuracy of the method.

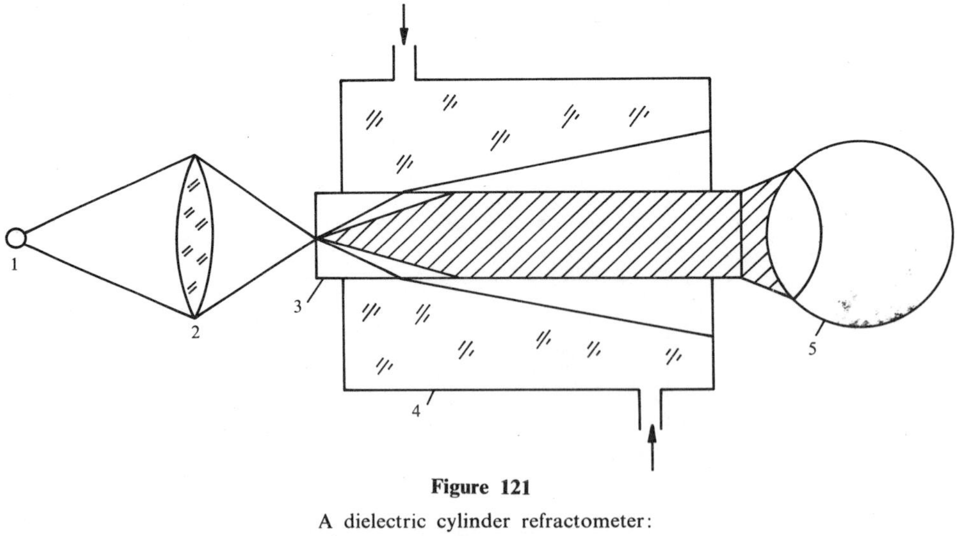

Figure 121

A dielectric cylinder refractometer:

1) light source, 2) microobjective, 3) glass cylinder, 4) cell with test liquid, 5) detector.

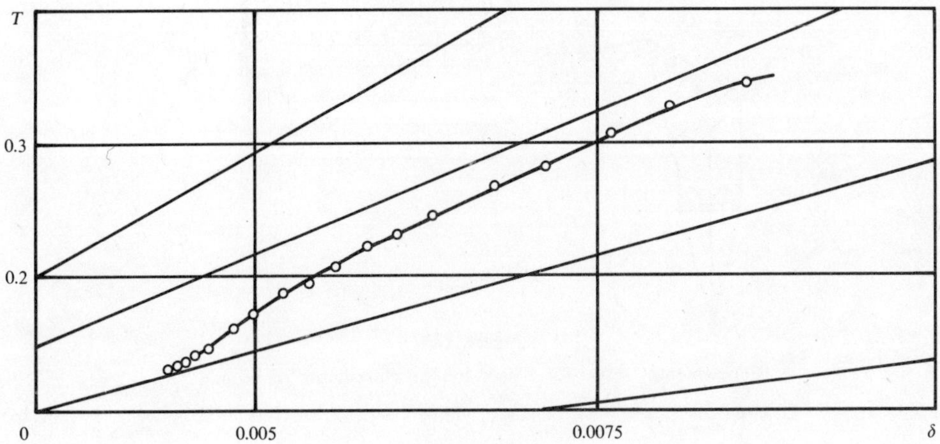

Figure 122

Light-transfer efficiency of the dielectric cylinder refractometer vs. δ: solid curves, theoretical; dots, experimental.

The dielectric cylinder refractometer can be used in automatic control processes for monitoring the continuous production of liquids. The observed changes in the refractive index are readily linked with the state of the liquid, its temperature, or concentration of the monitored component. This instrument has its obvious uses in a research laboratory as well, e.g., for dispersion measurements. A diagram of the setup for measuring the dispersion of liquids is shown in Figure 123.

For a fixed setting of the monochromator drum, the light cone is alternately passed through cylinders with gradually decreasing refractive indices, and the detector signal is measured. At first, the cylinders are optically denser than the liquid, and the difference δ is positive. As δ decreases, the detector signal should also diminish. Finally, the signal reaches a minimum for one of the cylinders (this does not mean, however, that δ will drop to zero, which happens only when n_1 and n_2 are equal). When the next rod is interposed, the signal increases, since there is no total internal reflection and ordinary reflection from the cylindrical glass–liquid interface will be observed.

There is a possibility that by changing from $n_1 > n_2$ to $n_1 < n_2$ we will actually reduce the light transfer, instead of increasing it. Nevertheless, experimental data and previous knowledge of the optical characteristics of the test object will permit selecting a suitable cylinder corresponding to minimum light transfer for $n_1 > n_2$. According to (317), the energy hitting the detector in this case is

$$E_1 = 2\pi E_0 n_1 \delta. \tag{323}$$

For the preceding cylinder, whose refractive index is $n_1 + \Delta$, we have

$$E_2 = 2\pi E_0 n_1 (\delta + \Delta). \tag{324}$$

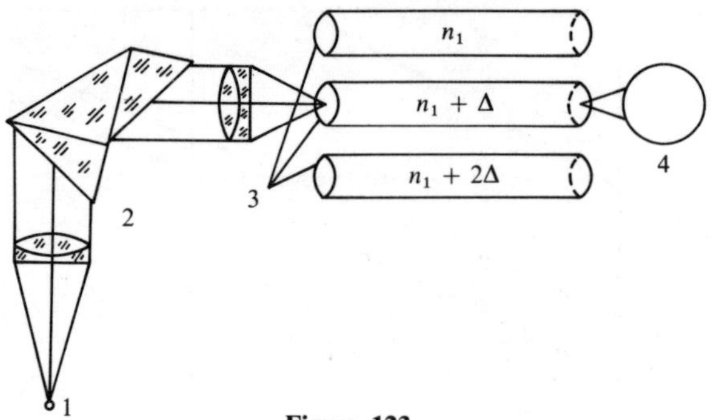

Figure 123

Experimental setup for dispersion measurements of liquids:

1) light source, 2) monochromator, 3) dielectric cylinders with various refractive indices, 4) detector.

Since all the cylinders are geometrically identical, and Δ is small, the energy distribution is the same in the light cones emerging from the two cylinders. This means that the factor E_0 is equal in (323) and (324) and these expressions can be applied to determine δ and hence the refractive index of the test liquid.

The expression for δ is readily obtained from (323) and (324):

$$\delta = \frac{E_1}{\Delta E} \Delta, \tag{325}$$

where $\Delta E = E_2 - E_1$.

By assumption, Δ is known in advance, whereas E_1 and E_2 are measured experimentally. Varying the monochromator output wavelength, we similarly find the refractive index of the test substance at another wavelength, and so on, until enough points for the entire dispersion curve are available.

7 Fiber Optics in Electron-Optical Systems

1. Introduction

The common electron-optical devices of automatic and remote control, such as cathode-ray tubes (CRT), electron-optical image converters (EOC), multistage image intensifiers, etc., are not free from numerous shorcomings. The information fed into these devices is displayed in the form of an optical image on a luminescent screen, which generally has a low resolving power and a low luminous yield.

Automation and computer applications require much higher resolving power and better image quality. It is particularly important to increase the resolving power of CRT in view of the wide scope of their uses.

The quantity of information displayed on the CRT screen is determined by the size of the glowing phosphor spot. The smaller the spot, the higher is the resolving power of the screen and the greater is the quantity of information that it can transmit.

The spot size depends on the electron focusing system, the electron beam shaping and control system, power stability, and quality of the luminescent layer. The resolving power of CRT markedly decreases when voltage fluctuations exceed 0.25% of the total voltage level, or when the focusing coil current changes by more than 0.1% [164]. Electron optics aberrations also diffuse the spot. Complex problems of electronic engineering have to be solved if we are to increase the resolving power of CRT. It is further necessary to improve the quality of the CRT screen, since the spot size largely depends on the quality of the luminescent material and the screen coating technology.

The screen of an ordinary CRT is a thick glass plate with a layer of settled phosphor on the inside surface. Diffusion of light in the phosphor layer markedly lowers the resolving power. The image quality is detrimentally affected by the aureole associated with total internal reflection from the surface of the substrate.

The CRT must be maintained under high vacuum. As a result, the tube should be made sufficiently strong to withstand the atmospheric pressure. Additional strength is ensured by making the screen concave. However, high curvature leads to inevitable optical distortions of the transmitted image. The screen is therefore generally made in the form of a thick slab capable of withstanding considerable loads. Because of the large thickness of the screen, the information read process is distorted by parallax effects. We thus see that the resolving power and the image quality cannot be improved without a radical change of the screen.

Fiber-element plates with one of the surfaces coated with a luminescent layer markedly improve the image quality of CRT screens, EOC, and image intensifiers. They also ensure a simple and efficient information read system.

2. Image transmission by fibers from a luminescent screen

Before we can proceed to assess the efficiency of optical-fiber luminescent screens, we have to determine the angular emission diagram of a luminescent phosphor and to analyze the factors influencing the light transfer of fibers in contact with a luminescent layer.

The angular distribution of a phosphor emission excited by an impacting electron beam depends on the particle size of the luminescent coat, the state of their optical contact with the screen, diffusion of electrons in the luminescent layer, light scattering in the substrate, and other factors. Theoretical analysis of this problem therefore involves considerable difficulties.

The glow uniformity of ordinary CRT tubes has been studied in [164]. The experimental setup (Figure 124) consists of a microobjective, a photomultiplier, and an oscillograph, intended for observing slow processes. The spot produced on the CRT screen by a defocused electron beam is projected with a 10× microobjective onto the plane of a rotating disk pierced with a hole 75 μ in diameter at a distance of 4 mm from the axis. As the disk turns, the hole scans different portions of the luminescent layer within the spot area. Synchronizing the oscillograph time base with the disk rotation, we observe on the oscillograph screen the light from different points of the luminous CRT spot.

The oscillogram of a conventional CRT screen is shown in Figure 125. The luminosity is clearly nonuniform for different points of the screen. The main reason for this is the granular structure of the luminescent material, i.e., the existence of sufficiently large grains which form when the screen is coated with the phosphor. Reducing the graininess we substantially improve the glow uniformity of the screen (Figure 125*b*).

If the phosphor is not in optical contact with the screen surface, the light will be reflected from the substrate. The lower the reflection, the higher is the phosphor efficiency and the image quality. The degree of optical contact with the substrate is therefore one of the basic factors which determine the efficiency of luminescent screens.

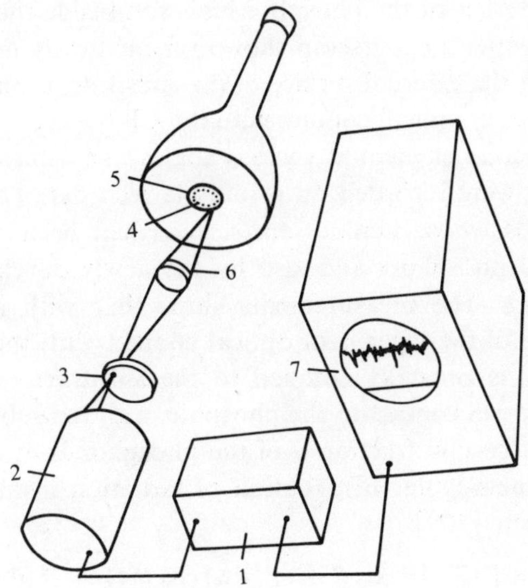

Figure 124

Experimental setup for checking the glow uniformity of a cathode ray tube screen:

1) amplifier, 2) photomultiplier, 3) rotatory disk, 4) scanning trajectory, 5) luminous spot, 6) microobjective, 7) oscillograph.

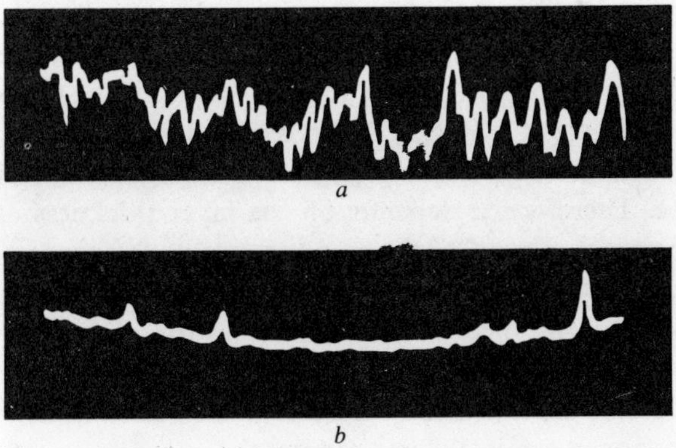

Figure 125

The oscillogram of the emission from an ordinary ZnS luminescent screen:

a coarse-grained settled phosphor, *b* fine-grained phosphor.

The angular distribution of the phosphor emission inside the substrate has been determined with the experimental setup shown schematically in Figure 126*a* [109]. To avoid refraction at the external surface of the substrate, a cemented semicircular glass element was used, in optical contact with the CRT screen. The radiation emerging from the semicircular element at various angles was detected by a photomultiplier, whose readings were recorded on a suitable recorder. The experiments were carried out with tubes whose luminescent screens had been coated by the usual coarse-grained settled phosphors and also by the newly developed low-graininess transparent phosphors. The measurements show that with a settled phosphor, only a minor fraction of the layer is in optical contact with the substrate, whereas the transparent layer is optically coupled to the substrate.

Since the actual state of contact of the phosphor with the substrate is unspecified, we will assume that a certain fraction α of the phosphor is in optical contact with the substrate. Then the angular distribution of radiation inside the substrate can be written in the form [109]

$$F(\theta) = F(\theta_0)\left[\alpha \cos\theta + (1-\alpha)\cos\theta' (1 - R(\theta'))\right], \tag{326}$$

where $F(\theta)$ is the phosphor emission function, θ' is the angle of the ray in the substrate, $R(\theta')$ is the reflection coefficient at the phosphor–substrate surface, $F(\theta_0)$ is the emission of the phosphor in the normal direction.

Figure 126*b* plots experimental and theoretical curves for various degrees of optical contact, calculated from (326) for various values of α. Various factors may be responsible for the lack of coincidence between the experimental and the theoretical curves. Nonuniform distribution of the emission centers, electron diffusion, and light scattering in the phosphor are some of the important factors.

With a low-graininess transparent phosphor, these effects are much less pronounced because of its greater homogeneity. The exciting electrons see this phosphor as a homogeneous absorbing layer. Therefore, for relatively low accelerating potentials (2–5 kV), these effects are comparatively inconspicuous, since the penetration depth of electrons of these energies is at most 1–2 μ. From this point of view, the parameter α may be regarded as the part of the phosphor emitting in immediate contact with the screen surface. Therefore, α depends on the layer thickness and the energy of the exciting electrons.

It is clear from the preceding that the parameter α has two slightly different interpretations. To simplify the theory and to estimate the efficacy of optical fibers in CRT screens, we may assume that only a minor fraction of a coarsegrained settled phosphor is in optical contact with the screen, whereas the entire surface of the transparent phosphor is optically bonded to the screen.

The emission of the settled phosphor (Figure 126*b*) is more directional and is almost entirely confined within a cone with an opening half-angle of not more than 45°. At the same time, the emission of the transparent phosphor almost fully corre-

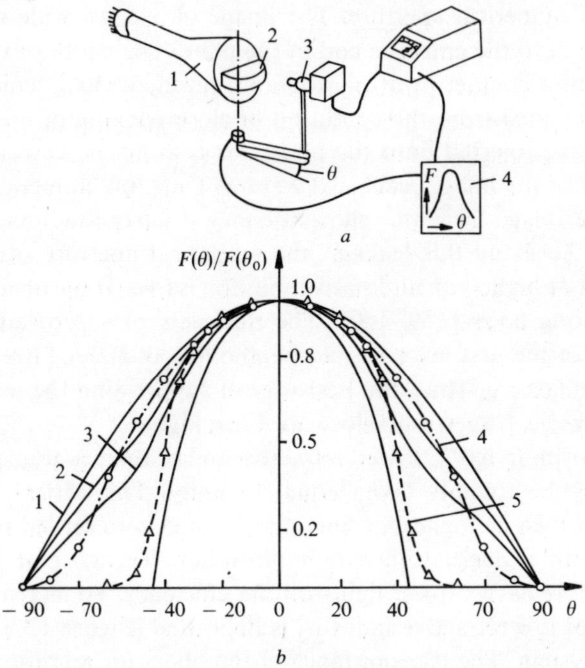

Figure 126

Experimental setup for measuring the angular distribution of phosphor emission in the substrate: *a* experimental setup: 1) CRT, 2) semicircular glass element, 3) goniometer-aligned photomultiplier, 4) electronic recorder; *b* theoretical and experimental curves for various degrees of optical contact: 1) $\alpha = 1$ (total optical contact), 2) $\alpha = 0.8$, 3) $\alpha = 0.6$, 4) experimental curve for a transparent phosphor, 5) ditto for a settled phosphor.

sponds to a Lambertian emitter in optical contact with the substrate. However, this does not signify that optical fibers of low numerical aperture should be used. For the settled phosphor, this may be partly justified because of the high directivity of radiation.

With the transparent phosphor, high numerical aperture fibers should be used, free from light leakage. If the numerical aperture is small, rays incident at a supercritical angle on the end face of the fiber will penetrate through the cylindrical surface. Meeting the surface of the adjacent fibers, they are partly refracted producing parasitic light fluxes inside the unfilled portions of the bundle. If the substrate thickness is small (5–10 mm), these fluxes experience no losses on their way to the exit face of the screen, thus creating a background which has a detrimental effect on image quality.

The effect of light leakage on image quality has been studied in [109]. The author used a plate 6 mm thick with an area of $20 \times 70\ \text{mm}^2$, made up of fibers 50 μ in diameter with 0.58 numerical aperture. The image of a 200 μ wide slit was projected with a 1:10 lens onto the entrance end of the plate. The width of the slit on the exit end, recorded on a contact print photograph, reached 350 μ, which is close to the 300 μ width calculated from the condition of close packing of the fibers. However, when the slit was projected onto the phosphor coating the entrance surface of the plate, the width of the image reached 1.25 mm. Thus low numerical aperture fibers transmitting the image from the phosphor give a fairly low image quality due to leakage effects. To avoid this leakage, the numerical aperture of the fibers should be increased to 1 or higher; an alternative solution is to coat the fibers in an additional optically insulating layer [159, 160]. The thickness of a protective metallic layer applied on top of the first layer should be about 300–500 Å [109]. Such fibers are difficult to manufacture. Thus, the best way of suppressing the leakage is by using fibers with numerical apertures close to 1 or higher.

A settled phosphor has a lower refractive index than a transparent phosphor. It may therefore be roughly taken equal to unity. The critical angle θ_{cr} defined by Eq. (132) will then be higher for fiber ends coated with settled phosphor than for those coated with a dense transparent phosphor. Transparent phosphor coated fibers will therefore have a lower light-transfer efficiency. To increase the light transfer, a thin film of low refractive index n_3 is deposited (Figure 127). This film acts as an immersion system. The transmittance of the fibers for meridional rays then can be written in the form [109]

$$T = 2\pi A \int_0^{\theta_{cr}} I_0 \cos\theta \,[1 - R_{13}(\theta)]\,[1 - R_{10}(\theta')]\, e^{-Kl\sec\theta'} \sin\theta \, d\theta, \qquad (327)$$

where I_0 is the phosphor emission intensity normal to the fiber end, θ is the incidence angle, θ' is the refraction angle inside the substrate, $R_{13}(\theta)$ and $R_{10}(\theta')$ are the Fresnel losses at the entrance and the exit ends, respectively, K is the absorption coefficient of the core material, l is the length of the fiber, A is a normalization coefficient defined by the equation

$$2\pi A \int_0^{\pi/2} I_0 \cos\theta \sin\theta \, d\theta = 1. \qquad (328)$$

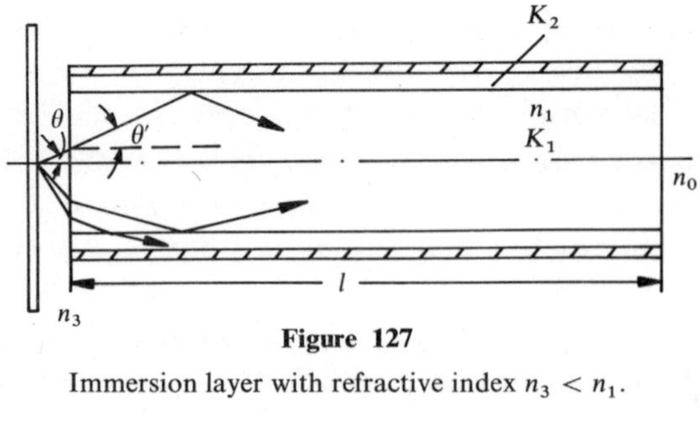

Figure 127

Immersion layer with refractive index $n_3 < n_1$.

T was calculated on a computer for particular numerical values of n_1, n_2, n_3 for a phosphor in full optical contact with the plate. The results are valid only for relatively short fibers, since Eq. (327) ignores reflection losses at the cylindrical surface. In calculating the transmittance T, the fiber length l in (327) was set equal to 25 mm, and the attenuation of the light beam due to absorption was taken as 1% [110].

Figure 128*a* shows the transmission curves of fibers with various n_1 for a settled phosphor as a function of the ratio n_1/n_2. For certain values of n_1/n_2, all the curves reach saturation, without quite attaining 100%, although the numerical aperture may approach unity and the fiber is therefore capable of collecting the entire phosphor emission. Moreover, the higher the core refractive index, the lower is the maximum value of T. For example, for $n_1 = 1.8$ and $n_2 = 1.5$ ($A \approx 1$), the light transmission of the fiber is about 87%, falling short of the expected 100%. This reduction of T is associated with Fresnel losses at the entrance and exit ends (see Chapter 2). When the core refractive index is decreased, these losses diminish, and a high transmittance can be attained for sufficiently high numerical aperture fibers. For $n_1 = 1.5$ and $n_2 = 1.1$, the transmittance reaches 95%.

Figure 128, *b* and *c*, shows the curves of $\sin \theta_{cr}$ and T of an optical fiber with $n_1 = 1.8$ as a function of n_1/n_2 for various values of the refractive index of the intermediate coating n_3 (a transparent phosphor). We see from the curves that T increases with a decrease in n_2, i.e., with the increase in the numerical aperture, and also with a decrease in n_3. However, in work with a transparent phosphor, T is somewhat lower than for a settled phosphor. For $n_1 = 1.8$ and $n_2 = 1.5$, the T for a transparent phosphor is about 40–60% (depending on the value of n_3), whereas for a settled phosphor $T \approx 85\%$. The poorer light transfer is attributed to the decrease in the critical incidence angle for the entrance end of a fiber in optical contact with a transparent phosphor, whose refractive index is substantially higher than the refractive index of the fiber core. As the critical angle decreases, the luminous energy that can be conveyed by internal reflections diminishes. This is also the reason for light leakage into adjacent fibers, leading to a deterioration of image quality.

When a low refractive index intermediate layer is used, θ_{cr} increases and a higher T is obtained. In parallel with the increase in θ_{cr}, light leakage is eliminated. This coating thus proved highly effective. For example, by coating the fiber ends with a thin film of CaF_2 or MgF_2 with a refractive index close to 1.3, $\sin \theta_{cr}$ can be increased to 0.78 for a fiber with $n_1 = 1.8$ and $n_2 = 1.5$ (Figure 128*b*). The light transfer (Figure 128*c*) will increase at the same time to 60%. It is nevertheless desirable to encase the fiber in another absorbing envelope to eliminate light leakage. However, the use of a fiber with a numerical aperture greater than unity considerably simplifies the problem.

Eq. (327) is written for the meridional rays only. Since radiation is also transferred by oblique rays, the actual values of T will be somewhat higher.

Fiber plates intended for CRT screens, EOC, and similar systems should be made

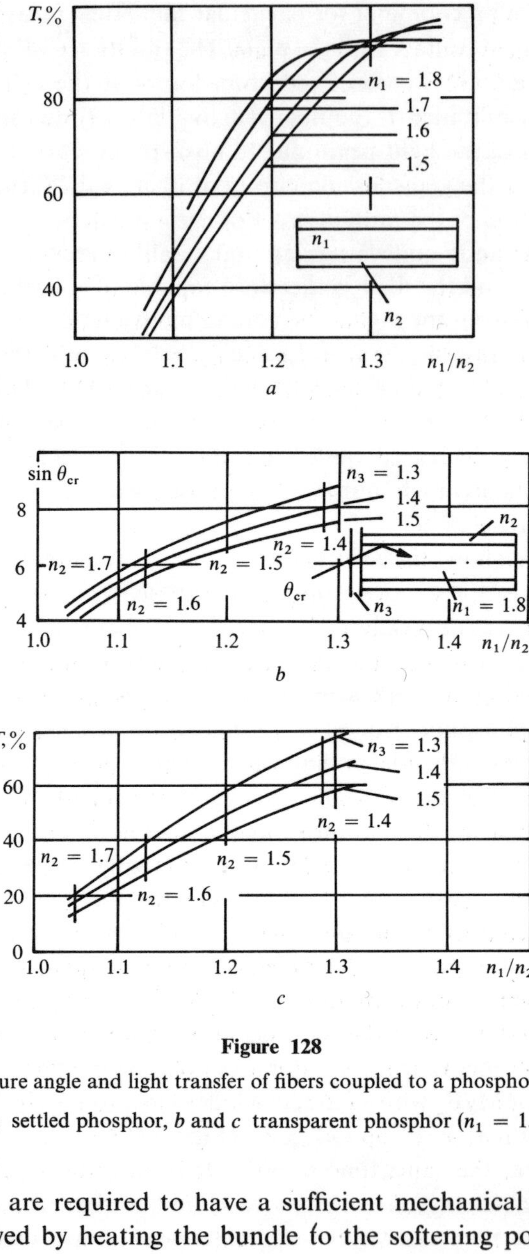

Figure 128

Aperture angle and light transfer of fibers coupled to a phosphor layer:

a settled phosphor, *b* and *c* transparent phosphor (n_1 = 1.8).

vacuum-tight and are required to have a sufficient mechanical strength. Vacuum-tightness is achieved by heating the bundle to the softening point and pressing it into a monolith slab. The fiber casings fill the gaps, forming a matrix in which the individual fibers are embedded. If such vacuum-tight plates are used in conjunction with a small-aperture source, the fiber aperture may be made small and the fiber casing may be quite thick.

When using Lambert-law sources, the fiber should have a double coat to suppress light leakage and the associated aureole. Since the fiber casing does not transmit useful information, the resolving power of the screen diminishes. Moreover, the fiber cores undergo a certain deformation during the heating and the compression stages; each fiber acquires an irregular polygonal cross section with a varying number of contact faces with the adjacent fibers. The presence of these contact faces enhances the leakage effect, so that additional coating is essential.

It follows from the preceding that fiber-optics luminescent screens should be made of optical fibers drawn from high refractive index heavy flints. Encasing the flint fiber in an envelope with a suitable refractive index, we can obtain a high numerical aperture and avoid the necessity of additional coating. Fibers with $n_1 = 1.9$ and $n_2 = 1.5$ have an aperture of 1.17 and do not show any light-leakage aureole when coupled to a settled phosphor.

3. CRT with a fiber-optics screen

Various reports on the uses of optical fibers in CRT screens have been published in the literature [52, 111, 116, 117, 121, 128, 138, 150, 153, 154]. These tubes have a number of advantages compared to ordinary CRT. The image they form is free from parallax and is favorably distinguished by high brightness and high resolution. Moreover, a flat image can be formed in this case, which greatly simplifies the information pickoff: a simple contact print can be readily obtained on paper.

The image on the fiber-optics CRT screen is more than 40 times brighter than the image on conventional screens. The emission from the external surface of a conventional screen is no more than 5–6% of the total phosphor emission, whereas in fiber-optics CRT it reaches 85–90% [52]. However, even if there is no need to increase the image brightness, fiber elements will improve other CRT parameters, e.g., they will increase the sweep rate, lower the anode voltage, reduce the power requirements of the deflection system, lower the electron beam current and the modulating voltage amplitude. If low brightness is acceptable, it is advisable to increase the contrast and the resolution by using a screen of optical fibers coated with a second absorbing layer. The light transfer somewhat decreases, but the image contrast becomes markedly higher; it retains its level even under conditions of strong external illumination. These CRT therefore may be used in devices functioning at high external illumination levels.

Screens are generally manufactured from fibers 10–20 μ in diameter. A screen of 10-μ fibers is capable of resolving more than 50 luminous points per millimeter. Therefore, the overall resolving power of the fiber-optics CRT is determined mainly by the diameter of the luminous spot on the screen, which in modern instruments can be reduced to 8 μ [141].

The number of fibers needed to make up the screen is very large. A tube 76 mm in diameter with a screen made of 8-μ fibers contains about 40 million individual elements [52]. Such screens are very difficult to manufacture. Nevertheless, the diameter of fiber-optics screens has been successfully brought up to 20 cm. The difficulties in the manufacture of CRT and other electron-optical devices are associated with the unusual thermal expansion properties of the inhomogeneous fiber slab. Special technology is required to fuse the screen into the glass tube. One of the possible methods of screen–tube matching is described in [77].

A Kovar or a similar alloy 2 is fused into the tube 1 (Figure 129). The edges 5 of the fiber slab 4 are silvered and are then bonded with AgCl to a silver ring 3, which is soldered with Kovar or another alloy. This technology ensures a high vacuum and an adequate mechanical strength of the device.

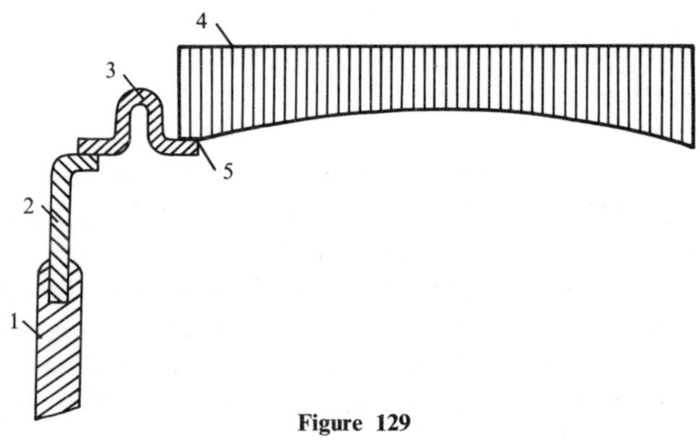

Figure 129

Bonding a fiber screen to a CRT glass envelope:

1) glass wall, 2) Kovar, 3) silver, 4) screen, 5) silver chloride.

4. Fiber elements in television

The use of fiber elements in camera tubes and picture tubes considerably widens the scope of television. The principal aim of any camera tube is to convert an optical image into a sequence of electrical signals.

As an example, let us consider the design and the operation of a vidicon with fast-electron scanning (Figure 130). A transparent conductive layer, known as the signal plate, is applied onto the inner surface of the entrance window. The signal plate is connected to the signal electrode in the shape of a ring fused into the glass tube of the vidicon. A thin coat of highly photoconductive semiconductor material is applied onto the signal plate. A load resistance and a d.c. voltage source are coupled into the plate circuit. Each element of the photoconductive target thus can be treated as an elementary capacitor shunted by a leakage load.

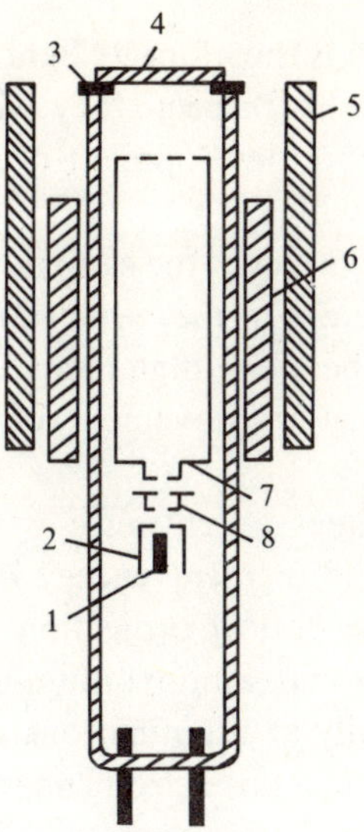

Figure 130

Vidicon:

1) cathode, 2) modulator, 3) signal electrode, 4) transparent conductive layer, 5) focusing coil, 6) deflecting coil, 7) collector, 8) anode.

In the absence of illumination and without a scanning electron beam, the two surfaces of the target are maintained at the same potential because of the dark conductivity of the semiconductor layer. When the target is exposed to the electron beam, it emits secondary electrons which are picked up by the collector. The target surface facing the electron beam acquires a positive potential, which rises to the anode potential. A certain potential difference is established between the two target surfaces. This process may be interpreted as the charging of elementary capacitors.

Between successive scannings, the capacitors are discharged owing to the dark conductivity of the semiconductor layer. When an image is projected onto the target, the conductivity of the illuminated portions markedly increases, and the charging current produced by the scanning beam is therefore higher. Thus, as the electron beam scans the target, charging current pulses are delivered from the elementary capacitors to the load resistance, creating a video signal voltage across the load.

In a camera tube of a different design, an image orthicon or a superorthicon, the optical image is projected onto a sensitive photocathode, which converts the optical image into an electron pattern. The electron pattern is transferred to the target, which is scanned by an electron beam to produce a video signal.

The conversion of the optical image into an electrical charge pattern may thus utilize both the internal and the external photoeffect. The main factor influencing the sensitivity of a camera tube is therefore its light yield efficiency. Image orthicons have the highest efficiency in this respect: they will transmit even under ordinary daylight illumination. The future development of television requires camera tubes of substantially higher sensitivities.

The image brightness on the target of the camera tube can be increased by improving the speed of the lens system. On the other hand, image quality considerations limit the objective diameter. Therefore, high image quality at low illumination levels is unattainable without radical improvement of the electron optics in modern camera tubes [49].

The application of fiber plates as entrance windows in camera tubes ensures more efficient utilization of the incident energy and smaller light losses, without any basic modifications of the optical projection system. A fiber-window vidicon will transmit a good image for end-window illumination of about 5 lux, and image orthicons will fucntion normally at illuminations as low as 0.2 lux [128].

Entrance windows are manufactured from fibers 10–20 μ in diameter, which is quite sufficient for television resolution purposes. Some television and cathode-ray tubes with a fiber screen are shown in Figure 131.

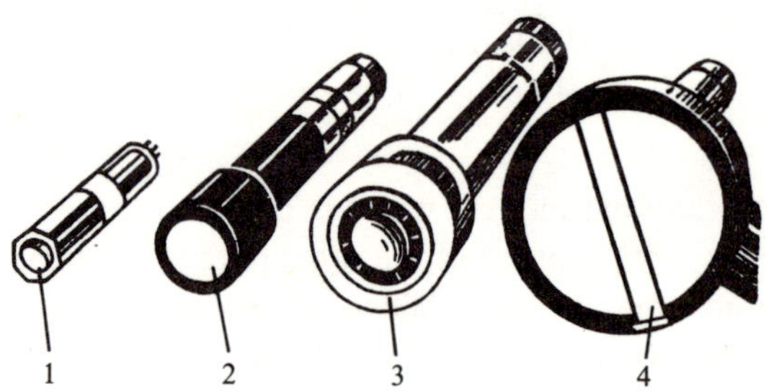

Figure 131

Television and cathode-ray tubes with a fiber-optics screen:

1) vidicon, 2) CRT, 3) image orthicon, 4) CRT with a fiber screen in the form of a single line for contact readoff of information.

The manufacture of fiber-optics image orthicons involves certain technological difficulties, associated with the chemical interaction between the photocathode material and the fiber. These difficulties, however, have been successfully overcome [91, 158].

The main advantage of a fiber-window vidicon or any other camera tube with a fiber-optics entrance window is that it can be easily matched with a CRT, an image converter, or an image intensifier, with screens also made of optical fibers [62, 63, 127, 130]. The entrance window in these cases is placed in direct contact with the luminescent screen of the corresponding electron-optical device. The resolving power of the combined system somewhat decreases, but sufficiently thin fibers will nevertheless ensure an adequate resolution.

Despite the drop in the resolving power, the phosphor emission is utilized much more efficiently. When the entrance window is in direct contact with the screen, more than 80% of the phosphor emission is transmitted to the vidicon target. Normal magnification lens projection systems, on the other hand, lose about 95% of the phosphor emission [62]. Moreover, the fiber-window camera tube is easily coupled to the image by a flexible bundle, which may be quite long. A cable 6 m long made up of 350,000 fibers 15 μ in diameter was used in [145]. Despite the significant light losses in light pipes of this length, the illumination was sufficient to produce an image of adequate quality.

Fiber optics also can be advantageously used in picture tubes (kinescopes). The fiber screen of the picture tube ensures a higher brightness and a higher image quality. Moreover, fiber elements permit designing considerably larger screens than normal. without fiber optics, the screen is increased by using ever larger picture tubes, and the increasing size naturally imposes progressively more exacting requirements on the mechanical strength of the tube. The television picture can be magnified with the aid of optical systems which project the picture from the tube screen onto a larger outside screen. Image brightness in the plane of the external screen is approximately described by the formula

$$B_s = B_t \tau \frac{S_t}{S_s} \frac{1}{4} \left(\frac{d}{f}\right)^2, \tag{330}$$

where B_t is the picture brightness on the screen of the picture tube, τ is the transparency of the objective, d/f is the objective relative aperture, S_t and S_s is the picture area on the tube and the screen, respectively.

If the image is projected by an objective with $d/f = 1{:}2$ and $\tau = 80\%$ without magnification, the ratio B_s/B_t is very small (about 0.05). Therefore conventional picture tubes cannot be used in projection systems of this kind. High screen brightness tubes are obviously needed.

The phosphor emission intensity is proportional to the density of the incident electron beam and the accelerating potential. The electron beam density is limited by considerations of beam defocusing as a result of Coulomb repulsion of the electrons. A considerable increase in brightness can be attained by stepping up the accelerating potential. Modern projection tubes operate at the maximum permissible electron beam density (still ensuring the required focusing) and high acceleration potentials (up to 80 kV). These operating conditions, however, inevitably lead to a rapid burnup of the phosphor and shorten the service life of the picture tube.

Fiber-optics screens not only produce pictures of sufficient brightness, but also markedly lower the accelerating voltage and the electron beam density. At lower electron beam densities the focusing is simpler and the image quality is correspondingly improved. Moreover, the picture on a fiber screen has a higher contrast and a higher resolution than the pictures obtained on an ordinary screen. Fiber elements make the operating conditions of the picture tubes less strenuous and give a simpler solution to the problem of building large screens: conical fiber bundles can be used to increase the screen area.

Figure 132 is a schematic drawing of a picture tube made of conical fibers [35]. The narrow end of a conical fiber bundle, matching in size the picture tube screen (60–100 mm in diameter), is fitted to the surface of a conventional picture tube. The surface of the screen and the entrance end of the conical bundle must be carefully polished to ensure perfect optical contact. A magnified image is obtained on the side exit end of the conical bundle, which is substantially larger than the entrance end. The fibers should be arranged in a definite order at the entrance and the exit ends only, and the system therefore can be made fairly compact and yet ensure an adequate magnification of the picture. The phosphor emission is utilized in these fiber systems with a much higher efficiency than in lens systems, and sufficient picture brightness is attained even with fibers about 1 m long.

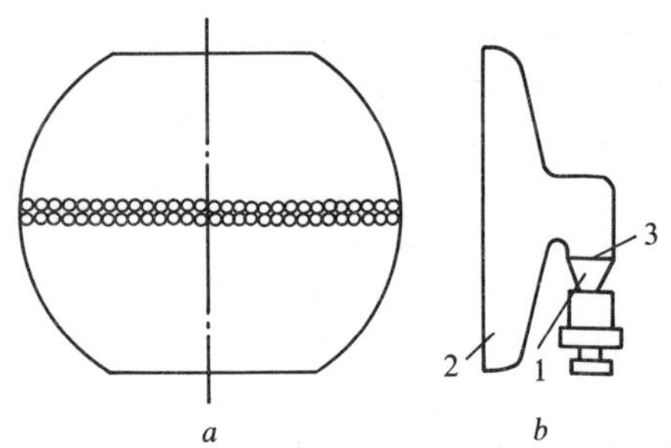

Figure 132

Picture tube screen made of conical fibers:

a front view, *b* side view: 1) projection picture tube, 2) exit end—viewing screen, 3) entrance end of conical bundle.

The image quality can be improved by introducing an additional absorbing coating. In this case, the image contrast at the exit end of the cone is mainly determined by the contrast of the picture on the tube screen. The number of component fibers in the bundle should be very high, to ensure adequate resolution of the television picture (about 400,000 fibers).

The maximum efficiency of phosphor emission utilization is attained when the picture tube screen is the entrance end of a conical fiber bundle. In practical applications, however, this system is not particularly convenient and it is difficult to manufacture, since the entrance end should be made vacuum-tight. It is better to use a separate fiber slab with the picture tube. To increase the brightness and to simplify the matching of the fiber slab to the screen, the picture tube screen also should be made of optical fibers. Although the resolution of the combined system in this case is somewhat lower, thin fibers will nevertheless provide adequate resolution.

Fiber optics also can be used in color television. This approach simplifies the problem of color superposition and lowers the cost of equipment.

A modern color picture tube with a three-color screen and a shadow mask is a complex electron system the cost of which is approximately three times the cost of a black-and-white TV set. The structure of a color picture tube is schematically shown in Figure 133*a*. The principal elements are three electron guns and a screen composed of at least 400,000 phosphor groups. Each group is made up of three grains of different phosphor, which emit red, blue, and green light; the phosphor grains are arranged in a strictly fixed pattern. The shadow mask mounted right behind the screen is perforated with a number of holes matching the number of phosphor groups. The mask is so mounted in relation to the screen that each electron beam modulated by one of the color components of the image passes through the mask perforations to hit a phosphor grain of the appropriate color. The deflecting system is common for all the three electron beams, so that the position of the electron beams relative to the different color phosphors is conserved in scanning. Since the phosphor grains are very small, the different color pictures appear to coincide.

The quality of the color picture on a screen of this design is very high, but the tube manufacturing technology is highly complex. The relative spatial position of the three electron beams, the mask, and the screen requires extremely high precision on the assembly line. The service life of a color picture tube does not exceed 500 hrs; it is mainly determined by the efficiency of the cathodes and the heaters. When one of the electron guns fails, the entire tube has to be replaced.

Figure 133*b* is a diagram of a screen of a basically different design—the three-color chromatrone [65] which can be built as a three-beam or a one-beam model. Let us consider the operation of the one-beam chromatrone, when the color control technique is similar to that in a shadow-mask color tube.

The phosphor dots are replaced in the chromatrone screen by three groups of vertical (or horizontal) stripes of different color phosphors. The number of green stripes is double the number of red and blue phosphor stripes, but their width is half the width of the other stripes. The different color phosphors are applied to the screen in a certain sequence, so that each red and blue stripe is flanked by two green stripes. A color separating grid, made of two groups of wires stretched taut along the red and the blue phosphor stripes, is mounted near the screen. Variable voltage is applied to each wire. If the potential difference between the two groups of wires is zero, a green raster is produced by the scanning electron beam. If a potential difference of a few hundred volts is applied between the two groups of wires, the grid field will deflect the electron beam toward the positive potential wires. The scanning electron beam will therefore hit only those color stripes which lie behind the positive

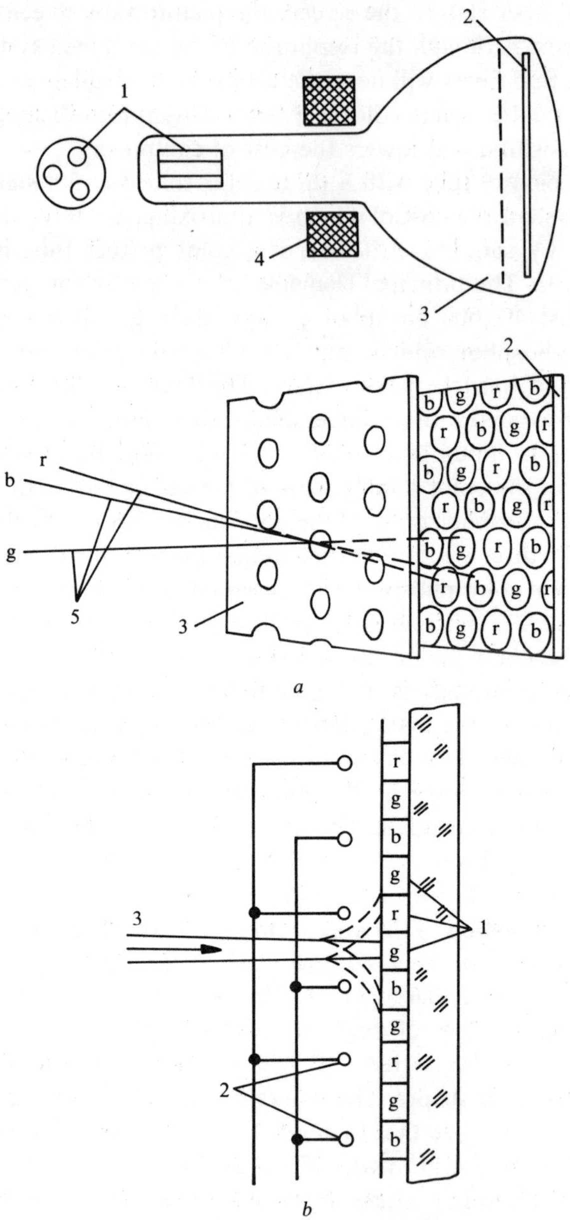

Figure 133

Color picture tubes:

a picture tube with mosaic screen and shadow mask: 1) three electron guns, 2) mosaic screen, 3) shadow mask, 4) deflecting system, 5) electron beam axes; *b* a striped chromatrone screen: 1) phosphor stripes, 2) control grids, 3) electron beam.

potential wires, and the raster will acquire a different, say red, color. By reversing the voltage polarity, a blue raster is obtained. In practice, the screen color between the two groups of wires is controlled by a sinusoidal voltage whose frequency is equal to the frequency of color changes.

The color video signals change in synchronism with the beam deflections. The video signals related to the modulating electrode circuit in such a way that when the beam hits a certain color component of the screen, the modulating electrode receives information about the corresponding color component of the video signal.

A chromatrone is as difficult to manufacture as a shadow-mask picture tube. One of the major problems is how to build a sufficiently large color-separating grid that will not deform when heated. Considerable difficulties are also associated with the generation of color control pulses owing to the parasitic capacitance (about 1000 pF) between the wires and the large amplitudes of voltage pulses required (up to 800 V). A one-beam chromatrone is therefore manufactured with its own special high-power generator. In a system with successive color transmission, when the colors change at a moderate frequency, the required voltage is easier to obtain.

A striped-screen picture tube is simpler to control and to maintain than a shadow-mask picture tube, but its cost is very high, whereas the service life is not much longer than that of a shadow-mask screen. Therefore, economically, it is better to use three different picture tubes with different color screens, combining the three pictures on a fourth screen. The picture brightness of each screen should be very high, because of the low efficiency of phosphor emission utilization. Optical fibers will greatly simplify the picture matching system and thus increase the efficiency.

Figure 134 shows how three color pictures are combined on a single screen with the aid of conical fibers [35], which at the same time may be used to magnify the picture. The wide ends of the conical fibers forming the viewing screen are arranged in rows in a certain pattern, so that each row of fibers transmitting red and blue images falls between two rows of green fibers. The fiber bundle transferring the green picture should thus have twice as many fibers as the red and the blue bundles. The total number of fibers needed to make up a picture screen is at least 1.6 million. The basic similarity between this screen and the original chromatrone screen is self-evident.

The narrow ends of the fibers transmitting one of the color components are joined together and then fitted to the corresponding picture tube screen. To avoid distortion in transmission, the three picture tubes should have the same rasters.

The advantages of this four-screen color system compared to ordinary mirror–lens systems is obvious.

The quality of the picture formed by this system is much higher than with optical systems, where the parasitic cross-illumination of the different screens lowers the contrast. It is, however, inferior to the image quality in a color picture tube with a mosaic screen and a shadow mask.

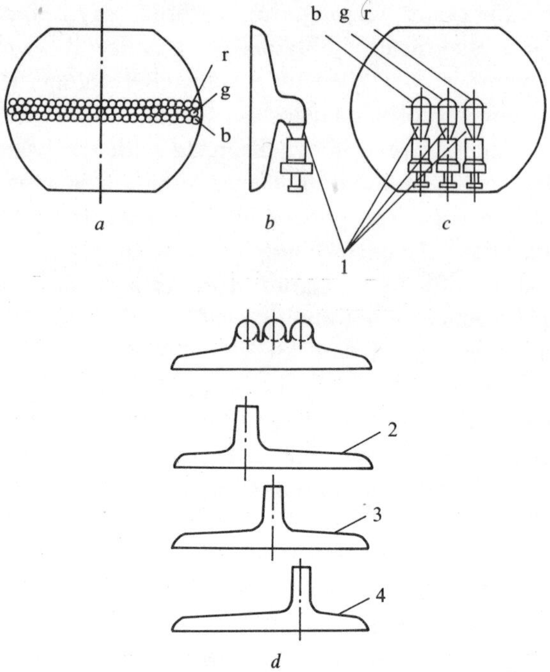

Figure 134

Superposition of color pictures on a single screen by conical optical fibers:

a front view; *b* side view; *c* rear view; *d* top view: 1) projecting picture tubes, 2, 3, 4) rows of fibers transmitting the red, the green, and the blue color pictures, respectively.

The picture formed by fiber bundles has a high contrast and a high brightness, so that it can be viewed in daylight. The service life of the entire system is substantially longer than that of a three-beam shadow-mask tube. When one of the picture tubes fails, it can be replaced without difficulty, and the adjustment of the screen relative to the entrance end of the corresponding fiber bundle does not take long. The ability of fibers to retain their high light-transfer efficiency when bent helps to considerably reduce the size of the TV sets.

5. Image converters and image intensifiers

The spectral response of the human eye is confined to wavelengths between 0.4 and 0.75 μ, and various electron-optical and infrared image converters are needed for other spectral regions.

One of the most effective infrared converters is apparently the evaporograph

[38] which even reveals objects illuminated by the far infrared. Unfortunately, these devices are still in the laboratory stage.

Electron-optical image converters functioning in the near infrared, ultraviolet, and X-ray regions of the spectrum are widely used. All these devices are based on the same principle. In near infrared and ultraviolet image converters, the invisible image is focused on the photocathode. In X-ray image converters, the image is focused on a phosphor whose emission then hits the photocathode. The photocathode emits electrons which are accelerated by an electrical field and are focused by electrostatic lenses onto a luminescent screen. The resolving power of these systems is not particularly high, since each small portion of the phosphor is a Lambert-law source emitting within a solid angle of 2π. Therefore, optical-fiber screens may substantially increase the resolving power of these systems [59, 69].

The focal surface of the focusing electrostatic system is not flat, and the image is therefore sharp only at the center of the screen. This effect is particularly noticeable in image converters with a large photocathode (5 cm in diameter and larger). To ensure satisfactory focusing on the luminescent screen, the photocathode surface should have a certain curvature. The focal surface of the projecting optical system should naturally have a matching curvature. An entrance window made of optical fibers can be readily shaped to a profile of desired curvature; the curved inner surface will then be coated with the photocathode material [59, 150]. The external surface may remain flat or have any other suitable shape.

Similar advantages are characteristic of fiber-optics image intensifiers. The common intensifiers have two or more intensification stages. In a multistage intensifier, the image from the luminescent screen of each stage should be transferred to the photocathode of the next stage. If lenses are used for image transmission, substantial intensification is difficult to achieve, since at most 10% of the phosphor emission is utilized in this case. To ensure high intensification, the luminescent screen of each stage is coupled to the next photocathode by a thin mica sheet, with a phosphor coat applied on one side and the photocathode on the other. Considerable image intensification is obtained owing to effective utilization of the phosphor emission. However, because of the finite thickness of the mica sheet, the resolving power somewhat decreases. The mica sheet thickness is limited from below by considerations of mechanical strength of the system.

The phosphor emission is even more efficiently utilized if the successive stages of the image intensifier are coupled by fiber elements (Figure 135*a*). Light losses in this case are determined mainly by absorption in the fiber core. Since the fibers are short, almost the entire emission is effectively transferred to the photocathode.

A succession of several stages will produce virtually any desired image brightness. The entire system can be assembled from separate bundles (Figure 135*b*) with flat entrance and exit ends [60]. The light-transfer efficiency from the luminescent screen to the photocathode somewhat decreases in this system (to approxi-

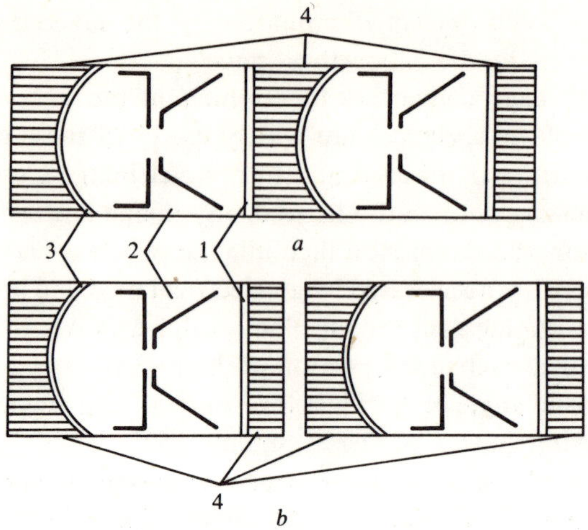

Figure 135

Optical fibers in image intensifiers:

a plates, *b* blocks: 1) phosphor, 2) electrostatic focusing, 3) photocathode, 4) fiber optics elements.

mately 80%) because of the relative mismatching of the fiber ends. For the same reason, the resolving power of the system somewhat diminishes.

In a single-stage intensifier with large fiber-optics windows, intensification corresponding to that of a multistage intensifier can be obtained by using flexible coherent bundles [60]. The operation of this device is shown in Figure 136. The external optical system focuses the primary image on a small part of the photocathode surface. The photoelectrons are accelerated and focused on part of the luminescent screen. The phosphor emission is conveyed by a flexible coherent bundle to the entrance end, illuminating the next portion of the photocathode. The number of the successive intensification stages is determined by the size of the photocathode and the screen.

The resolution of the final image is naturally much poorer than in single-stage intensification. This is associated with the reduction of the effective working surfaces of the photocathode and the screen and with imperfect matching of the flexible bundle and the fiber windows. In a multistage image intensifier assembled of individual bundles, additional resolving power losses are caused by the mismatching of the fibers in the adjoining ends.

The intensification obtained from a single stage with flexible fiber bundles may reach several hundreds and even thousands. The intensification after a single transmission is determined by the accelerating potential (it is approximately equal to 20). If only one of every four photons transmitted by the flexible bundle hits the photocathode, the intensity of the secondary image will be five times the intensity of the

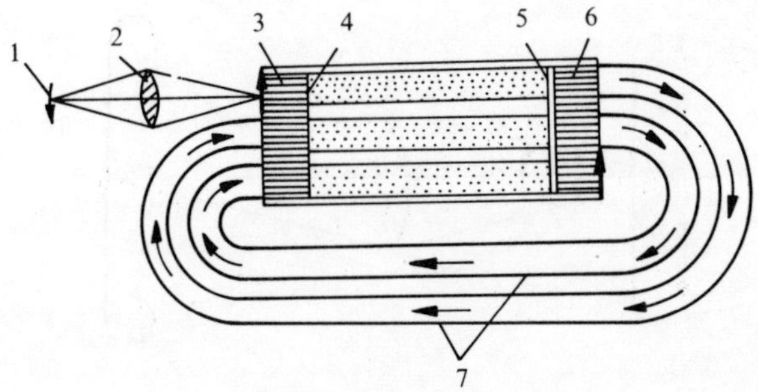

Figure 136

Image intensifier functioning as a multistage intensifier:

1) object, 2) focusing lens, 3) intensifier entrance window, 4) photocathode, 5) phosphor, 6) exit window, 7) coherent fiber bundles.

primary image. The image intensity produced by the device shown in Figure 136 is therefore 500 times the intensity of the primary image.

Application of image intensifiers with fiber optics is particularly promising in electron-optical systems. Such an intensifier coupled to a television tube will transmit the picture under very low illumination conditions. Sufficiently thin fibers will ensure complete resolution of the television raster.

6. Information pickoff and transmission

The photographic recording of information displayed by the electron-optical systems on a luminescent screen has a number of shortcomings, associated with the low image quality on the screen and further deterioration of the image due to diffusion of electrons in the phosphor, light scattering in the substrate, and halation.

A conventional photographic recording system is shown in Figure 137. Let us estimate its photometric efficiency. Since the phosphor emits into the substrate within a solid angle of 2π, part of the rays hit the substrate–air interface at a supercritical angle and experience total internal reflection. This produces halation associated with multiple reflections of these rays from the screen surface and their scattering near the phosphor–substrate interface. The halo leads to a marked deterioration of the image quality.

The relative amount of luminous energy reaching the point P' is expressed by the equality

$$E_{P'} = \frac{\int_0^u F(\theta')\,[1 - R(\theta')]\,\tau(\theta') \sin\theta'\, d\theta'}{\int_0^{\pi/2} F(\theta') \sin\theta'\, d\theta'}, \tag{331}$$

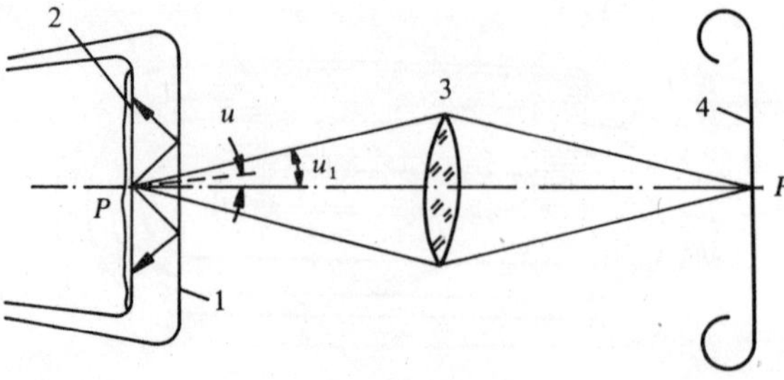

Figure 137

A photographic system for information pickoff from CRT screen:

1) exit end of CRT, 2) phosphor, 3) objective, 4) photographic film.

where $F(\theta')$ is the phosphor emission function at the point P, $R(\theta')$ is the Fresnel reflection coefficient at the substrate–air interface, $\tau(\theta')$ is the transmittance of the projecting system, u is the maximum angle of rays inside the substrate which still reach the point P'.

For a normal magnification optical system, the aperture angle is $u_1 \approx 14°$. If the substrate refractive index is 1.5, we have $u = 9°$. If moreover $R(\theta') = 0.04$ and $\tau = 0.7$, we find from (331) $E_{P'} = 0.018$, i.e., about 2% of the energy emitted by the phosphor at the point P reaches the point P' of the detector. The halation energy can be estimated from the expression

$$E_{\text{halo}} = \frac{\int_{\theta_{\text{cr}}}^{\pi/2} F(\theta') \sin \theta' \, d\theta'}{\int_0^{\pi/2} F(\theta') \sin \theta' \, d\theta'}. \tag{331a}$$

For a substrate with $n = 1.5$, θ_{cr} is approximately 42°. Inserting this figure in (331a), we find that the halo accounts for 56% of the phosphor emission at the point P. Because of this exceptionally low photometric efficiency, photographic recording systems will require long exposure times, and the information pickoff speed will diminish correspondingly.

Contact prints taken through a glass plate 25–30 μ thick were used in [129]. Although this method substantially decreases the exposure, it is not free from shortcomings either (brittle plate, low resolution).

The pickoff situation markedly improves if the substrate of the luminescent screen is made from optical fibers. The information then can be transferred onto photographic paper or film by contact printing.

By appropriately choosing the values of n_1 and n_2, one can ensure a sufficiently high light-transfer efficiency of the fibers (over 80%) and markedly increase the proportion of light energy reaching the photographic film. If the fiber numerical aperture is 0.49, the light flux increases by a factor of 20–40; for a numerical aperture of 1.2, it increases more than 60-fold compared to the results obtained with the recording system shown in Figure 137 [109].

The increase of the photometric efficiency of a fiber-optics screen has been verified experimentally [117]. A CRT screen comprised 100,000 fibers 75 μ in diameter, and its numerical aperture was 0.86. The photometric efficiency was inferred from the photographs shown in Figure 138. The first photograph (Figure 138*a*) was taken with an ordinary photographic objective of 1:2 relative aperture; the second photograph (Figure 138*b*) was taken by placing the photographic material in direct contact with the tube screen. In both cases, the film speed, the exposure time, the brightness of the ray trace on the screen, and other parameters were identical. Because of the high optical efficiency of contact prints picked off the fiber screen, the photograph in Figure 138*b* is clearly overexposed. To reduce the photographic density to the same level as in Figure 138*a*, a neutral-gray filter with an attenuation ratio of 50 had to be interposed between the photographic material and the screen.

This example clearly proves the high photometric efficiency of a fiber-optics screen. Systems of this kind can be successfully applied to information pickoff in various computer configurations.

The printout speeds are currently much lower than the output speeds of modern computers. Indeed, the existing printers have reached speeds of about 2,000 characters per second, and this record will hardly be exceeded. On the other hand, the output speeds of computers reach 100,000 characters per second and more [117].

It is clear that radically new methods and systems are required for increasing the rates of information pickoff and recording. Fiber-optics CRT again provide a promising solution in this direction. A computer may display its output in the form of alphanumerical characters on a CRT screen. This information then can be picked off by a conventional contact print from a fiber-optics element.

The information can be recorded on a photosensitive layer or on photoconductive-coated paper. In either case, the pickoff speed may reach 20,000 characters per second

a *b* *c*

Figure 138

Photographs of a singal displayed on a fiber luminescent screen:

a formed by a 1:2 objective, *b* with the film held in direct contact with the screen, *c* taken with a neutral filter attenuating the flux intensity to 1/50.

[138]. As an example, consider the technique described in [130]. A paper tape coated with a photoconductive ZnO layer is first charged to negative electric charge in a corona discharge. When the paper tape is drawn in front of a screen displaying an image, the potential of the illuminated portions changes (owing to the photoconductivity of ZnO), and the magnitude of this change depends on the illumination intensity. A potential profile of the image is thus imprinted into the paper tape. This profile is developed by dusting the tape with positively charged powder. The image is then fixed thermally.

A high-speed information recording system of this kind is shown in Figure 139. Figure 140 shows special CRT intended for high-speed contact printing of alphanumerical data [40, 72]. The screen is an optical-fiber element in the form of a single line 260 mm long and 13 mm wide. Because of the high image contrast and the high optical efficiency of the screen, the printing rate may reach almost 20,000 characters per second.

The recording speeds can be further increased using the system shown in Figure 141. Fiber components of a certain shape convey the image from the CRT screens to the

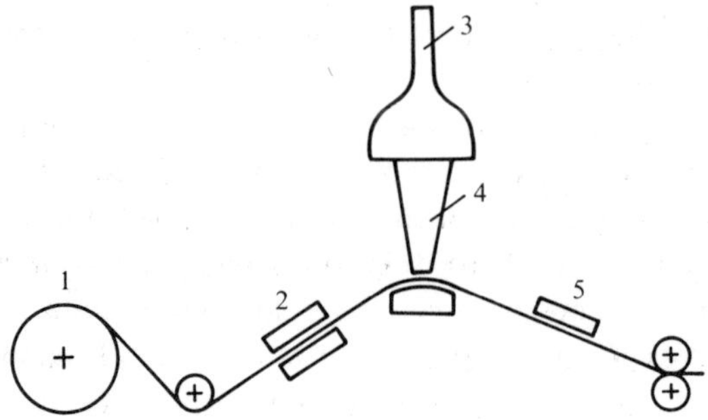

Figure 139

High-speed photographic information recording system:

1) drum (cassette), 2) discharge gap, 3) CRT, 4) fiber image shape converter, 5) dusting nozzle.

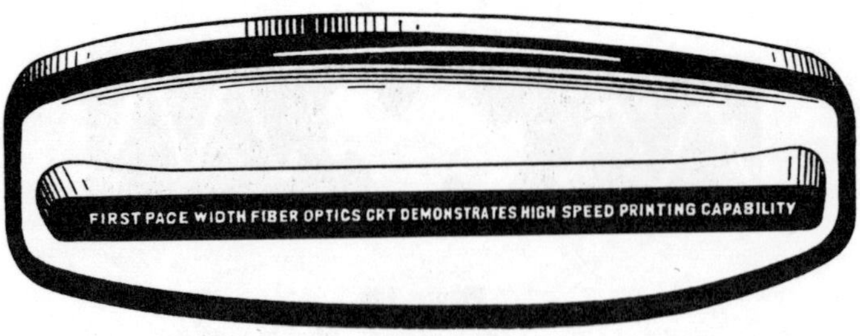

Figure 140

A photograph of the CRT for high-speed alphanumerical printing.

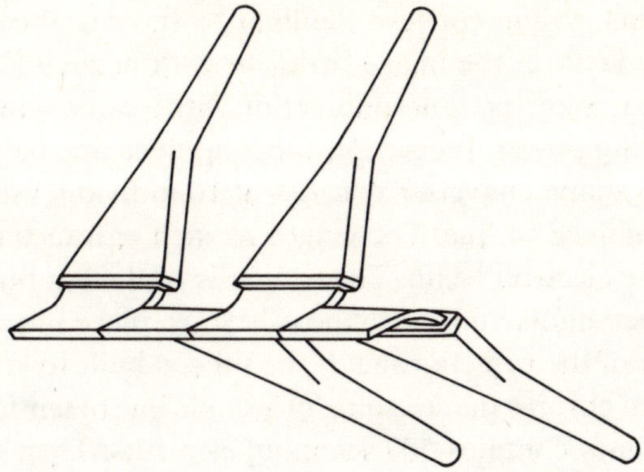

Figure 141
A system of tubes for increasing the information printing rate.

exit ends of the individual bundles which are arranged in line. The recording rate is increased by a factor equal to the number of CRT screens used (in this particular example, it reaches 80,000 characters per second).

A characteristic feature of fiber optics is that it enables any image to be converted into any desired shape [26, 64, 112, 121]. The shape of the CRT screen is determined by the electron-optical system. A square screen provides for optimum electron-beam scanning, and most screens are therefore of this shape.

Figure 142 shows the screen of an image orthicon fitted with a fiber image shape converter: the entrance end is stretched into a long narrow line onto which the objective projects the image. The image shape converter consists of several fiber bundles whose exit ends are stacked one on top of the other, forming a square or a rectangle. The exit end of the converter is placed in contact with the entrance end of the image orthicon, which is also made up of optical fibers. Since the light transfer of the fibers

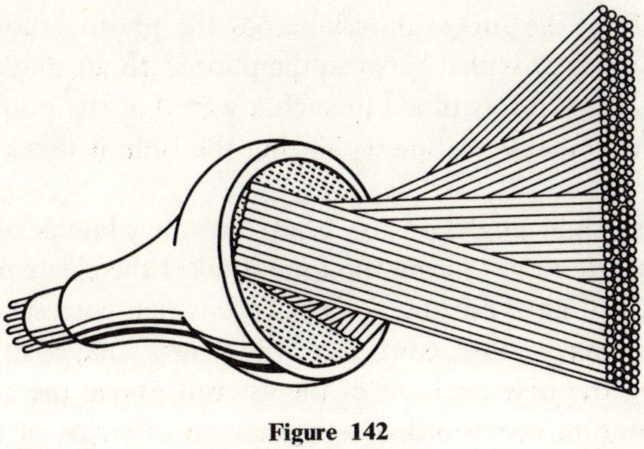

Figure 142
An image shape converter.

is sufficiently high, this system conveys the light to the entrance window and then to the photosensitive layer of the image orthicon with negligible losses.

Using this image shape converter in conjunction with a camera tube, we can markedly increase the resolving power. Indeed, let the long fiber line be made up of 10 segments, i.e., the image shape converter consists of 10 individual fiber bundles whose entrance ends are arranged in line. The length of each entrance end is equal to the scanning length of the electron beam. The exit ends of the ten bundles are arranged into a square or a rectangular image shape. They are placed in direct contact with the entrance window of the camera tube. If the tube is built to scan successively ten lines, this arrangement ensures the scanning of a single line of ten-fold length. Suppose that each of the ten lines contains 500 scanning elements. Then the single line contains 5,000 elements, which exceeds the line resolving power of the existing picture tubes. Up to forty scanning lines have been successfully combined into a single line, giving 20,000 scanning elements [124]. This is far beyond the ability of modern TV equipment.

The image shape converter described above clearly can operate in the reverse direction, by stretching the image on a CRT screen into a single long line, thus ensuring a high resolution in high-speed information recording systems (see Figure 142).

Fiber optics enables us to capitalize in full on the various advantages of TV equipment in air reconnaissance. Wide-angle TV cameras can be used with image shape converters. Because of the high line resolution, even the smallest objects become visible.

An electronic fiber-optics air reconnaissance system (Figure 143) operates as follows [117]. Part of the reconnaissance area is projected onto the long-line entrance end of the image shape converter. The exit end of the converter is square or rectangular, placed in contact with the entrance end of a fiber-optics image orthicon. The image formed on the orthicon target is scanned by an electron beam, which produces a video signal. The signal is transmitted directly to an earthbound receiver or is recorded on a video tape.

As the aircraft flies, the image moves across the photocathode surface. A suitable electronic device is provided between the photocathode and the image orthicon target which scans the photocathode in such a way that the potential profile of the image on the target remains stationary during the time it takes the electron beam to scan the target.

Field scanning at right angles to the beam scanning line is obtained as a direct result of the aircraft motion. During the time it takes the electron beam to scan the target, the aircraft will have advanced a distance which corresponds on the surface to the width of the image shape converter line. A new image is then formed on the target, representing the new position of the aircraft above the surface. The system thus provides a continuous record of a succession of strips of the reconnaissance area.

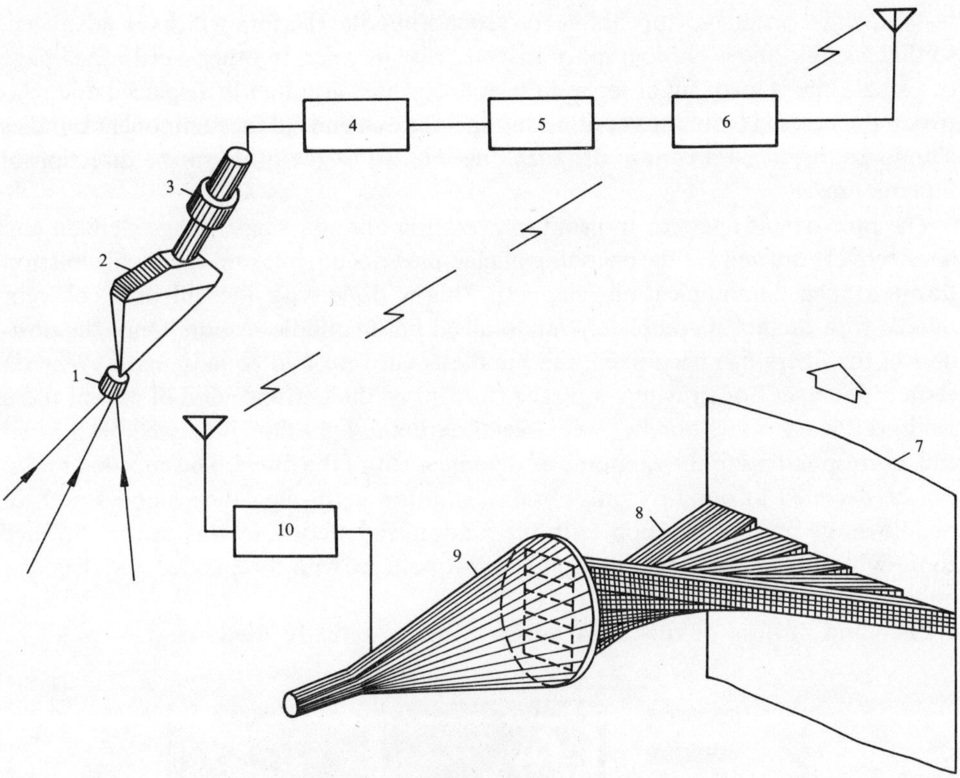

Figure 143

Aerial reconnaissance system using fiber optics:

1) aerial objective, 2) image shape converter, 3) fiber-optics image orthicon, 4) video signal generator, 5) amplifier, 6) transmitter, 7) film, 8) image shape converter (analogous to 2), 9) receiving tube, 10) receiver.

The earthbound receiver has a CRT with a fiber-optics screen and an image shape converter, analogous to that in the transmitting equipment. The electron ray scanning the CRT screen along horizontal lines from top to bottom reconstitutes the image formed on the target of the camera tube. The fiber-optics image shape converter converts the image into a long line, thereby reproducing a portion of the reconnaissance area caught by the objective of the transmitting equipment. A photographic film moves synchronously with the motion of the aircraft across the exit end of the image shape converter.

Since the entrance ends of the individual fiber bundles comprising the image shape converter are stacked in a square or a rectangular pattern, a finite time is required for the electron beam to scan the picture tube screen. Therefore, when the film is exposed to the first bundle, there is still no image in the second bundle. At the next instant, when an image appears in the second bundle, the film will have advanced, so that a single line is photographed in a stepwise manner. In other words, the image of a single line is made up of separate segments which are slightly displaced one relative to the other. To obtain a continuous line, the exit ends of the component bundles should be displaced a certain distance one relative to the other in the direction of film motion.

The property of fibers to transmit information about a single image element can be effectively utilized for the encoding of classified documents and their transmission through open communication channels. This is done with the aid of a coherent bundle with the fibers completely randomized in the middle section. Once the position of the fibers has been fixed, the bundle is cut into two equal lengths. When a certain document or drawing is placed in front of the entrance end of one of these sections (the encoding bundle), a jumble of dark and light dots is formed on the exit end, corresponding to the randomized arrangement of the fibers. The encoded image can be decoded in one way only: by transmitting it through the second length of bundle in the reverse direction, with the randomized section serving as the entrance end. Figure 144 shows a photograph of a test pattern, with its encoded and decoded images.

Encoding devices of this kind are now being actually used.

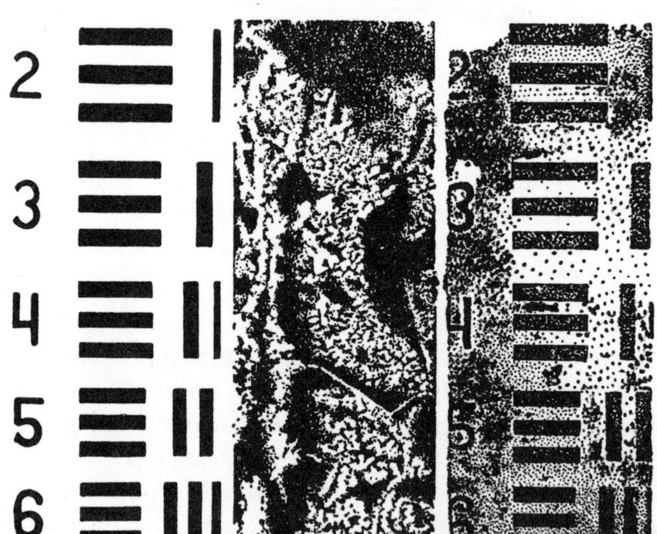

a *b* *c*

Figure 144

Encoding of information with fiber optics:

a test pattern, *b* encoded image, *c* transmitted image.

7. Scanning and control systems

Image shape conversion techniques open new horizons for the application of optical fibers in various scanning devices which scan different elements of the image field in succession along a certain trajectory and pick off information about each of the image elements.

We distinguish between electronic and mechanical scanning systems. Mechanical scanning devices are relatively simple, stable in operation, and have a long service life. However, their scanning speeds are substantially lower than those of electronic scanners, and they can execute only very simple scanning trajectories. The oldest mechanical scanning device is Nipkov's disk used at the dawn of the television era to break the picture into separate elements and to produce a video signal.

A picture transmission system using Nipkov's disk is shown in Figure 145. The disk is perforated with small square holes which are spaced at equal angular intervals in the sense of rotation and are displaced by one hole width in the radial direction. The picture is projected onto the disk. As the disk turns, each hole traces an arc within the picture frame, and this trace can be interpreted as one line of the scan. A photocell set behind the disk responds to the changes in illumination as the hole moves along the line, generating a proportional signal current. The photocurrent is determined at any time by the illumination of the scanned element. The current through the load produces a video signal. During one revolution, the holes of the disk scan the entire picture area. The picture is thus broken down into scanning elements and a video signal is produced.

The receiver in this system should be equipped with a similar disk rotating syn-

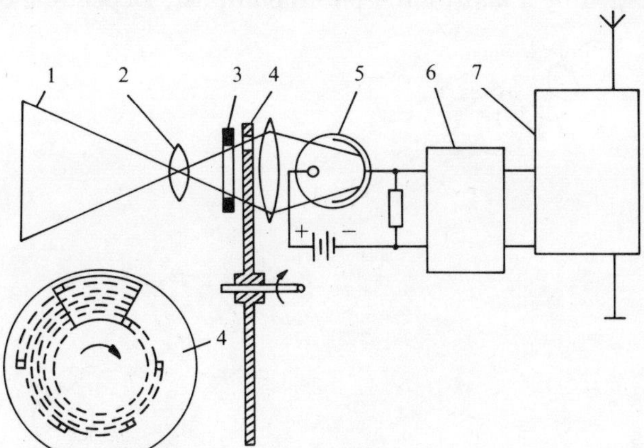

Figure 145

A television tansmitter with Nipkov's disk:

1) object, 2) objective, 3) picture-limiting frame, 4) Nipkov's perforated disk, 5) photocell, 6) amplifier, 7) transmitter.

chronously and in phase with the scanning disk. Since the area of each hole is small, only a small fraction of the total luminous flux reaches the photosensitive layer. Because of low resolving power and low efficiency, the quality of the image produced by this system is very low. With further development of television, the mechanical scanners were replaced with more efficient electronic devices.

The main advantage of electron beam scanning is the high scanning speed. Consider a 60-line picture to be scanned by a mechanical scanner. If the motor turns at a speed of 3600 rpm, scanning a single line of the raster in one revolution, a whole second will be needed to scan the entire picture, i.e., the pictures will follow one another at a frequency of 1 Hz. Electronic scanning, on the other hand, ensures frequencies of over 200 Hz. In general, the picture repetition frequency is chosen so as to ensure smooth viewing, without blinking, and it may vary depending on the detector. For the human eye, frequencies of about 20 Hz are quite sufficient.

Besides their high speed, electronic systems also have a high resolving power, e.g., the television picture can be broken up into 400,000 scanning elements. However, despite the advantages of these systems, they are not free from certain shortcomings, such as the need for a special electronic beam deflection unit, focusing fields, synchronizing elements. A hot cathode is used to produce the electron beam, and the service life of the device is correspondingly reduced.

Application of fiber optics to mechanical systems may substantially increase the scanning speeds and broaden the uses of mechanical scanners. One of such systems is shown in Figure 146 [64]. The image is projected onto the entrance end of an image shape converter with a ring-shaped exit configuration. A radial slit cut in the disk alternately scans the exit ends of all the fibers as the disk rotates. The light from each fiber is focused by a lens onto a photomultiplier. Depending on the slit width

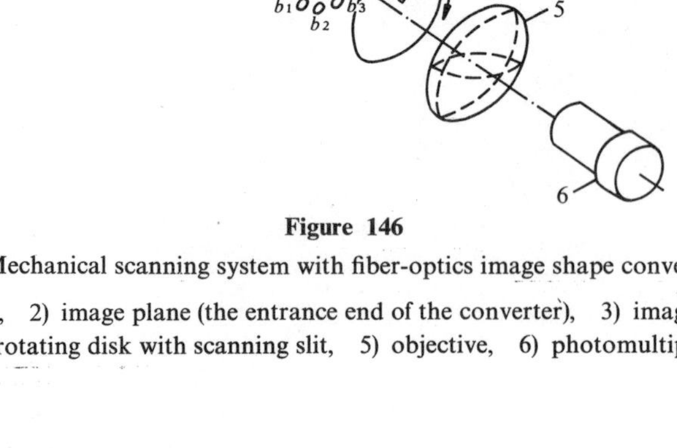

Figure 146

Mechanical scanning system with fiber-optics image shape converter:

1) projecting lens, 2) image plane (the entrance end of the converter), 3) image shape converter, 4) rotating disk with scanning slit, 5) objective, 6) photomultiplier.

and the fiber spacing at the exit end, the photomultiplier signal will present either an electrical analog of the illumination distribution over the ring or a series of discrete pulses with amplitudes corresponding to the illumination of the individual fibers. A fiber image shape converter thus breaks the image into separate elements and ensures optical scanning by means of simple mechanical motion. During a single revolution of the disk, all the image elements, whose number is determined by the number of component fibers, are scanned. A high number of small-diameter fibers and a scanning slit of suitable width will ensure a satisfactory resolution and a high scanning frequency (60 Hz at 3600 rpm).

Mechanical scanning systems with fiber optics have a number of distinct advantages which make them particularly suitable for various servo and control systems. Their design is simple, they are cheap, stable, with a long service life, have a small size, and contain no complex electronic components; moreover, they ensure a high resolving power and a high scanning frequency, and any image shape can be scanned along a variety of trajectories—lines, circles, spirals.

Figure 147 shows an information pickoff system from a paper tape chart, which may be used in various automatic data analysis systems [64]. The device is equipped with two interchangeable band-shaped fiber elements whose entrance ends are straight or arched for perfect matching with the coordinate system marked on the charting paper. The exit ends are shaped into an arc of an appropriate radius of curvature, so that when the disk rotates, each of the lenses in the disk scans the exit ends of the fibers.

If the disk has four scanning lenses, the image is projected four times onto the photomultiplier window during each revolution of the disk (240 Hz scanning frequency at 3600 rpm). The transport speed of the chart is therefore considerably increased, and the time needed for chart analysis is correspondingly reduced.

The time of a single scanning cycle is one of the principal characteristics of scanning systems. The cycle length determines the time required to scan all the points of the field. Using fiber elements, the search time for a certain object in the scanning field can be reduced by up to one order of magnitude [18]. For a square screen, this is accomplished by scanning simultaneously in two perpendicular directions.

Suppose that the control field is projected onto a square screen made up of 100 scanning elements, 10 lines with 10 elements each. Light pipes are coupled to each element of one line, and then combined into a bundle. If the same operation is repeated for all the ten lines, we obtain ten bundles whose exit ends are arranged along the scanning trajectory. Similar fiber configurations are provided for the scanning elements in each vertical line. The original 100-element field is thus broken into two 10-element fields. The two fields are scanned synchronously by two scanners, in perpendicular directions. If the scanning time of a single element is Δt, a total time of 100 Δt will be needed to scan the entire 100-element field. By using the fiber-optical configuration described above, the scanning time can be reduced to 1/10

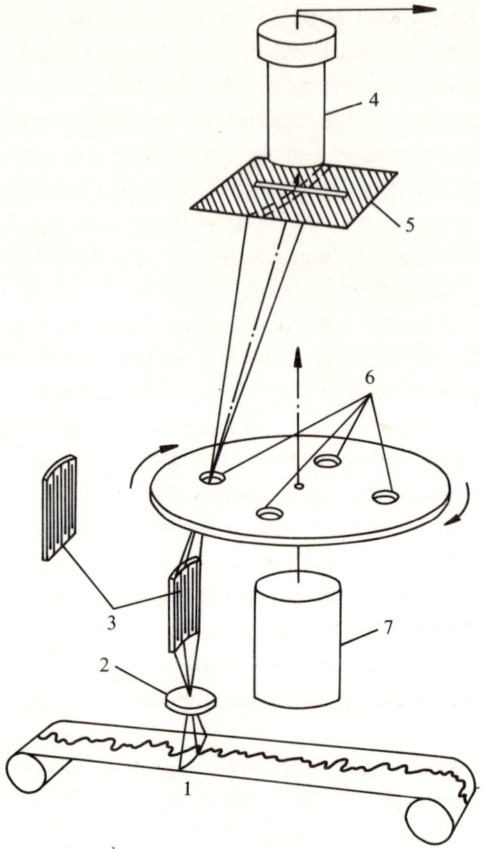

Figure 147

A scanning system for high-speed automatic analysis of tape-charted data:

1) chart tape, 2) projection objective, 3) fiber optics, 4) photomultiplier, 5) aperture, 6) scanning lenses, 7) motor.

of this value, i.e., for a square scanning field with N elements, fiber optics reduces the scanning time by a factor of $\sqrt{N}$.

A similar scanning device is described in [19] (Figure 148). The fiber converter is made from noncoherent conical fibers, whose total number is equal to the total number of scanning elements in the field. The fibers are packed in a regular pattern, so that they form a single common bundle with a square entrance end. This square acts as a screen on which the scanning field is projected.

A diagram of the screen and its coupling to the scanning device is shown in Figure 149. For simplicity, the drawing shows only 36 scanning elements (six horizontal and six vertical lines). Each bundle, with its entrance end corresponding to a single field element, is split into two threads, connected to the horizontal and vertical scan-

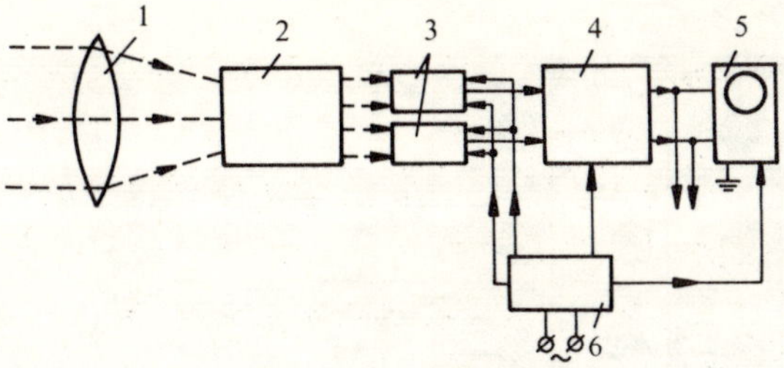

Figure 148

A fiber-optics search device:

1) optical system focusing the image of the controlled field onto the entrance end of the fiber element, 2) fiber converter, 3) photoelectronic disectors for scanning, 4) circuit forming control signals, 5) electron-beam indicator, 6) power supply.

ners. One thread from each of the fiber bundles in the same line is taken to form a single bundle with a circular exit end. The number of the resulting bundles is equal to the number of horizontal lines. The exit ends of these bundles are arranged along the scanning trajectory (along a circle, as shown in Figure 150).

The other half of the bundles whose entrance ends fall along the same vertical line of the screen are also tied into a single bundle. A similar operation is repeated for all the vertical lines. We thus obtain a second group of bundles, whose number is equal to the number of vertical lines. The exit ends of these bundles are arranged along a second circle (Figure 149), which constitutes an additional scanning trajectory.

The fiber converter in its finished form has a square exit end with N horizontal and N vertical lines and two circular entrance ends, each formed by N bundles arranged along a circle. The exit ends of the fiber converter are scanned by disectors 3, and the field is therefore scanned simultaneously in two mutually perpendicular directions.

The disectors use photomultipliers, each provided with two deflecting coils fed by sinusoidal voltages with a $\pi/2$ phase shift. The resulting magnetic field rotates the electronic image so that each point describes a constant radius circle. A radial slit intercepts the electron beam moving to the anode. When the exit ends of the bundles are projected onto the entrance end of one of the disectors, the electrons from all the portions of the photocathode in contact with the exit ends of the bundles pass successively across the slit as the electron image is rotated.

The relevant information about an object in the scanned field is conveyed by the appropriate light pipes to the entrance ends of the two disectors. Comparing the disector outputs with the reference pulses delivered to the scanning device, we can determine the coordinates of the object in the field.

It is clear from the above description that single-valued position fixing is possible only when a single object lies in the scanned field. In this case, one of the disectors defines the horizontal line of the object, and the other the vertical column. If the

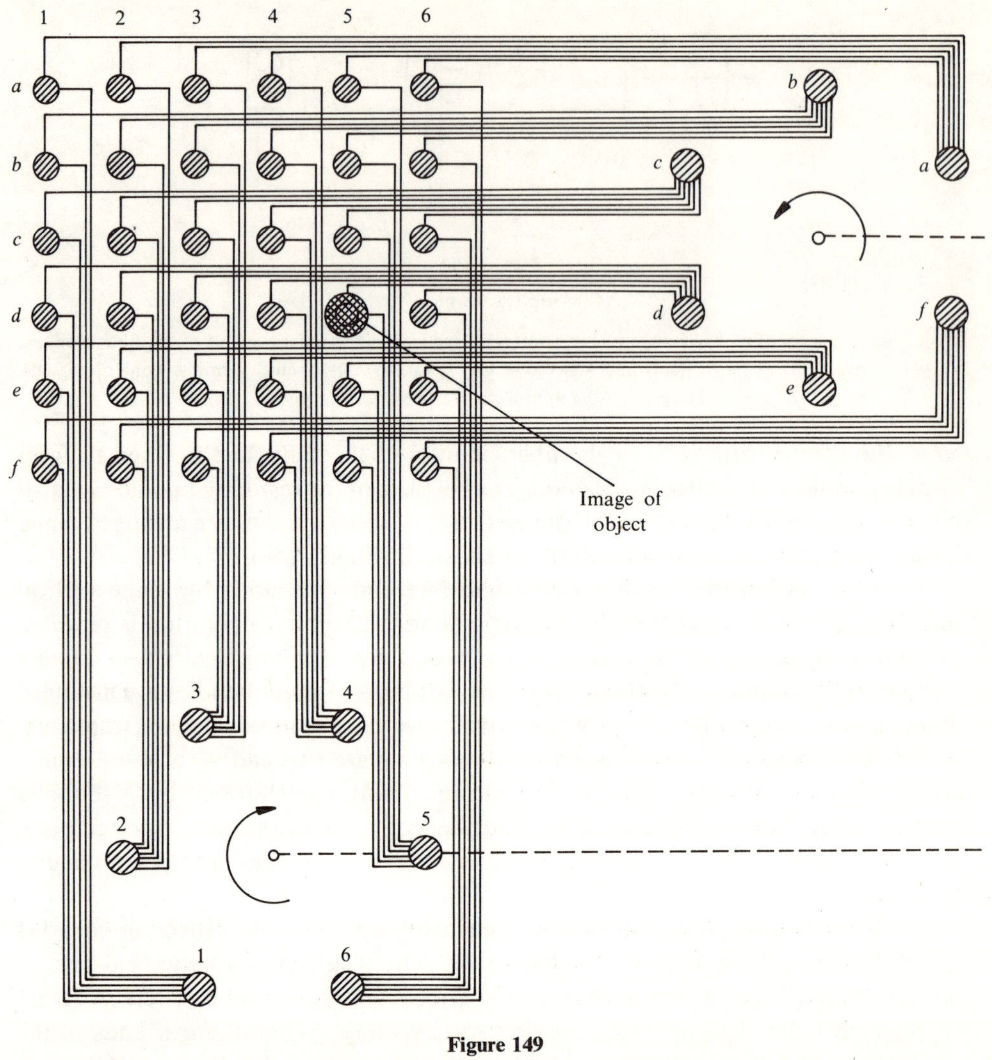

Figure 149

Diagram showing the screen design of the search system and its coupling to the scanning devices.

field contains two objects, no conclusive information can be obtained on their exact position. For example, consider two objects at points *a*2 and *c*3 (Figure 149). The disector output signals identify the presence of objects in rows *a*, *c* and columns 2, 3. However, the exact position of the objects cannot be determined: there are in fact two alternative interpretations of the data: either *a*2, *c*3 or *a*3, *c*2. The problem naturally grows in complexity as the number of objects increases.

If only one moving object is sought in the scanned field, the disector output can be fed to a computer to determine the original position. The computer output is delivered to a servo system which shifts the optical axis. In this way, continuous tracking of a moving object is effected. The advantage of this system compared with ordinary systems is that the search time is reduced and the coordinates are determined very rapidly, without loss in the resolving power.

This device also can be manufactured with a circular screen [18]. In this case, scanning is done simultaneously along radii and concentric circles. A diagram of this setup is shown in Figure 150. As in the previous case, each of the bundles corresponding to the various scanning elements of the field is split into two. One branch from each of the bundles lying along a constant radius circle is tied into a common bundle. The number of these combined bundles is equal to the number of concentric circles comprising the scanned field. The remaining branches of all the bundles lying along the same radius are then tied up into a common bundle. The number of combined radial bundles is equal to the number of radial lines of the scanned field. (This can be done if the diameter of the fibers at the center of the screen is less than the diameter of the fibers along the periphery.)

The two groups of bundles formed in this way couple the screen and the field projection with two photosensitive scanning devices. When the scanners operate in synchronism, the entire field is scanned simultaneously along the radii and the concentric circles, so that the search time is greatly reduced.

Since the light transfer of optical fibers does not change when the fibers are bent or twisted, fibers can be freely used in various control systems, light-sensitive relay devices, etc.

A diagram of a system for programmed control of a servo motor axis is shown in Figure 151 [81]. The principal elements ensuring the exceptionally high accuracy of the control device are the fiber converters *A* (the stator) and *B* (the rotor). Con-

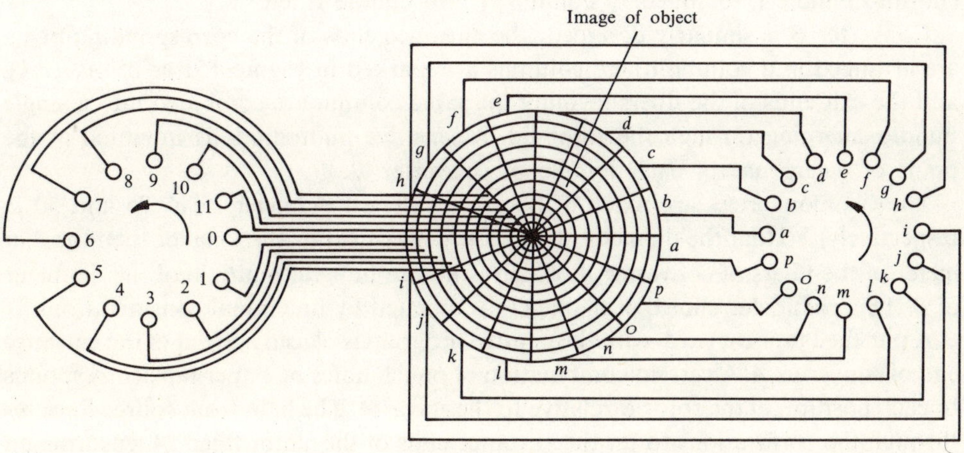

Figure 150

Diagram showing the design of a circular screen and its coupling to the scanning devices.

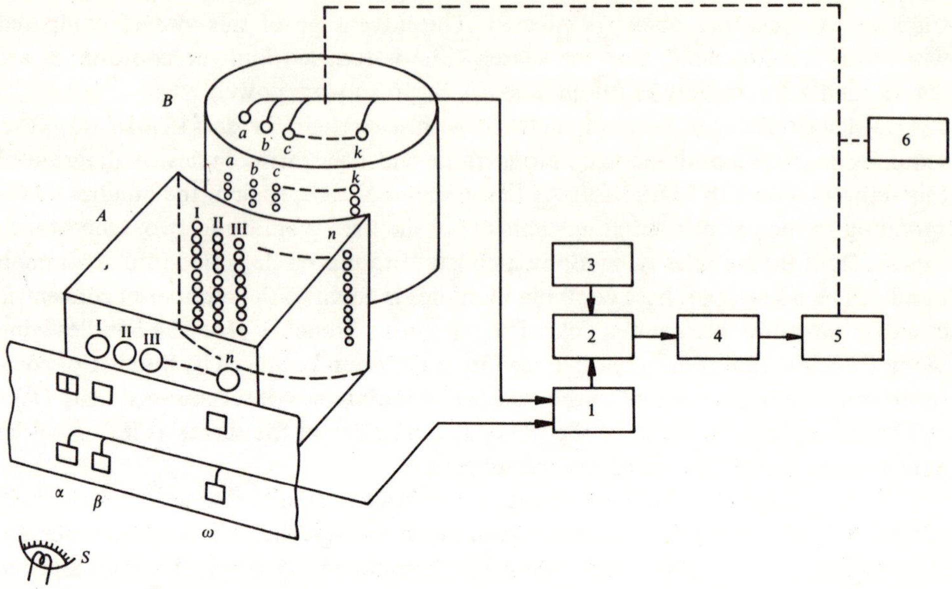

Figure 151

Programmed control system for a servo motor.

verter A consists of separate noncoherent bundles whose fibers are packed in a circular arrangement at the entrance end. At the exit end, the fibers are packed in a vertical column of one-diameter width. All the fibers of column I are joined at the entrance end into bundle I, the fibers of column II into bundle II, etc.

Converter B is similarly designed: the entrance ends of the corresponding fibers are arranged in a column (these columns are marked in Figure 151 as $a, b, c, \ldots, k$), and the exit ends of the fibers forming the same column are combined into a single bundle emerging through the rotor lid. A sensitive photodiode is mounted at the point of emergence of each bundle of converter B.

The two converters are made of fibers of different diameters (e.g., 75 and 50 μ, respectively), so that the diameter of the fibers of converter B is smaller than the diameter of the fibers of converter A, which is not an integral multiple of the diameter of B. This particular choice of diameters is dictated by functional considerations.

A punched tape (or card) with the control program is placed in front of the entrance end of converter A. A certain combination of punch holes on paper tape corresponds to each position of the rotor B relative to the stator A. The light from source S passes through the perforations to hit the entrance ends of the stator fibers A, ensuring an exact definition of the light–shadow boundary on the rotor B. For example, if the light–shadow boundary is to fall between columns III and IV of converter A, the punch holes should be arranged so that the light from the source reaches the entrance ends of bundles I, II, III of converter *A*, whereas bundles from IV to *n* remain unilluminated.

The light–shadow boundary can be set between any pair of columns corresponding to a certain combination of punch holes on paper tape, i.e., its position can be programmed beforehand.

The motor axis 5 which receives the unbalance signal is coupled to the rotor and the axis of the control device 6. The logic circuit 1 selects a certain column of converter *B* and generates an unbalance signal representing the displacement of this column relative to the light–shadow boundary on converter *A*. The selected column corresponds to a certain combination of punch holes on the tape, and the column is therefore selected automatically by the logic circuit for each position of the paper tape. The hole combinations on the paper tape controlling the column selection are identified as $\alpha, \beta, \ldots, \omega$ in the figure. The unbalance signal is delivered to a calculator unit 2 and after amplification by amplifier 4 it actuates the motor 5. When the fiber column of rotor *B* selected by the logic circuit coincides with the light–shadow boundary on the stator, the unbalance signal vanishes and motor rotation ceases.

This control system will align the axis with an accuracy better than one fiber diameter of the rotor *B*. For example, for fiber diameters of 75 and 50 μ, the aiming accuracy is 25 μ.

A schematic diagram illustrating the operation of the programmed control system is shown in Figure 152. The plane of the drawing represents part of the horizontal section of the converters *A* and *B* (view from above of the system shown in Figure 151). The numbered lines correspond to the edges of the converter columns. The initial position of the rotor relative to the stator is selected so that the left edge of column I of the stator coincides with the left edge of the column *a* of the rotor (identified as 0 in Figure 152). This position is fixed by a special electronic circuit, and the selection by the logic circuit is done according to column *a* of rotor *B*.

Suppose rotor *B* should be given a tangential displacement of 25 μ relative to stator *A*. This displacement is equal to half the diameter of the fibers in converter *B* and to one third of the fiber diameter in converter *A*. To obtain the required displacement, the paper tape position should be changed so that the light–shadow boundary shifts to position I on stator *A*, i.e., column I is illuminated and column II remains in the dark (Figure 152*b*). The logic circuit receives the appropriate signal from the table and the selection is done according to column *b* of the rotor *B*. The system will continue generating a signal as long as the left edge of column *b* does not coincide with the light–shadow boundary (Figure 152*b*).

It is clear from the last figure that rotor *B* shifts by 25 μ relative to the initial position shown in Figure 152*a*. To obtain a displacement of 50 μ, the tape should again be moved until the light–shadow boundary occupies position 2 of stator *A*; columns

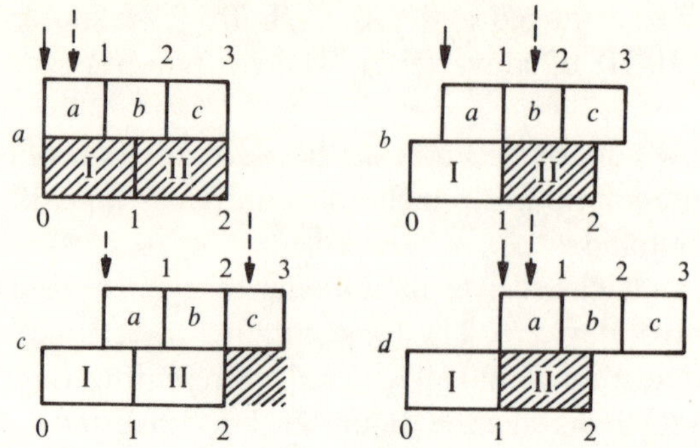

Figure 152

Schematic diagram illustrating the operation of a programmed control system (the solid arrow marks the zero point, the dashed arrow identifies the selected column).

I and II are then illuminated, and column III is in the dark (Figure 152*c*). At the same time, the logic circuit receives a signal instructing to proceed with selection according to column *c* of rotor *B*. We see from the figure that rotor *B* shifts by 50 μ relative to its initial position.

If the light–shadow line is in position I (Figure 152*d*), i.e., column I is illuminated and the selection is done according to column *a* of rotor *B*, the rotor is displaced 75 μ relative to its initial position. The controlled axis thus can be aimed in any desired direction with a fairly high accuracy. The aiming accuracy is determined by the ratio of the fiber diameter of converters *A* and *B*. If this ratio is 9:10, the rotor position can be fixed to within 1/10 of the fiber diameter of stator *A*.

Absence of friction contacts and the high accuracy of the system distinguish it favorably from analogous electron-mechanical systems. The system can be modified to convert the information about the axis position into digital or coded form, for direct computer input. This version of the system can be used for measuring the altitude above sea level, as an aid in sea and air navigation, or for determining the roll angle of a ship or an aircraft relative to the horizon.

Optical fibers in servo and tracking systems greatly simplify the design of the rotatory optical heads. Figure 153 shows an infrared fiber-optics servo system [16].

The optical head consists of a projection objective and three fiber elements: a sphere and two lens-like components. The first lens transfers the projected image to the sphere, and the second conveys it from the sphere to the vidicon. The gaps between the sphere and the adjoining elements are filled with an organosilicon lubricant having the same refractive index as the fiber material. This minimizes the light losses when the image is transferred to the photosensitive layer of the vidicon. The light transfer of this system is markedly higher than that of ordinary servo systems, since fiber elements replace a complex system of rotation-compensating rotatory prisms and correcting lenses.

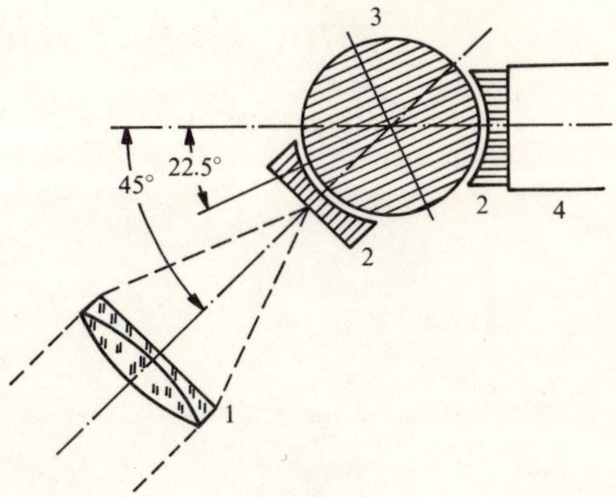

Figure 153

An optical head for a fiber-optics servo system:

1) objective, 2) lens-shaped fiber elements, 3) fiber-optics sphere, 4) vidicon.

This device is small, simple, and convenient in practical use. It scans in 45° angles in any direction, while the fiber sphere is turned only through half the total deflection angle by a differential spring transmission. The fibers used are 10 μ in diameter, with 0.66 numerical aperture. Nevertheless, the resolving power of the entire system is about 11 lines per mm, i.e., quite acceptable for most applications.

8 *Other Applications of Fiber Elements*

1. Observation of inaccessible cavities

Widely used fiber-optics elements are fiberscopes, which constitute flexible fiber bundles intended for observation of inaccessible cavities and transmission of images over curved paths [88, 145, 161].

A fiberscope is a coherent bundle of regularly packed fibers which are fixed at the two ends only. In the middle part, the fibers are free, so that the bundle is sufficiently flexible. Since light transfer is not affected by fiber bending, the quality of the transferred picture remains uniformly adequate. To protect the fibers from mechanical damage and contamination, the entire fiber bundle is encased in a protective sheath in the form of an elastic stainless-steel spring or a plastic envelope.

Fiberscopes may reach a few meters in length [117], with up to 1 million fibers per 1 cm^2 of cross sectional area [146]. A general view of a fiberscope is shown in Figure 154. An objective fixed at the entrance end of the instrument projects the picture. The transmitted picture is observed at the exit end through an eyepiece fitted in the fiberscope sheath.

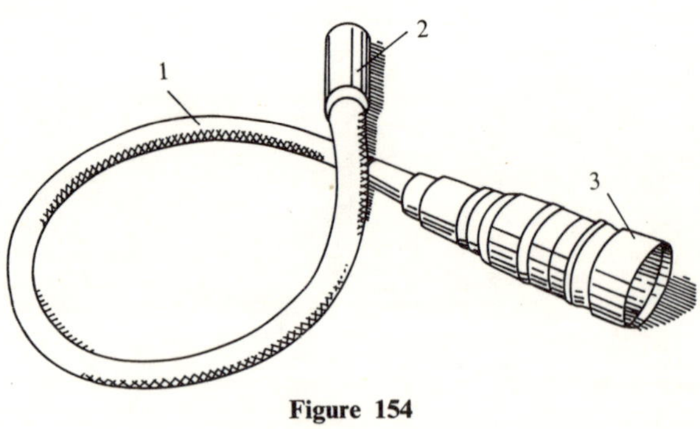

Figure 154

General view of a fiberscope:

1) fiber bundle, 2) objective, 3) eyepiece.

Flexible fiberscopes are used in observations of various processes and in monitoring the operation of equipment in chemical reaction chambers or in premises which are hazardous to the personnel. They are also effective in various control systems, e.g., in fire-hazardous premises and cavities, such as jet-engine nozzles. In the latter case, the picture of the nozzle may be transmitted directly to a control panel in the cockpit [146]. Such a warning system enables the pilot to rapidly assess the situation in an emergency and to take appropriate steps.

Flexible bundles may also be used for monitoring various inaccessible cavities, e.g., jet engines, fuel tanks, etc. The problem of illumination can be solved with the aid of the same fiberscope: the peripheral fibers are combined into a single bundle which acts as a conventional light pipe, transmitting light from an external source to the object surface. The picture of the illuminated object is then transferred through the fibers in the central part of the bundle.

Fiberscope diameters may be fairly small (a few millimeters), with up to 40,000 fibers. Fiberscopes therefore may be inserted through very narrow channels, e.g., gun barrels and jet engine ducts.

Because of the high elasticity, excellent light transfer, high resolution, and small diameter, fiberscopes are becoming increasingly popular in medicine. The old dream of physicians, physiologists, and biologists, to observe directly the internal organs and cavities of the human body (the stomach, the bronchi, the heart, etc.) is thus coming true thanks to fiber optics [33, 50].

Prototypes of such instruments, known in medical use as endoscopes, have been used for quite some time now in diagnosis of internal diseases. These endoscopes, however, do not ensure a high quality picture and have much more restricted uses than fiber systems. They incorporate prisms, numerous lenses and mirrors, so that the overall light intensity is markedly reduced. The illumination system leaves much to be desired also. Endoscopes do not have sufficient elasticity and mobility, and in some cases they are therefore not usable.

Fiber optics can markedly improve the endoscope performance [50]. The mechanical and optical characteristics of fiber-optical endoscopes [61, 83, 86, 87] are markedly better than the corresponding characteristics of the conventional instruments.

Figure 155 shows an endoscope described in [87]. It consists of 150,000 fibers, each 14 μ in diameter and 92 cm long. Thicker fibers arranged along the perimeter of the bundle are intended to provide object illumination. An incandescent bulb is fixed in the viewing head, near the high-magnification eyepiece, and its light is projected onto the peripheral fibers. The entrance end, which is inserted into the organ, has a projecting lens. The entire fiberscope is encased in a smooth flexible plastic sheath not exceeding 11 mm in diameter.

Clinical observations of the stomach cavity conducted with this fiberscope established its adequate performance.

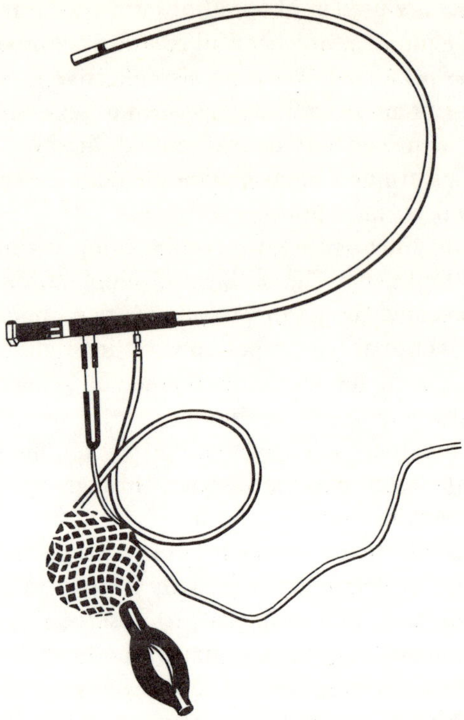

Figure 155
Fiber-optics endoscope.

The potential uses of fiber optics in medicine are not restricted to observation of internal cavities in the human body. Fiber-optical elements may also be applied to observe processes which take place in the subcutaneous tissues [54, 120].

Theoretical and experimental studies [95] show that bundles consisting of isolated fibers 1 μ in dia. have adequate light transfer and faithfully transmit colors. Using such thin fibers, probe bundles of sufficiently small diameter can be prepared for subcutaneous observations. For example, a probe 0.5 mm in diameter contains some 200,000 fibers of 1 μ diameter each, thus ensuring a resolving power of up to 500 details per 1 mm [54].

Projection optics are relatively difficult to manufacture and assemble for the inserted end of very thin fiber bundles. These instruments therefore use no projection optics whatsoever, and only those tissues which come in direct contact with the entrance end of the bundle can be viewed.*

The use of very thin bundles also presents a problem with regard to object illum-

* The entrance end of the bundle is the surface facing the object of study.

ination. One of the possible solutions to the problem of illumination of subcutaneous tissues [55] is illustrated in Figure 156*a*. Light from an extraneous source is projected into a fiber bundle, which transmits it to the object surface; the light is partly reflected by the tissue, permitting direct viewing. This system is not easy to implement in practice, since light is also reflected from the exit end of the bundle, thus markedly lowering the picture contrast. Dark-field observations are therefore more advantageous.

In dark-field observations, the end faces of the bundle must be plane and strictly perpendicular to the optical axis of the instrument. We see from Figure 156*b* that the information-carrying signal propagates inside a hollow light cone which accommodates the rays directly reflected by the exit end of the bundle. To satisfy this condition, the numerical apertures of the fibers and the condenser lens must carefully match. This matching can be readily ensured by selecting an appropriate lens for fibers of given aperture. In certain cases, a bundle can be replaced by a single multifilar fiber with a sufficient number of light-carrying elements.

The best resolution conditions are attained with direct and immediate contact between the entrance end of the bundle and the object surface. Often, however, no direct contact can be established, so that each fiber receives light reflected from a surface element whose area is larger than the entrance end area of the corresponding fiber. Information relating to a single surface element is thus transmitted by several fibers simultaneously, and detail resolution deteriorates.

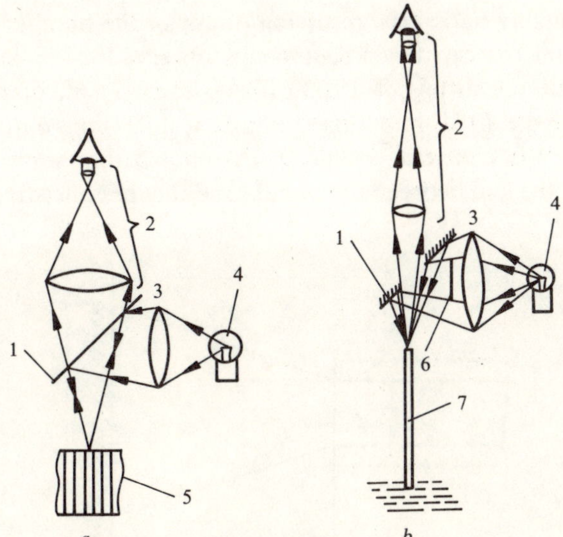

Figure 156

A vertical system for the illumination of subcutaneous tissues:

a observations on a light field, *b* observations on a dark background; 1) mirror, 2) microscope, 3) condenser, 4) light source, 5) fiber element, 6) screen, 7) fiber bundle.

The diameter d of the surface element scanned by a single fiber element depends on the fiber diameter D, the numerical aperture of the fiber, and the gap t between the entrance end and the object surface. We see from Figure 157 that

$$d = D + 2t \tan \theta. \tag{332}$$

Since the numerical aperture of the entrance end of a fiber is

$$n_3 \sin \theta = (n_1^2 - n_2^2)^{1/2},$$

we have

$$\operatorname{tg} \theta = \frac{\sin \theta}{(1 - \sin^2 \theta)^{1/2}} = \frac{a}{n_3 \left(1 - \dfrac{a^2}{n_3^2}\right)^{1/2}} = \frac{a}{(n_3^2 - a^2)^{1/2}}, \tag{333}$$

where

$$a = (n_1^2 - n_2^2)^{1/2}.$$

Inserting (333) in (332), we find

$$d = D + \frac{2t}{\sqrt{\left(\dfrac{n_3}{a}\right)^2 - 1}}. \tag{332a}$$

Let $n_3 = 1.4$, $a = 0.5$; then for $D = 10\ \mu$ and $t = 13\ \mu$, relation (332a) gives $d = 20\ \mu$, i.e., a $13\ \mu$ gap roughly halves the resolving power of the bundle.

A fiber-optics microscope for subcutaneous observations is described in [120]. A coherent bundle consisting of 10,000 fibers, each $5\ \mu$ in diameter, is inserted into a syringe needle (0.81 mm inner diameter, 1.07 mm outer diameter). The outer diameter of the coherent bundle is 0.5 mm, so that some forty 50-μ fibers can be inserted in the gap between the bundle and the inner needle wall for purposes of illumination.

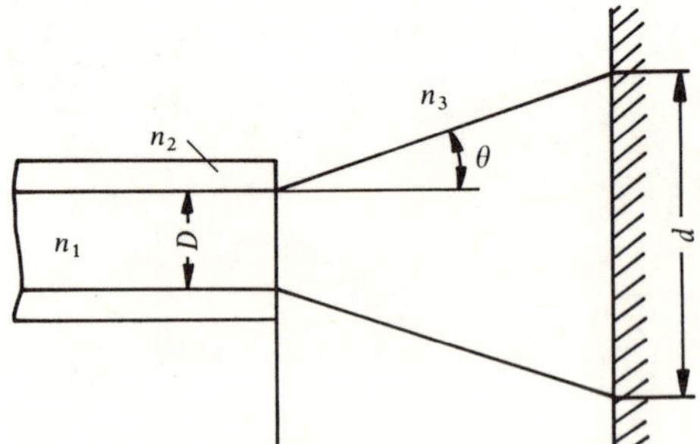

Figure 157

Determining the resolution of a fiber held at some distance from the scanned surface.

The entrance end of the needle and the bundle is cut at 45° to the axis, and the observation field is therefore elliptical. The observation system incorporates two interchangeable objectives with 5 × and 10 × magnification and a special distortion-correcting eyepiece, needed because of the oblique cut of the entrance end. Distortion correction is done by special cylindrical lenses which ensure the required magnification ratio along the major and the minor axes of the ellipse.

A cross section of the subcutaneous microscope is shown in Figure 158. The resolving power is different along the major and the minor axes of the ellipse. According to the authors' estimates, the resolution limit is 14 μ along the major axis and 10 μ along the minor axis. Dynamic scanning in the direction of the minor axis will resolve details up to 5 μ across (fiber diameter).

According to the authors, the main shortcoming of visual observations of subcutaneous tissues is the low picture contrast, since the different tissue components do not differ significantly when illuminated with white or monochromatic light. Before conducting the visual observations, certain pigments should be injected subcutaneously in order to stain the colorless structures, e.g., nuclei, cytoplasm, and

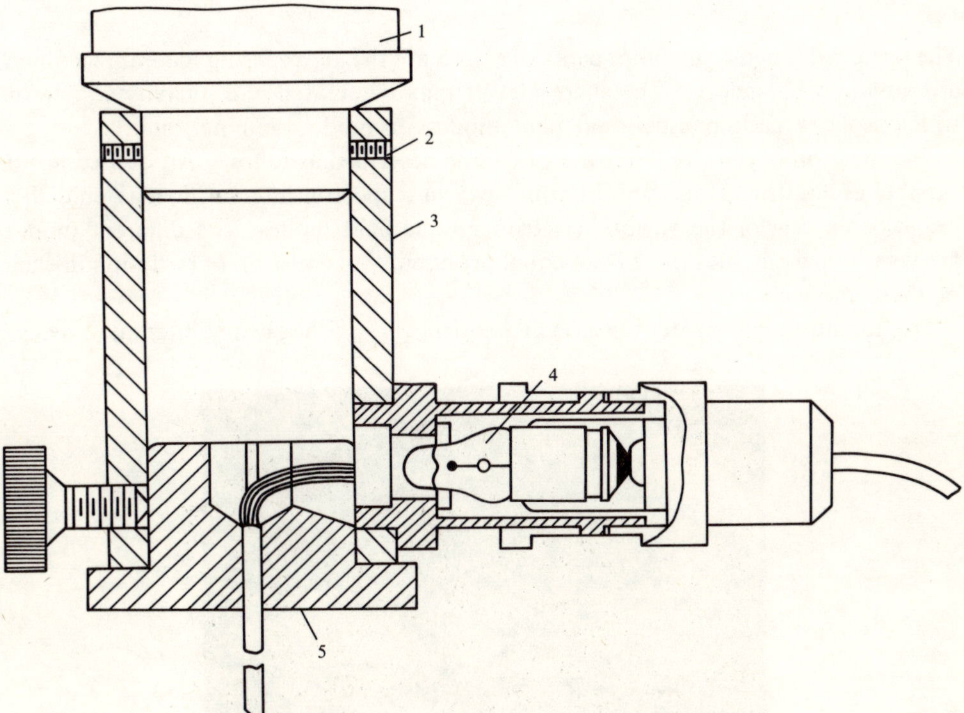

Figure 158

Cross section through a microscope for subcutaneous observations:

1) microobjective, 2) focusing screw, 3) tube, 4) light source, 5) needle holder.

other components of the living cell. The pigments should be so chosen as to ensure selective staining of the different tissue components.

Figure 159 reproduces a photograph of the microscope field. The visible area is $500 \times 700\ \mu^2$. The dark strip $90\ \mu$ wide is a blood vessel. The elliptical field is the result of applying a cylindrical lens for distortion correction. These instruments can be applied to diagnose diseases of internal organs and subcutaneous tumors. In clinical conditions, they can be effectively used for monitoring the effect of drugs and for observing the processes which take place in tissues and blood vessels in reaction to extraneous agents.

Fibers transparent in the ultraviolet [90] are used in fluorescence observations, extending the scope of study of the physicochemical processes in the living organism.

Combination of fiber optics and lasers provides a basically new means for the illumination of subcutaneous tissues and may be applied to investigate the effect of high light fluxes on living tissue.

2. Application of fibers in lasers

The principal functional components of a laser are the active lasing medium in which a population inversion of the energy levels may be created and a cavity resonator which isolates certain individual field modes or mode combinations.

The function of the cavity resonator can be described as follows. An excited atom capable of emitting a photon of frequency ν in a spectral interval $d\nu$ may emit this radiation in one of the various electromagnetic field modes. The different modes are very numerous, and there is an equal probability of emission of radiation in each of these modes.

The radiation energy density $u(\nu)$ at frequency ν in a frequency interval $d\nu$ is ex-

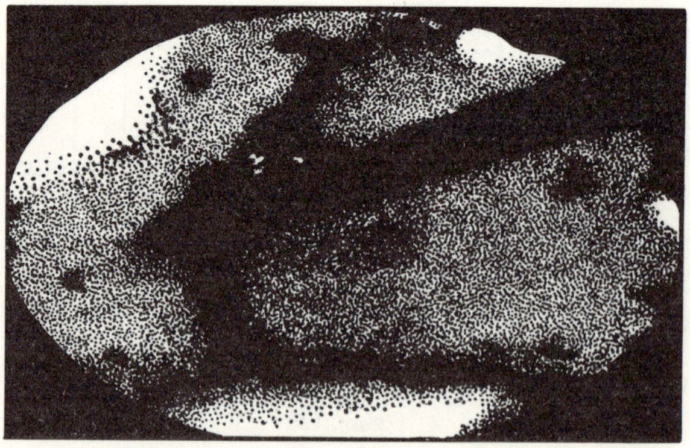

Figure 159

The field of view of a subcutaneous microscope.

pressed by Planck's formula

$$u(\nu)d\nu = \frac{8\pi\nu^2}{c^3}\,\frac{h\nu}{e^{h\nu/kT}-1}\,d\nu, \tag{334}$$

where k is Boltzmann's constant, T is the absolute temperature of the emitter.

The factor $(8\pi\nu^2/c^3)d\nu$ determines the number of the degrees of freedom or the number of oscillatory modes per unit volume at frequencies between ν and $\nu + d\nu$. The factor

$$\frac{h\nu}{e^{h\nu/kT}-1}$$

expresses the average energy per degree of freedom.

The number of different oscillatory modes in a volume V for a spectral line of wavelength λ with a Lorentzian contour and a line width $\Delta\lambda$ is given by [143]

$$N(\lambda)\,d\lambda = \frac{8\pi^2 V\Delta\lambda}{\lambda^4}. \tag{335}$$

For $V = 1\ \text{cm}^3$, $\lambda = 6000$ Å, and $\Delta\lambda = 10^{-2}$ Å, this relation gives 10^9. The radiation emitted by an ensemble of excited atoms in a volume $V = 1\ \text{cm}^3$ will thus be equally distributed between 10^9 modes. A cavity resonator selecting certain oscillatory modes is necessary in order to isolate the desired field components.

Mode selection in the microwave region is accomplished by cavity resonators whose geometrical dimensions are comparable with the wavelength. In the visible spectrum, no such resonance cavities can be built because of the short wavelengths involved. Lasers therefore use Fabry–Perot interferometers for this purpose. The simplest interferometer of this type consists of two plane mirrors separated by a certain gap L, with the active medium filling the gap. A resonator of this type selects only the axial modes, i.e., those modes which propagate perpendicularly to the mirrors, satisfying the condition

$$L = m\frac{\lambda}{2}, \tag{336}$$

where m is an integer.

We see from (336) that the modes selected by this resonator are standing waves.

According to the geometrical optics treatment, the modes transferred inside a fiber correspond to the light waves which hit the core–envelope interface at an angle greater than the total inner reflection angle. To confine all the modes inside the fiber, the index of refraction of the core should be infinitely large compared to the index of refraction of the envelope. In practice, however, these refractive indices differ by no more than a few percent, so that only a small number of modes remain confined in the fiber. Tests made with cylindrical dielectric waveguides [96, 118, 148, 149] bear out this conclusion. Similar tests were carried out with Pyrex-coated ruby waveguides of rectangular cross section [51]. The core cross section reached 80 × 120 μ^2 and the length was 2.5 cm. The results of both passive (without generation in the waveguide) and active tests established the existence of well-defined field modes in the waveguide; these were mostly high-order modes.

The use of fibers as active resonators for stimulated radiation was investigated in [47, 82]. If the population inversion of the relevant energy levels is such that the top level has a population excess of N particles, the condition of generation in a fiber of volume V may be written in the form [143]

$$N \geqq \frac{Vh\Delta\nu}{4\pi\mu^2 Q}, \tag{337}$$

where $\Delta\nu$ is the line halfwidth of the radiation corresponding to transitions between the relevant levels, μ is the transition matrix element.

The Q factor of the fiber resonator is determined by radiation absorption in the core and by the reflection coefficient of the fiber ends. For fibers a few centimeters long made of high-quality optical glasses, the absorption is negligible. If the reflection coefficient of the fiber ends is $R = 0.9$–0.98, we may approximately write [147]

$$Q = 2\pi \left(\frac{L}{\lambda}\right)(1 - R)^{-1}. \tag{338}$$

The minimum pumping power required to create a population inversion of N particles in the fiber is

$$P = \frac{Nh\nu}{\tau}, \tag{339}$$

where τ is the lifetime of the particles on the top level.

Since N is a function of fiber volume V, both N and P diminish with decreasing fiber diameter. This is not unexpected, since the operation of a laser is mainly determined by the pumping density and the density of the active centers; it is independent of the absolute values of the parameters.

The principal conclusion reached in [147] is that fibers ensure a much better mode selection and a much stronger mode coupling than a Fabry–Perot interferometer. It is apparently the strong coupling leading to significant energy redistribution between the modes that lowers the lasing threshold. These advantages of a fiber-optics laser become prominent only if we can ensure high-efficiency utilization of the pumping power: pumping a small-diameter cylinder is not an easy undertaking.

One of the possible ways of overcoming this difficulty was proposed in [147]. Active material of refractive index n_1 is machined into a core of sufficiently small radius r_1, so that only one or few modes can propagate through it. The core is coated with an envelope of radius r_2 and refractive index $n_2 < n_1$; this envelope is then topped with another envelope of refractive index n_3. The two envelopes may be manufactured from glasses having different dispersions; the glasses should be selected so that $n_2 > n_3$ for the pumping radiation and $n_2 < n_3$ for the laser radiation. This choice of the refractive indices promotes the penetration of the pumping energy into the core and markedly suppresses its leakage from the fiber. The fiber acts as an efficient light pipe of radius r_2 for the pumping radiation. When generation begins, modes for which the condition of total inner reflection at the core–envelope boundary is not satisfied freely leak out of the fiber.

The various solids used in laser technology are very difficult to draw into fibers. The most promising materials for this purpose are glasses and rare-earth doped plastics.

In one of the experiments stimulated radiation was generated in glass rods with neodymium impurity [30]. The bundle consisted of 2000 fibers 20 μ in diameter, with 40 μ separation between the fibers. The stimulated radiation had a wavelength of 1.06 μ. Each fiber thus acted as a waveguide with a diameter almost 20 times the generated wavelength, so that numerous modes could be sustained in the fibers.

Generation was observed at pumping energies of 40 J and higher. The individual fibers generated independently, so that the radiation emitted by the entire bundle was noncoherent.

In [67] stimulated radiation was obtained with polymethylacrylate plastic fibers activated with chelate molecules containing Eu(III) atoms. The fibers were 50 cm long and 1 mm in diameter. At liquid nitrogen point, they generated monochromatic radiation at 6130 Å. The pumping was done with ultraviolet pulses (3400 Å) of 250 μsec duration.

Chelate compounds have pincer-shaped molecules in which the central rare-earth ion is covalently bonded with organic groups. Figure 160 shows the structural formula of the chelate compound europium tribenzoylacetonate.

The distinguishing feature of chelate compounds is that their fluorescence spectrum is mainly determined by the central rare-earth ion, whereas the absorption

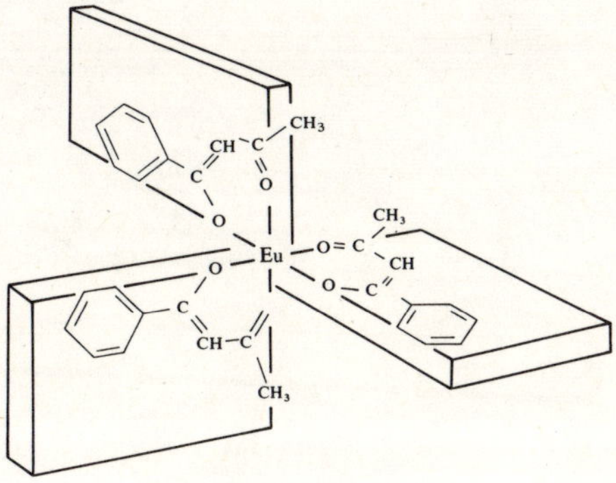

Figure 160

Molecular structure of europium tribenzoylacetonate.

is governed by the organic radicals. The absorbed energy is transmitted through intramolecular interactions to the rare-earth ion, which becomes excited. This is a highly significant factor, since the narrow absorption lines of the rare-earth ions require very high pumping powers in order to create population inversion. Organic molecules, on the other hand, are characterized by strong absorption in a wide range of wavelengths in the ultraviolet, sometimes even encompassing a substantial part of the visible spectrum, so that chelate compounds utilize the light of flash lamps with fairly high efficiency.

Another valuable feature of these compounds is their high solubility in various organic solvents and high-molecular compounds from which hard plastics can be prepared. This simplifies the production of laser fibers.

Since each fiber generates independently, fiber-optics laser bundles should preferably be used for light amplification, and not for generation of coherent radiation. An amplification of this kind is described in [114], and its principal components are shown in Figure 161. The active medium is a neodymium-doped glass fiber coated with an envelope having a lower refractive index. The fiber is 1 m long, the core is 10 μ in diameter, and the diameter of the envelope ranges from 0.75 to 1.5 mm. While still in the hot state, the fiber was shaped into a spiral, whose inner diameter exceeded the diameter of the flash lamp. The light signal for amplification was generated by an ordinary neodymium laser and focused on the entrance end of the fiber. The core was protected from stray light by 0.1 mm apertures at the two ends of the fiber. The flash lamp was triggered in synchronism with the beginning of generation of the input signal. The amplification of this device was determined as the ratio of the output signals for pumped and unpumped fiber. The measurement results are plotted schematically in Figure 162, which shows that the amplification increases with increasing pumping energy E_p, reaching saturation at a certain E_p. This saturation corresponds to the point where the fiber starts generating. Genera-

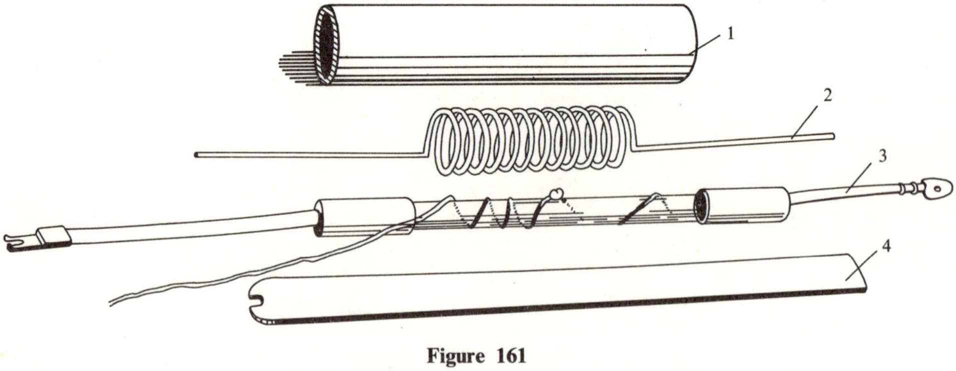

Figure 161

Basic constituent elements of a fiber laser:

1) reflector, 2) fiber, 3) flash lamp, 4) graduated rule.

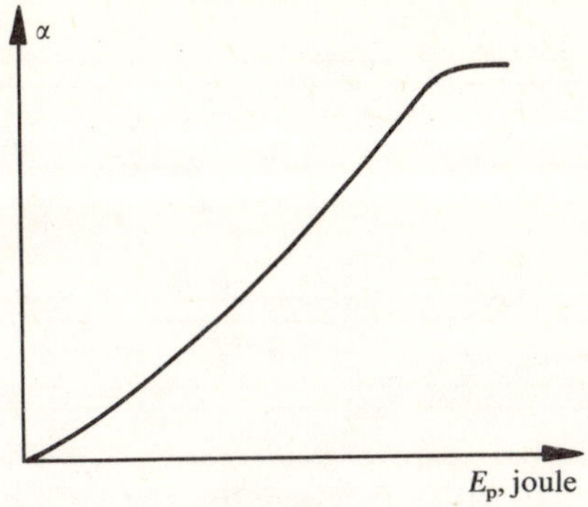

Figure 162

Amplification α as a function of the pumping energy E_p.

tion in the fiber is caused by feedback which is triggered at a certain population inversion as a result of reflection from the fiber ends.

Different methods are available for reducing end reflection, e.g., coated optics or immersion liquids. It is very difficult, however, to bring this reflection down to 0.1 % in a sufficiently wide range of incidence angles. Therefore, to reduce the feedback, the authors chose to alter the fiber geometry. The fiber ends were cut at an angle ϕ to the axis; this angle was selected from the condition

$$\cos^{-1}\frac{n_2}{n_1} < \phi < \cos^{-1}\frac{1}{n_1} - \cos^{-1}\frac{n_2}{n_1}. \tag{340}$$

Relation (340) is obtained from the following considerations. Let the extreme meridional rays of the light cone transferred inside the fiber make an angle θ' with the axis. Then for $\phi = \theta'$ the lower extreme ray will be normally incident on the end face (Figure 163*a*). Meridional rays confined inside the light cone will hit the end face at nonzero angles and, after refraction, leak out through the lateral surface of the fiber. Part of the incident energy will naturally be reflected back into the fiber, and this fraction will increase as the angle of incidence on the fiber end increases. The largest angle of incidence corresponds to the topmost ray of the light cone in the fiber. Therefore, the maximum angle ϕ is selected from the condition that this ray should not experience total inner reflection at the fiber end (Figure 163*b*), i.e., from the condition

$$n_1 \sin(\phi + \theta') < 1.$$

The angle ϕ should thus be taken between these extremes, i.e., between the limits (340). For example, for $n_1 = 1.5230$ and $n_2 = 1.5082$, we have $80° > \phi > 33°$. The feedback is thus markedly suppressed, although it is not quite eliminated. For $\phi = 0$, generation in a fiber with these values of n_1 and n_2 begins at pumping energy

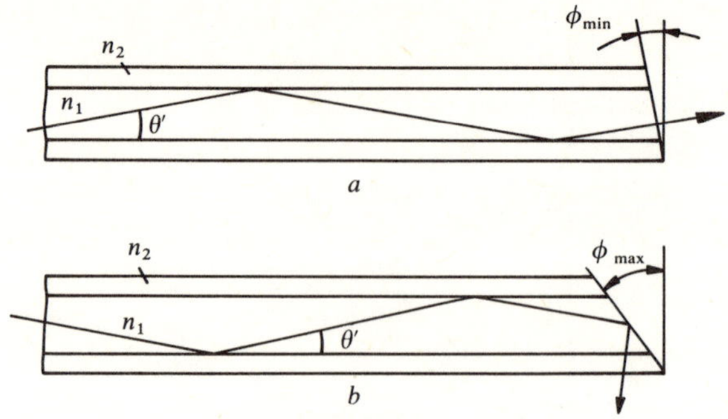

Figure 163

Choosing the limit values of the angle ϕ to reduce feedback.

of 40 J, whereas for $\phi = 10°$ it begins at 370 J. Increasing the threshold in this way, we can easily achieve fairly high amplification.

Thus good mode selection and strong mode coupling make fibers potentially suitable for use as laser resonators. The advantage of these elements is that they are characterized by a fairly high surface-to-volume ratio, and the cooling problems for the laser can therefore be solved relatively easily. Moreover, fibers can be manufactured in a variety of shapes, simplifying the design of various special structures.

3. Scintillators

Luminescence chambers are used in nuclear physics [13] for recording particle tracks and for particle energy measurements. The main constituent element in such chambers is a scintillator crystal which glows under the impact of ionizing particles.

The glow intensity of a luminescence chamber is low, and it is therefore generally used in conjunction with a brightness intensifier whose light signal is fed to a photomultiplier cathode or to a photosensitive layer.

The efficiency of a luminescence chamber is mainly determined by the luminous yield of the crystal and by its transparency to the luminescence wavelengths. Currently used scintillator crystals incude Na I (Tl), Cs I (Tl), anthracene, tolan, stilbene, and others.

Consider the efficiency of a chamber using a thallium-activated cesium iodide crystal [13]. Suppose a singly charged relativistic particle enters the chamber and loses 5 MeV of its energy over each centimeter traversed in the crystal. Since 35 eV are required to generate a single photon, the particle will generate $1.4 \cdot 10^5$ photons in a 4π angle along a path length of 1 cm. The number of quanta reaching the brightness intensifier photocathode is determined by the objective angle and satisfies the relation

$$N' = \frac{N}{16}\left(\frac{d}{f}\right)^2 \frac{M^2}{(1+M)^2} g, \tag{341}$$

where N is the number of photons generated, d/f is the relative aperture of the objective, M is the amplification, g is the ratio of the corresponding solid angles in the crystal and in air (this ratio is approximately equal to $1/n^2$, and for CsI $n = 1.79$).

If we take $d/f = 1/2$, $M = 1$, we find $N' = 170$ photons/cm. In other words, the brightness intensifier photocathode collects about 0.1% of the total radiation generated in the scintillator. The scintillator efficiency can be increased by using fiber optics [28].

The design of a fiber-optics scintillator is schematically shown in Figure 164. The luminescing fibers are packed in alternating layers with a mutually perpendicular configuration [3, 85]. The entrance ends of the fibers may be maintained in direct contact with a photographic plate or with the brightness intensifier photocathode. The fraction of the luminous flux generated by the ionizing particle in the fibers which satisfies the condition of total inner reflection on the lateral fiber surfaces propagates in two opposite directions along the fiber axis. The radiation reaching the fiber ends exposes the photosensitive layer, which registers the projection of the particle trajectory on the end-face plane of the fibers. The perpendicular packing of the fibers permits recording the particle trajectory in two mutually perpendicular planes, and the three-dimensional trajectory thus can be reconstructed without difficulty.

Scintillators of this kind can be manufactured from various plastic fibers using appropriate luminescing impurity elements, or thin-walled small-bore glass tubes filled with liquid solutions of luminescing composition. Glass tubes have certain

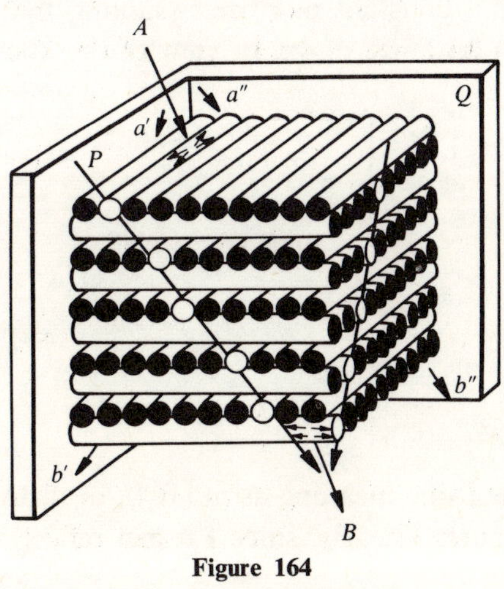

Figure 164

Fiber-optics scintillator:

A–B particle trajectory, *a′–b′* projection of the trajectory on plane *P*, *a″–b″* projection on plane *Q*.

distinct advantages compared to the plastic fibers. These include high uniformity and small inner reflection losses because of the higher surface quality. Glass-tube scintillators, however, have a large inactive volume—this is the volume occupied by the glass walls and by the air gaps between the tubes. In densely packed plastic scintillators the inactive volume accounts for 9.3%, whereas in glass-tube scintillators with tube diameters 10 times the tube wall thickness the inactive volume is as high as 42% [135]. For tube diameters 20 times the wall thickness, the inactive volume is still 27%.

The reduction of the light flux from liquid scintillators is partly offset by the higher luminous yield, since the energy conversion efficiency of some solutions is 30% higher than for plastics. The choice of a scintillator is therefore largely determined by the particular problem being considered.

The light flux emerging from a scintillation fiber and its angular distribution may be determined from equations (164) and (171), respectively. It should be remembered, however, that the light is not introduced into the fiber from an external source: it is generated inside the fiber and thus propagates within an angle of 4π. If the emitting center is located on the axis, the luminous fluxes propagating in the two opposite directions in the fiber will be confined inside a right cone with the linear half-angle

$$\theta'_{cr} = \cos^{-1}(n_1/n_2), \tag{342}$$

which is the angle between the axis and the extreme ray for which the condition of internal inner reflection from the cylindrical surface is just satisfied. Therefore the function $F(\theta')$ in (164) should be replaced by another function $G(\theta')$ which describes the angular distribution of the radiation generated in the fiber. To fair approximation, $G(\theta')$ may be treated as a constant number, assuming that the entire radiation is uniformly distributed in an angle of 4π. In view of the above, we have

$$\begin{aligned} \Phi = 2\pi D^2 \int_{\theta'=0}^{\theta'_{cr}} \int_{\gamma=0}^{\pi/2} & tG(\theta')\, \rho^{l \tan\theta'/D\cos\gamma} \exp(-Kl\sec\theta') \times \\ & \times \sin\theta' \cos^2\gamma d\gamma d\theta' + \\ + 2\pi D^2 \int_{\theta'=\theta'_{cr}}^{\theta'=\theta'_{lim}} \int_{\gamma=\gamma_{min}}^{\pi/2} & tG(\theta')\, \rho^{l \tan\theta'/D\cos\gamma} \exp(-Kl\sec\theta') \times \\ & \times \sin\theta' \cos^2\gamma d\gamma d\theta'. \end{aligned} \tag{343}$$

All the symbols have the same meaning as in (164); in distinction from (164), however, the coefficient t figures linearly, since Fresnel reflection needs be considered for the exit end only.

In accordance with (171), the angular distribution of the radiation emerging from a scintillation fiber may be written as

$$f_1(\theta) = D^2 t G(\theta') \frac{n_0(n_0^2 - n_1^2 \sin^2 \theta')^{1/2}}{n_1^2 \cos \theta'} \exp(-Kl \sec \theta') \times$$

$$\times \left[\int_{\gamma=0}^{\pi/2} \rho^{l \tan \theta'/D \cos \gamma} \cos^2 \gamma \, d\gamma + \int_{\gamma=\gamma_{min}}^{\pi/2} \rho^{l \tan \theta'/D \cos \gamma} \cos^2 \gamma \, d\gamma. \right]. \tag{344}$$

The transparency of plastic scintillator fibers to luminescence and the angular distribution of the outgoing light flux were studied experimentally in [131]. The measurements were carried out on a setup which is shown schematically in Figure 165. When processing the measurement results, it was assumed that constant-intensity ultraviolet radiation causes different points in the fiber to luminesce with the same intensity. Fibers without encapsulating envelopes were used, so that special measures had to be taken to protect the surface from dust and harmful mechanical and chemical factors.

In outgoing-light intensity measurements, the effective, and not the real, fiber length was taken into consideration; the effective length is the distance from the excitation point to the recording device maintained in contact with the exit end of the fiber. An integrating sphere with a photomultiplier was used as a detector.

Strictly speaking, the measurements should have been carried out for real-length fibers, ensuring excitation at the entrance end. The author correspondingly conducted two series of experiments. First the outgoing light flux was measured for various positions of the ultraviolet source, i.e., with the luminescence excited at different points in the fiber. After that, analogous measurements were repeated for end-face

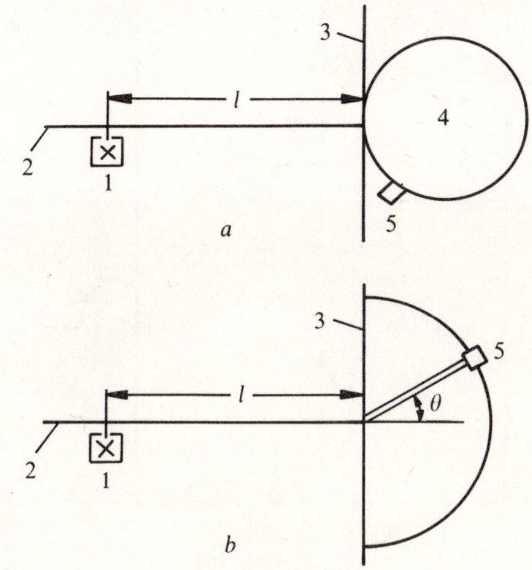

Figure 165

System for measuring the light flux emerging from scintillator fibers (a) and its angular distribution (b): 1) ultraviolet lamp, 2) scintillator fiber, 3) screen, 4) integrating sphere, 5) photomultiplier.

excitation, using truncated fibers for the purpose. Experiments showed that the difference between the two series was less than 10%.

The measurements were performed with fibers 1 m long and 1 and 0.5 mm in dia. The length of the excitation region was 2 mm in the axial direction, so that the detector in all probability responded to the luminescence flux only.

The theoretical calculations were based on relation (343) for fibers 1 mm in diameter for different values of K and ρ. The normalized curves are plotted in a semilogarithmic scale in Figure 166 (solid curves). The normalization condition assumed a unit outgoing light flux for a fiber of zero length. The experimental points fell between the two dashed lines. The results were applied to estimate K and ρ. Figure 166 gives $K = 0.012\ \text{cm}^{-1}$ and $\rho = 0.995$.

The angular distribution of the outgoing light flux was measured with a photomultiplier mounted on a goniometer (Figure 165*b*). The numerical aperture of the detector was about 10^{-3}, which is much higher than the requisite resolving power. The experimental findings are plotted in Figure 167 for various effective fiber lengths; the curves are normalized in relation to the axial position of the detector. Each curve is an average of numerous measurements carried out with several fibers.

Comparison of the theoretical data obtained from (344) with the experimental findings gives 0.018 cm^{-1} and 0.991 for the absorption coefficient and the total inner reflection coefficient, respectively.

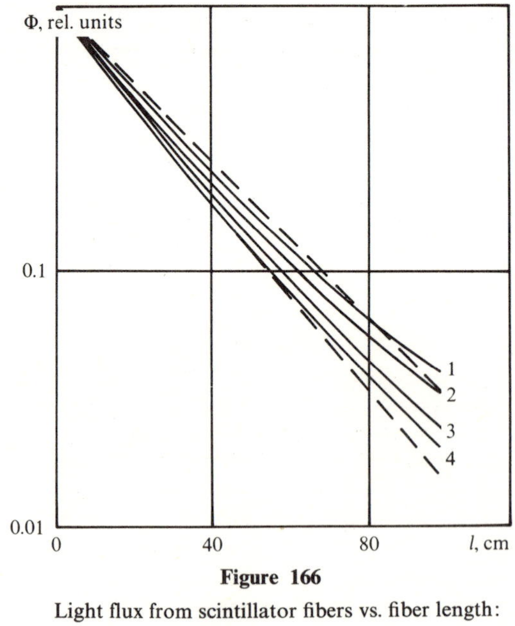

Figure 166

Light flux from scintillator fibers vs. fiber length:

1) $\rho = 0.995$, $K = 0.01\ \text{cm}^{-1}$, 2) $\rho = 0.995$, $K = 0.012\ \text{cm}^{-1}$, 3) $\rho = 0.995$, $K = 0.015\ \text{cm}^{-1}$, 4) $\rho = 0.996$, $K = 0.02\ \text{cm}^{-1}$.

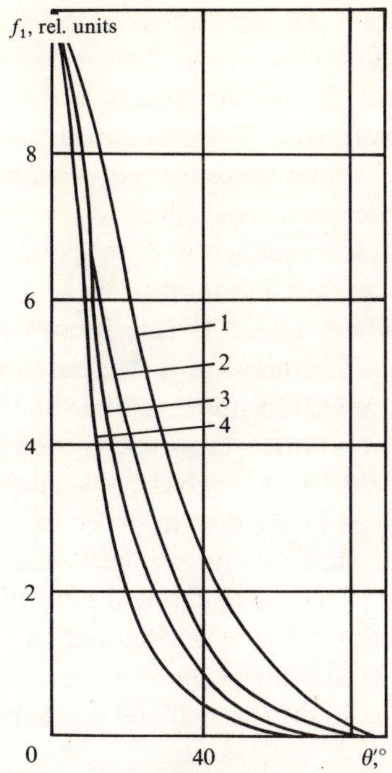

Figure 167

Angular distribution of outgoing light flux for various fiber lengths:

1) $l = 20$ cm, 2) $l = 30$ cm, 3) $l = 40$ cm, 4) $l = 50$ cm.

Similar tests were carried out with fibers 0.5 mm in dia. The average results give for plastic scintillator fibers $\rho = 0.993 \pm 0.002$ and $K = 0.015 \pm 0.003$ cm^{-1}.

Results of similar tests with plastic fibers 1 mm in diameter and uviol glass tubes (outer diameter 1 mm, wall thickness 0.075 mm) filled with liquid scintillators are given in [135]. The total inner reflection coefficient for tubes was found to be 0.984 and for fibers 0.96–0.98 (i.e., somewhat lower). However, the same parameter was somewhat higher for the fibers used in [131]. Unfortunately, the author did not know what material the fibers were made from, and it is therefore impossible to compare the optical properties of these fibers.

The values of K and ρ from [131, 135] should be treated as average figures, since they are based on light transfer measurements, which are largely determined by the number of reflections inside the fiber and the geometrical path of the ray. Scattering effects associated with local variations in the index of refraction were ignored in theoretical computations. Since volume scattering inevitably lowers the intensity of the outgoing radiation, the neglect of scattering effects artificially raised the absorption coefficient. Thus when measuring the absorption coefficient of the plastic scintillator samples from [131], the authors of [133] established that it was somewhat lower than the absorption coefficient obtained in fiber experiments. This seems to prove that fibers involve light losses which do not occur in large specimens. This result is quite understandable, since considerable internal stresses arise in plastic fibers during their manufacture (this is known from microscopic investigations), and these stresses lead to an additional reduction of light transfer and hence to an apparent increase of the absorption coefficient.

One of the possible reasons for a reduction of the total inner reflection coefficient is the scattering of light by surface inhomogeneities or extraneous particles, e.g., dust. The surface of plastic fibers markedly deteriorates with time under the effect of the surrounding medium, i.e., the fibers age. If the fibers are coated with an envelope, the conditions of total inner reflection improve and the lateral surface is protected from outside damage. This treatment, however, somewhat lowers the numerical aperture and increases light leakage through the side surfaces.

Theoretical calculations show [131] that in order to ensure values of ρ close to unity (0.999) and to maintain the light transfer characteristic of nonisolated fibers, the index of refraction of the envelope material should be 1.46. Optical fibers with this refractive index are currently being manufactured on an industrial scale, but the production of glass-encased plastic fibers involves considerable technological difficulties in view of the differences in the thermal and mechanical properties of the glass and the plastic.

Despite significant light losses, plastic scintillators have quite adequate performance characteristics [32, 136, 156]. A scintillator made from polystyrene or polyvinyltoluene base fibers with various luminescent impurities and measuring $5 \times 5 \times 5$ cm^3 is described in [32]. The external view of this scintillator is shown in Figure 168. Its adequacy was established by investigating the trajectory of a proton beam (3 mm in dia., 660 MeV).

In some nuclear physics experiments, e.g., when studying particle tracks in magnetic fields, the luminescence chamber and the brightness intensifier should be separated by a certain distance, so as to prevent the magnetic field from interfering with the proper operation of the intensifier. Since the luminescence intensity is low, the separation of the two components may reduce the incident light intensity to below the response threshold of the intensifier photocathode. Therefore high speed optics should be used to step up the light flux.

Methods of optical coupling between a luminescence chamber consisting of scintillator fibers 0.5 mm in diameter ($\rho = 0.993$, $K = 0.015$ cm^{-1}) and a brightness intensifier at a distance of some 60 cm are analyzed in [134]. The different coupling modes are shown in Figure 169.

The first coupling arrangement was tried using a special high-speed objective with a 1:1.5 relative aperture and focal distance of 14.92 cm. The experimental results show that normal magnification of the exit end image ensures that 4.3% of the radiation emerging from the scintillator hits the intensifier photocathode.

Figure 168

General view of a plastic fiber scintillator.

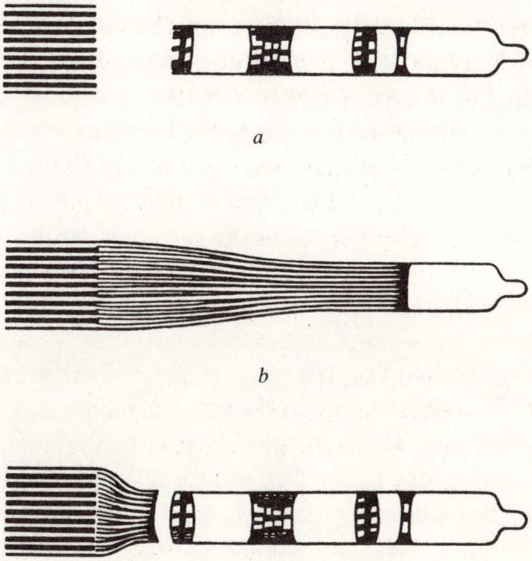

Figure 169

Different modes of optical coupling between luminescence chamber and brightness intensifier:

a by means of an objective, *b* by means of optical fibers, *c* combination method.

The use of glass fibers 0.5 mm in dia., with 0.65 numerical aperture, increases the effective light flux to 10.2%. A combination method which uses an objective and optical fibers gives 9.8%. Fiber optics may thus increase the effective light flux by a factor of 2–3 compared with very high-speed objectives. If we further remember that the introduction of fibers essentially simplifies the problems of photocathode surface curvature, the advantages of fiber-optical coupling become quite obvious.

4. High-speed photography

The main requirements from high-speed photographic cameras are high time and space resolution combined with a sufficient number of information-carrying film frames and high-speed optics. Since in some cases high-speed photography is applied to events which take up a relatively long time, a compromise solution has to be sought with regard to choice of time resolution and total number of film frames.

Modern high-speed cameras adequately satisfy these requirements, but they use shutters, scanning systems, and high-speed optics of highly complex design. The use of fiber elements will simplify the camera design and markedly reduce the production costs, as simpler optical-mechanical systems will be involved. Fiber optics resolves the image into individual elements, retaining high time and space resolution and ensuring faithful recording of relatively prolonged events [75, 112, 166].

One of the possible versions of fiber-optics high-speed photography is shown in Figure 170*a* [112]. The main element in this system is the fiber image shape converter 3, consisting of 1500 fibers 75 μ in dia. each. The converter is shaped in the form of a 5×1.8 mm^2 rectangle with an entrance end in the shape of a 116 mm $\times$ 75 μ strip. The photographed object is a magnesium filament 1 (0.127 mm in dia.) whose image is projected onto the entrance end of the converter by means of the objective 2 (relative aperture 1:2.9, focal distance 35 mm). The exit end is projected onto film 6 by means of objective 4 (1:5.6, $f = 90$ mm) and two plane mirrors 5 set at an angle of 45° to each other.

A film 610 mm long is spread on the inner surface of the drum 7, which is driven by an electric motor. The motor speed in these experiments was 10,000 rpm.

Since each fiber transmits a single image element, the film registers a series of dots which are stretched into tracks as the film moves. The number of tracks is equal to the number of fibers. The blackening density in each track corresponds to the variable illumination of the exit end of the corresponding fiber. The film thus presents a continuous record of a time-dependent event, and not a series of discrete photographs. The size of each "frame" may be taken equal to the height of the image of a single row obtained on film. In these experiments, 89 mm of film were used up, which corresponds to 9800 frames photographed in 0.87 msec, i.e., the exposure time of each frame was 90 nsec.

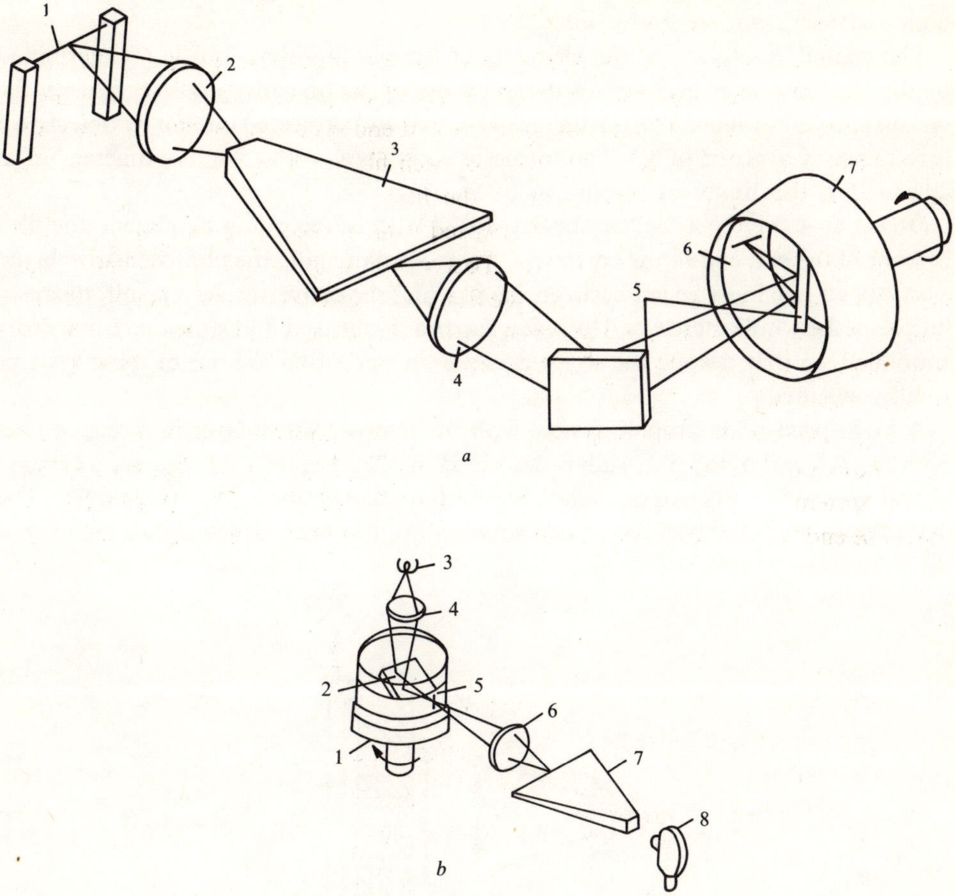

Figure 170
High-speed photographic recording (*a*) and reproduction (*b*) systems.

To reproduce the photographic information recorded on film, the film should be fed in the reverse direction through a similar fiber converter. A diagram of the device is shown schematically in Figure 170*b*. The treated film 5 is fixed on a transparent rotating drum 1. The illuminating system consists of a light source 3, a condenser 4, and a rotatory mirror 2. The recorded image is projected by objective 6 onto the entrance end of converter 7, shaped like a single long line. The image formed on the exit end of the converter is photographed by an ordinary cinecamera 8.

During the reproduction stage, the rotation speed of the drum is substantially lower than during recording, and all the details of the process may therefore be examined. To ensure faithful reproduction, the converter and the transparent drum should be carefully aligned. It is therefore advisable to use the same converter for both recording and reproduction.

The spatial resolution of the elements of the photographic image is determined by the fiber diameter and the resolving power of the objective and the film. In the system shown in Figure 170, the image of the exit end projected on film by objective 5 is reduced by a factor of 8.3. The image of each fiber is thus 9 μ in diameter, i.e., it falls within the limits of resolution of the film.

Objective 4 (Figure 170*a*) can be dispensed with in recording by placing the film in front of the exit end of the converter. To avoid damaging the photosensitive layer, a certain air gap must be left between the film and the converter. As a result, the resolution on the whole increases. However, camera design and adjustment are markedly simplified, so that despite the slight decrease in resolution the use of these systems is fully justified.

A high-speed photographic system with the photosensitive layer in direct contact with the exit end of the fiber slab is described in [73] (Figure 171). The main element of this system is a fiber-optics slab consisting of 32,000 fibers 25 μ in diameter. The entrance ends of the fibers are coated with a phosphor layer which allows high-speed

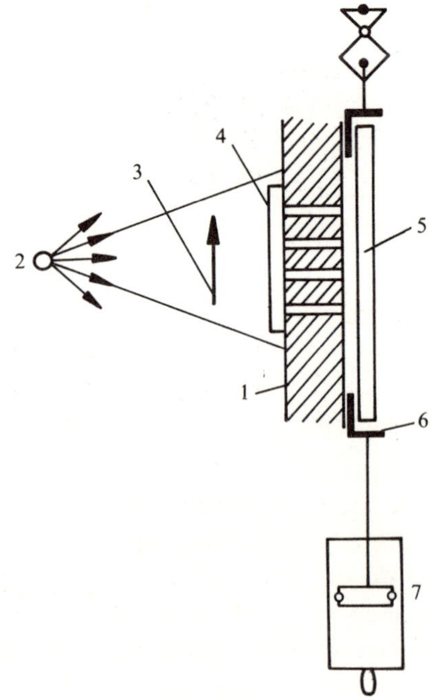

Figure 171

A system for high-speed photography:

1) fiber matrix, 2) light source, 3) object, 4) phosphor, 5) photographic plate, 6) guide rails, 7) pneumatic drive.

photography to be conducted in X-rays. The entrance end area of the slab is 18 cm^2, the length of the fibers is 25 mm. Each fiber is encapsulated in two envelopes, the topmost acting as an absorbing skin.

When the fibers have been packed in a bundle, the gaps are filled with absorbing epoxy resin. The fiber slab is thus treated as a monolith matrix with interior fiber light pipes.

The photographic plate was clamped in a frame which moved in preselected directions. Under conditions of continuous illumination, each fiber exposes an uninterrupted strip on the moving plate, the blackening density corresponding to the variable illumination of the given image element.

In order to obtain a series of frames, the plate should be displaced by a special pneumatic device in the vertical direction over a distance equal to one fiber diameter; a different system drives the frame with the plate in the horizontal direction with a velocity of some 10 m/sec.

For image reproduction, the photographic record is transmitted through the same fiber slab in the reverse direction. More detailed information about the process can be obtained by moving the plate at a slower speed. The picture can be observed visually or photographed with an ordinary cinecamera.

It is clear from the above that the fibers in the epoxy matrix should be packed in layers which are separated from one another by distances considerably greater than the fiber diameter, as otherwise successive photographs will overlap. The resolving power of this system is therefore substantially lower than that of the system shown in Figure 170*a*. Nevertheless, the fiber slab is markedly simpler and more convenient for practical uses. It may be applied to record fast processes both in the X-ray region and in the visible spectrum. In particular, it may be useful for studying laser emission and other fast processes.

Application of fiber optics to high-speed photography thus substantially simplifies camera design, eliminates high-speed shutters, and requires much less sophisticated optics, while retaining sufficiently high space and time resolution. Estimates [112] show that fiber optics will attain a resolution of 10^{-8} sec with a total of up to 250,000 photographs. Fiber elements may also be used for motion picture printing [125].

5. Coupling of electric circuits

Because of the rapid development of microelectronics and the steadily increasing number of elements in unit volume, the problem of suppression of spurious coupling between integrated circuits and individual elements has gained importance. Fiber optics provides an effective solution of this problem [78, 165].

Optical coupling between integrated circuits is made possible by *p-n* junctions which transform the electric current energy into recombination radiation and also effect reverse conversion of radiation into electric current. In the simplest case, the coupling is established with the aid of two *p-n* junctions one of which emits radiation and the other detects the radiation propagating in the light pipe (Figure 172).

The coupling efficiency is mainly determined by the conversion conditions from electric current to radiation and by the light transfer properties of the fibers. The conversion of electric current into radiation by diodes of gallium arsenide and other $A_{III}B_{V}$ compounds is substantially more efficient than the conversion by means of silicon diodes. For example, the recombination radiation yield in a silicon *p-n* junction is $5 \cdot 10^{-4}$ photons/electron [56], whereas the yield in a gallium arsenide *p-n* junction is 0.1 photon/electron [113].

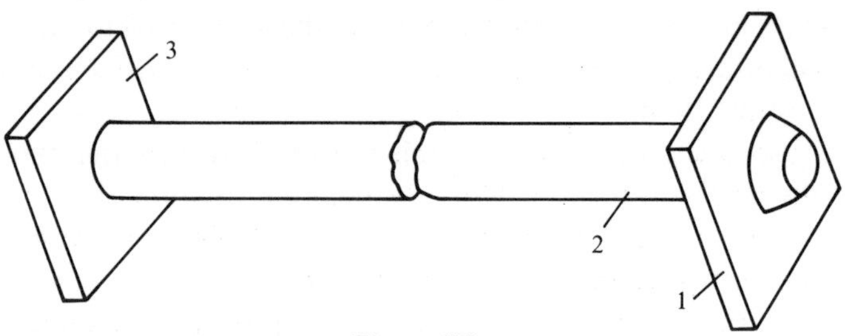

Figure 172

Optical coupling with a light pipe:

1) transmitter diode, 2) light pipe, 3) detector diode.

Despite the high inner yield of recombination radiation, good optical coupling between the light source and the detector is difficult to ensure because of the high indices of refraction of semiconductors. The index of refraction of gallium arsenide [126] and silicon [53] at the relevant wavelength is approximately 3.6, so that total inner reflection is observed at angles as large at 16°. As a result, most of the generated radiation is trapped inside the material (Figure 173*a*). The outgoing light flux may be substantially increased by using spherical diodices (Figure 173*b*). Since in this case the rays are always perpendicular to the surface, the radiation yield is maximal. Unfortunately, the methods of manufacture of spherical emitters are incompatible with the microcircuit technology. The radiation yield therefore cannot be increased substantially without extensive experimental studies and new technological development.

The most promising technique for increasing the yield calls for the use of immersion liquids and optical fibers with high refractive indices, approaching the refractive index of the semiconductor. Among the currently available materials, the most suitable one is As_2S_3, with a refractive index of 2.47 at $\lambda = 1\ \mu$. This material can readily be drawn into fibers with an exceptionally smooth surface, so that the inner reflection becomes maximally effective.

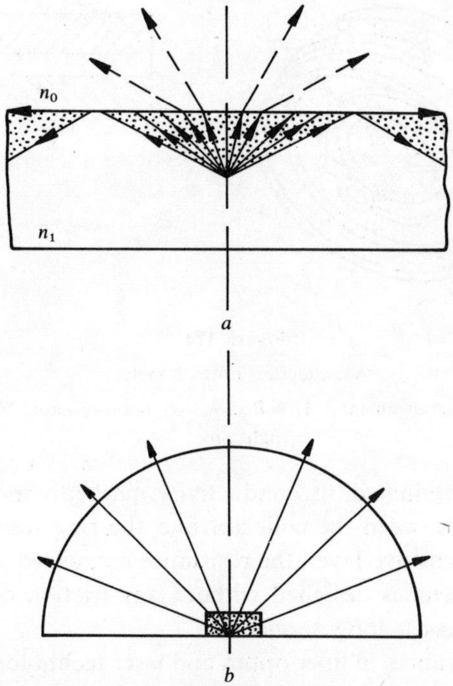

Figure 173
Emerging radiation from a semiconductor with a high refractive index:
a **plane surface,** ***b*** **spherical surface.**

Currently available optical coupling devices use gallium arsenide diodes as light sources and silicon diodes as detectors. As_2S_3 fibers are employed as light pipes. The desired mechanical strength of the diode-to-light pipe connections is ensured by using glasses containing arsenic, sulfur, and iodine in approximately equal weight quantities [74]. Since good bonding is assured for gallium arsenide only, whereas silicon and arsenic trisulfide display notoriously poor bonding, the semiconductor glasses should preferably be used in conjunction with a high-refraction viscous liquid thoroughly wetting all the components. A solution containing arsenic trisulfide, arsenic tribromide, antimony bromide, and selenium is currently used for this purpose. When heated, the volatile components diffuse into the As_2S_3 light pipe (or possibly react chemically with the material) and, partially evaporating, render the connection mechanically stronger [78].

The performance of electric circuits can be improved by using fiber-optical coupling. Figure 174 shows a potentiometer in which the sliding contact is replaced by a single fiber [128]. A photoresistor sandwiched between the ring resistor and the collector is illuminated by light from the exit end of the fiber attached to a rotating beam. When

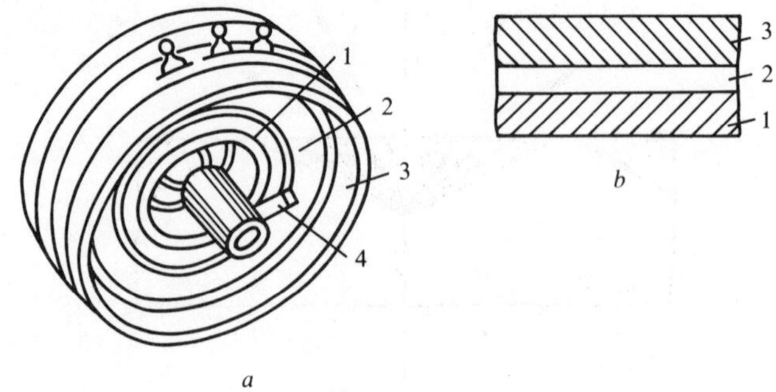

Figure 174

A contactless potentiometer:

a external view, *b* layer arrangement; 1) collector, 2) photoresistor, 3) ring resistor, 4) fiber light pipe.

the photoresistor is illuminated, its conductivity markedly increases and an ohmic contact is established between the collector and the ring resistor. If the light spot moves over the photosensitive layer, the resistance connected in the collector circuit varies. This potentiometer is designed without any friction contact, and it is thus free from noise and has a long service life.

The considerable advances in fiber optics and laser technology open new horizons for the design of optical computers. The main obstacle to increasing the speed of computers is posed by restrictions associated with the switching times of the active element and the time for the signal to propagate between any two elements.

The signal propagation time in an optical fiber is about $2 \cdot 10^{10}$ cm/sec, i.e., close to the propagation velocity of electric signals in a coaxial cable. Optical fibers therefore constitute a remarkable element for connections between electric circuits. They also effectively eliminate any undesirable coupling or interference between the active elements.

The light transfer of optical fibers is generally high. However, even if the transmitted signal is damped, this shortcoming can be overcome using fiber lasers, which may double as intensifiers offsetting the signal damping in the transmission line.

Coincidence logic circuits in optical computers may be set up using light beams of different polarization. Electric pulses are converted into light pulses in Kerr cells or lasers. Figure 175 shows a NOR logic circuit described in [68]. It comprises a vacuum element with a semiconductor anode connected in the excitation circuit of a diode laser. The potentials and the signal levels are selected so that the circuit functions as follows. The laser is excited by master pulses and emits light pulses if the photocathode of the vacuum photocell is not illuminated. When the photocathode is illuminated and photocurrent starts flowing, the laser is quenched, despite the presence of master pulses. Preliminary estimates show that this logic circuit operates with a speed of 2–10 nsec/oper.

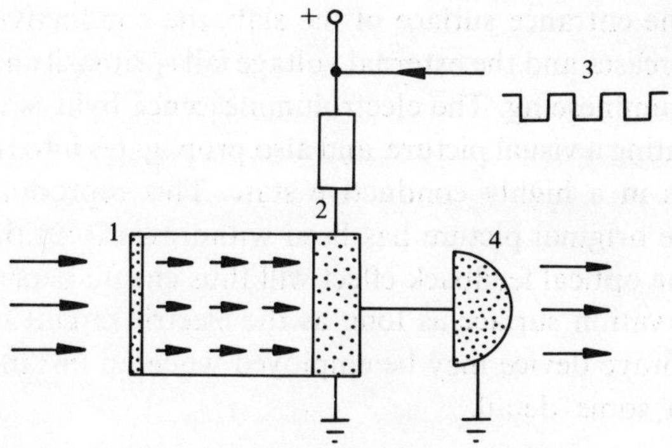

Figure 175

Optical NOR logic circuit:

1) vacuum photocell photocathode, 2) semiconductor anode, 3) master pulses, 4) laser diode.

The whole topic of optical computers is still largely theoretical, and yet the rapid development of laser and semiconductor technology and the recent advances in fiber optics give some hope for future feasibility of these systems.

An original storage device based on fiber optics is described in [71]: it is capable of storing an instantaneous light signal or a picture for a long time. The main constituent element of this storage is a fiber-optics slab (Figure 176). The lateral surface of each fiber is coated with a photoconductive selenium layer. A transparent conducting layer is deposited on the entrance end of the fiber slab, whereas the exit end is coated with an electroluminescent compound and a transparent conducting layer on top. The conducting layers are connected to an a.c. voltage source.

If the entrance end is not illuminated, almost the entire voltage falls across the high-resistance selenium layers and the electrophosphor does not emit. If a picture

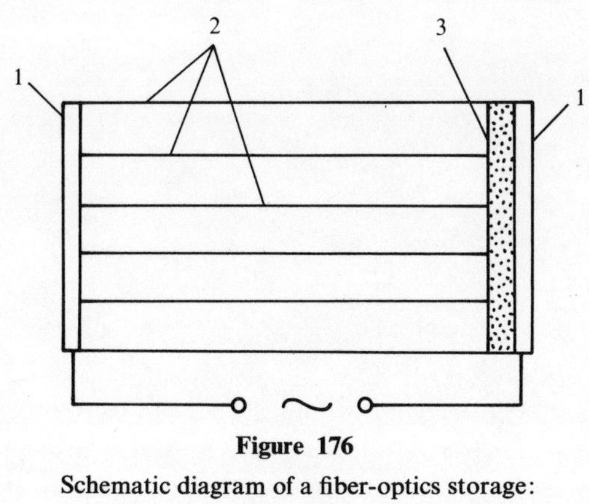

Figure 176

Schematic diagram of a fiber-optics storage:

1) transparent conducting layers, 2) photoconductive selenium layers, 3) electrophosphor.

is projected onto the entrance surface of the slab, the conductivity of the selenium layers markedly increases and the external voltage falls primarily across the phosphor layer which starts luminescing. The electroluminescence light is emitted in the outward direction, creating a visual picture, and also propagates into the fibers sustaining the selenium layers in a highly conductive state. This reproduction phenomenon will persist after the original picture has been withdrawn from the entrance surface of the fiber slab. The optical feedback effect will thus ensure persistence of the initial image on the observation surface as long as the electric circuit is connected to the a.c. source. This storage device may be employed when an instantaneous image has to be examined in some detail.

Appendix
Some Properties of Cylindrical Functions

The Appendix briefly describes some properties of Bessel and Hankel functions which are essential for the theory of dielectric waveguides (Chapter 3). A systematic treatment of the theory of these functions may be found in books by Watson* and Lebedev.**

Cylindrical functions are defined as solutions of the second-order linear differential equation

$$\frac{d^2v}{dr^2} + \frac{1}{r}\frac{dv}{dr} + \left(1 - \frac{n^2}{r^2}\right)v = 0, \tag{A.1}$$

where n is a constant.

Equation (A.1) is often referred to as Bessel's equation. A particular solution of this equation is sought in the form of an infinite power series

$$v = r^s \sum_{\lambda=0}^{\infty} a_\lambda r^\lambda \qquad (a_0 \neq 0). \tag{A.2}$$

Inserting v from (A.2) into (A.1), we obtain a recursive formula for the individual even-index coefficients

$$a_{2\lambda} = (-1)^\lambda \frac{a_0}{2^{2\lambda}(n+1)(n+2)\dots(n+\lambda)\,\lambda!} \quad (\lambda = 1, 2, 3, \dots). \tag{A.3}$$

All the odd-index coefficients of the series are zero:

$$a_{2\lambda+1} = 0 \qquad (\lambda = 0, 1, 2, \dots).$$

The parameter s may take on two values

$$s = \pm n. \tag{A.4}$$

* Watson, G.N. *A Treatise on the Theory of Bessel Functions.* Cambridge University Press. 1944.

** Lebedev, N.N. *Special Functions and Applications* [in Russian]. Moscow–Leningrad. Fizmatgiz. 1963.

To completely define the sought function, we choose the coefficient a_0 in the form

$$a_0 = \frac{1}{2^n \Gamma(n+1)}, \tag{A.5}$$

where $\Gamma(n)$ is the gamma function which for positive n is defined by the integral

$$\Gamma(n) = \int_0^\infty e^{-x} x^{n-1} dx.$$

Inserting these expressions for the coefficients in (A.2) and making some simplifications, we obtain for $s = +n$ a particular solution for equation (A.1) in the form

$$J_n(r) = \sum_{\lambda=0}^{\infty} \frac{(-1)^\lambda \left(\frac{r}{2}\right)^{2\lambda+n}}{\lambda!\,\Gamma(n+\lambda+1)}. \tag{A.6}$$

Relation (A.6) defines Bessel's function of the first kind of order n. Note that the series in (A.6) converges for all r.

Another particular solution is similarly obtained for $s = -n$:

$$J_{-n}(r) = \sum_{\lambda=0}^{\infty} \frac{(-1)^\lambda \left(\frac{r}{2}\right)^{2\lambda-n}}{\lambda!\,\Gamma(-n+\lambda+1)}. \tag{A.7}$$

In the particular cases $n = 0$ and $n = 1$ we obtain Bessel's functions of first kind of zero and first order, respectively. These functions have a fairly simple explicit form

$$J_0(r) = 1 - \frac{1}{(1!)^2}\left(\frac{r}{2}\right)^2 + \frac{1}{(2!)^2}\left(\frac{r}{2}\right)^4 - \frac{1}{(3!)^2}\left(\frac{r}{2}\right)^6 + \ldots,$$

$$J_1(r) = \frac{r}{2} - \frac{1}{1!2!}\left(\frac{r}{2}\right)^3 + \frac{1}{2!3!}\left(\frac{r}{2}\right)^5 - \ldots.$$

If n is not an integer, solutions (A.6) and (A.7) are linearly independent and the general solution of equation (A.1) takes the form

$$Z_n(r) = C_1 J_n(r) + C_2 J_{-n}(r), \tag{A.8}$$

where C_1 and C_2 are constants.

If n is an integer, J_n and J_{-n} are linearly dependent, and to construct a general solution, we need another linearly independent solution besides $J_n(r)$, i.e., Bessel's function of the second kind. This function has the form

$$Y_n(r) = \frac{J_n(r)\cos n\pi - J_{-n}(r)}{\sin n\pi}. {}^{*} \tag{A.9}$$

* Function (A.9) is sometimes called Neumann's function; the corresponding symbol is $N_n(r)$.

The general solution of the cylindrical equation is thus written in the form

$$Z_n(r) = C_1 J_n(r) + C_2 Y_n(r). \tag{A.10}$$

Since in the theory of directional waves n may only take integral values in (A.1), we should concentrate on solutions of the form (A.10). It is significant that because of physical considerations the coefficient C_2 should be taken equal to zero. Indeed, while function $J_n(r)$ remains finite for any positive n at the point $r = 0$ (the waveguide axis), $Y_n(r) \to -\infty$ as $r \to 0$.

Since in all problems of interest for our purposes n is an integer, the expression for Bessel's function can be simplified. In this case $\Gamma(n+1) = n!$ and the series in (A.6) may be rewritten as

$$J_n(r) = \sum_{\lambda=0}^{\infty} \frac{(-1)^\lambda \left(\frac{r}{2}\right)^{2\lambda+n}}{\lambda!\,(n+\lambda)!}. \tag{A.6a}$$

Together with Bessel's functions $J_n(r)$, we should also consider Hankel's functions, which are defined by the equality

$$H_n(r) = J_n(r) \pm iY_n(r). \tag{A.11}$$

The plus sign corresponds to Hankel's function of the first kind and the minus sign to Hankel's function of the second kind.

Finally, the propagation of an electromagnetic wave outside a waveguide is described in terms of MacDonald's functions (modified Hankel's functions), which are defined by the equality

$$K_n(r) = \frac{\pi}{2} i^{n+1} H_n(ir). \tag{A.12}$$

The solution of applied problems requires knowledge of the asymptotic behavior of these functions for very large or very small values of the argument. For small r,

$$J_0(r) \approx 1; \tag{A.13}$$

$$J_n(r) \approx \frac{r^n}{2^n \Gamma(n+1)}; \tag{A.14}$$

$$Y_n(r) \approx \frac{-2^n (n-1)!}{\pi r^n}. \tag{A.15}$$

For large r, the linear independent solutions of the cylindrical equation may be represented in the from

$$J_n(r) \approx \sqrt{\frac{2}{\pi r}} \cos\left(r - \frac{\pi}{4} - n\frac{\pi}{2}\right), \tag{A.16}$$

$$Y_n(r) \approx \sqrt{\frac{2}{\pi r}} \sin\left(r - \frac{\pi}{4} - n\frac{\pi}{2}\right). \tag{A.17}$$

Thus, for large values of the argument, Bessel's functions are expressed as a cosine and a sine with decreasing amplitudes. In other words, their behavior is analogous to functions which describe damped oscillations.

The modified Hankel's functions have a substantially different asymptotic behavior. For large values of the argument it may be approximated by a damped exponential,

$$K_n(r) \approx \sqrt{\frac{\pi}{2r}} e^{-r}. \tag{A.18}$$

For small r,

$$K_0(r) \approx \ln \frac{2}{\gamma r}, \tag{A.19}$$

where $\gamma = 1.781$ is Euler's constant, and

$$\ln \gamma = \lim_{n \to \infty} \left(1 + \frac{1}{2} + \frac{1}{3} + \ldots + \frac{1}{n} - \ln n\right) = 0.577, \tag{A.20}$$

$$K_n(r) \approx (n-1)!\, 2^{n-1} r^{-2}, \quad n \neq 0.$$

In conclusion, we give a number of useful relations between cylindrical functions of different orders. Thus,

$$\frac{d}{dr} J_0(r) = \frac{d}{dr}\left[1 - \left(\frac{r}{2}\right)^2 + \frac{1}{(2!)^2}\left(\frac{r}{2}\right)^4 - \frac{1}{(3!)^2}\left(\frac{r}{2}\right)^6 + \ldots\right] =$$

$$= -\left[\frac{r}{2} - \frac{1}{1!2!}\left(\frac{r}{2}\right)^3 + \frac{1}{2!3!}\left(\frac{r}{2}\right)^5 - \ldots\right] = -J_1(r),$$

i.e.,

$$J_0'(r) = -J_1(r). \tag{A.21}$$

Similarly, direct differentiation of the series in (A.6) may establish the following recursive formulas:

$$\frac{2n}{r} J_n'(r) = J_{n-1}(r) + J_{n+1}(r), \quad n = 1, 2, 3, \ldots. \tag{A.22}$$

$$2J_n'(r) = J_{n-1}(r) - J_{n+1}(r), \quad n = 1, 2, 3, \ldots. \tag{A.23}$$

$$J_{-n}(r) = (-1)^n J_n(r). \tag{A.24}$$

The first of these relations permits expressing a function of any order n in terms of the functions of zero and first order.

Similarly simple recursive relations apply to modified Hankel's functions:

$$-\frac{2n}{r}K_n(r) = K_{n-1}(r) - K_{n+1}(r)\,; \tag{A.25}$$

$$-K_n'(r) = K_{n-1}(r) + K_{n+1}(r)\,; \tag{A.26}$$

$$K_{-n}(r) = K_n(r)\,. \tag{A.27}$$

Relations (A.22)–(A.27) directly lead to expressions which are used for the analysis of directed waves in cylindrical dielectric waveguides. For small r we have

$$\frac{K_0(r)}{rK_1(r)} = \ln\frac{2}{\gamma r}\,; \tag{A.28}$$

$$\frac{K_{n-1}(r)}{rK_n(r)} = [2(n-1)]^{-1} \quad \text{for} \quad n \geqq 2. \tag{A.29}$$

For large r,

$$\frac{K_{n-1}(r)}{rK_n(r)} = \frac{1}{r}\,. \tag{A.30}$$

Bibliography

1. Born, M. *Optik*, Berlin, Springer, 1933.
2. Brillouin, L. *Wave Propagation in Periodic Structures,* N.Y., Dover, 1953.

*3. Veinberg, V.B., Bogatyreva, V.V., Ivanov, L.N. Image transmission by multifiber light pipes, *Optiko-Mekhanicheskaya Promyshlennost'*, 1960, No. 7.

*4. Veinberg, V.B., Sattarov, D.K. Polar diagrams of scattering and depolarization of fiber light pipes, *Optiko-Mekhanicheskaya Promyshlennost'*, 1961, No. 9.

*5. Veinberg, V.B., Sattarov, D.K. Propagation of light through a straight fiber, *Optiko-Mekhanicheskaya Promyshlennost'*, 1961, No. 12.

*6. Veinberg, V.B., Ispiryan, N.V., Belkina, L.A. Light pipes, *Optiko-Mekhanicheskaya Promyshlennost'*, 1962, No. 4.

*7. Veinberg, V.B., Ivanova, L.N., Sattarov, D.K. Pipes for radiant energy, *Svetotekhnika*, 1962, No. 7.

*8. Veinberg, V.B., Sattarov, D.K. Transmission of light by transparent light pipes, *Optiko-Mekhanicheskaya Promyshlennost'*, 1963, No. 2.

*9. Vlasov, A.G. Reflection and transmission of light by a system of thin films, *Optiko-Mekhanicheskaya Promyshlennost'*, 1946, No. 2.

*10. Grebenshchikov, I.V., Vlasov, A.G., Neporent, V.S., Suikovskaya, N.V. *Coated Optics*, Moscow–Leningrad, Gostekhteoretizdat, 1946.

*11. Daich, A.R., Tsirlin, Yu.A., Pargamanik, L.E. Light transmission through light pipes, *Optika i Spektroskopiya*, 1960, **8**, No. 5.

12. Drude, P. *The Theory of Optics*, New York, Dover, 1959.

*13. Zavoiskii, E.K., Butslov, M.M., Plakhov, A.G., Smolkin, E.G. Luminescence chamber, *Atomnaya Energiya*, 1956, No. 4.

*14. Zakhar'evskii, A.N. Determination of small differences in refraction and dispersion, *Trudy Gosudarstvennogo Opticheskogo Instituta*, 1929, **5**, No. 47.

*15. Zakhar'evskii, A.N. *Interferometers*, Oborongiz, 1952.

*16. IR tracking system with fiber optics, *Elektronika*, 1963, No. 46.

*17. Kard, P.G. New formulas for multilayer film computations, *Optika i Spektroskopiya*, 1963, **14**, No. 2.

* References preceded by an asterisk are in Russian.

*18. KATYS, G.P. *Searching and Tracking Photoelectric Scanning Devices*, Moscow, Nauka, 1964.

*19. KATYS, G.P., KARYAGIN, O.I. *A method of scanning and a scanning device*. Soviet Patent Claim No. 822313/26–10, 1963.

*20. KOROBKO–STEFANOV, A.A. Total internal reflection of waves, *Uspekhi Fizicheskikh Nauk*, 1950, **42**, No. 3.

*21. KOSHLYAKOV, N.S., GLINERG, E.B., SMIRNOV, M.M. *Fundamental Differential Equations of Mathematical Physics*, Moscow, Fizmatgiz, 1962.

*22. KAPANY, N.S., MILLER, B. Application of fiber optics in electron-optical systems, *Zarubezhnaya Radioelektronika*, 1961, No. 4.

*23. LANDAU, L.D., LIFSHITS, E.M. *Electrodynamics of Continuous Media*, Moscow, Fizmatgiz, 1959.

*24. LANDSBERG, G.S. *Optics*, Moscow, Gostekhteorizdat, 1952.

*25. LISITSA, M.P. Reflection and transmission of light by a system of thin and thick dielectric layers, *Zhurnal Tekhnicheskoi Fiziki*, 1954, **24**, No. 10.

*26. MALYSHEV, G.M., RYSKIN, A.I. Application of fibers optics in a system employing a Fabry-Perot interferometer and an electron-optical converter, *Optika i Spektroskopiya*, 1964, **17**, No. 5.

*27. MANDELSTAM, L.I. *Complete Collection of Works*, Moscow–Leningrad, USSR Acad. Sci., Vol. 1, 261, 1948.

*28. NIKANOROV, V.I., PISAREV, A.F., SAVCHENKO, O.V. *Report Laboratory Nuclear Problems JINS*, May, 1957.

*29. OBREIMOV, I.V. *Application of Fresnel Diffraction to Physical and Technical Measurements*, Moscow–Leningrad, USSR Acad. Sci., 1954.

*30. Optical generation in glass fibers and rods, *Elektronika*, 1962, No. 35.

*31. ROZENBERG, G.V. *Optics of Thin-Layer Coatings*, Moscow, Fizmatgiz, 1958.

*32. SAVCHENKO, O.V. A fiber scintillator for a luminescent chamber, *Pribory i Tekhnika Eksperimenta*, 1959, No. 4.

*33. SAZONTOVA, V.M., TSEPELEV, YU.A. Light pipes, *Novosti Meditsinskoi Tekhniki*, 1961, No. 6.

34. STRATTON, J.A. *Electromagnetic Theory*, McGraw-Hill, 1941.

*35. SUDRAVSKII, D.D. Reproduction of pictures with fibers optics, *Tekhnika Kino i Televideniya*, 1961, No. 5.

*36. TAMM, I.E. *Elements of the Theory of Electricity*, Moscow, Gostekhteorizdat, 1954.

*37. TUDOROVSKII, A.I. *Theory of Optical Instruments*, Part I, Moscow–Leningrad, USSR Acad. Sci., 1948.

*38. FAERMAN, G.P., SINTSOV, V.N., POPOVA, K.B. Evaporography, *Optiko-Mekhanicheskaya Promyshlennost'*, 1962, No. 11.

*39. FEDOROV, F.I. *Optics of Anisotropic Media*, Minsk, Belorussian Acad. Sci., 1958.

*40. Charactron with a fiber screen, *Elektronika*, 1961, No. 34.

41. VON HIPPEL, A.R. *Dielectrics and Waves*, Chapman & Hall, 1954.

42. SIMONYI, K. *Foundations of Electrical Engineering*, New York, Macmillan, 1963.

*43. SHKLYAREVSKII, I.N. Phase relations at a dielectric—metal interface, *Zhurnal Tekhnicheskoi Fiziki*, 1956, **26**, No. 2.

*44. SHCHEGLOV, P.V. A fiber field corrector for optical and electron-optical instruments, *Soobshcheniya GAISh*, 1961, No. 117.

*45. EICHENWALD, A.A. *Zhurnal Russkogo Fiziko-Khimicheskogo Obshchestva* (Phys. Ser.), 1909, **41**.

46. JAHNKE, E., EMDE, F. *Tables of Functions with Formulae and Curves*, New York, Dover, 1945.

47. BAUMEISTER, P.W., STONE J.M. Broad-band multilayer film for Fabry-Perot interferometer, *J. Opt. Soc. of America*, 1956, **46**, No. 3.

48. BEAM, R.E., ACTRAHAN, M.M., JAKES, W.C., WACHOWSKI, H.M., FIRESTONE, W.Z. *Northwestern Univ. Report*, ATI 94929, 1949.

49. BEURLE, R.L. Low light level television, *Proc. Inst. Electr.*, 1963, **110**, No. 10.

50. BIRD, E.W. Fibre optics, a current appraisal, *Med. and. Biol. Illustr.*, 1962, **12**, No. 13.

51. BIRD, V.R., CARPENTER, D.R., MCDERMOTT, P.S., POWELL, R.L. Rectangular optical dielectric waveguides as laser, *Laser and Applications*, ed. by W.S.C. CHANG, The Ohio State University, Columbus, 1963.

52. BORKMANN I. Die Faseroptik und ihre Anwendung in der Technik, *Radio und Fernsehen*, 1963, No. 5.

53. BRIGGS, H.B. Optical effects in bulk silicon and germanium, *Phys. Rev.*, 1950, **77,** No. 2.

54. CAPELLARO, D.F., KAPANY, N.S. A hypodermic probe using fiber optics, *Nature*, 1961, **191**, No. 4791.

55. CAPELLARO, D.F. Optical systems using fiber optics, *Light and Heat Sensing*, Pergamon Press, Oxford, London, N.Y., Paris, 1963.

56. CHYNOWETH, A.G., MCKAY, K.G. Photon emission from avalanche breakdown in silicon, *Phys. Rev.*, 1965, **102**, No. 2.

57. COCHE, A., LAUSTRAIT, G. Etude de conditions d'utilisation des scintillateurs liquides, *J. Phys. Rad.*, 1958, **19,** No. 6.

58. COOPER, B., NAJAC, H.W., SHAMOS, M.H., BREITJELLER, J.M., JACOBSON, J.H. Use of optical fibers for determination of irradiance at the retinal plane, *Acta ophtalmol.*, 1963, Suppl. 76.

59. COURTNEY-PRATT, J.S. *Improvements in or relating to electronic image forming tubes*, U.K. Patent, cl. 39(1), 97(1), No. 841200, 1960.

60. COURTNEY-PRATT, J.S. Image converter tube photography, *J. SMPTE*, 1962, **71**, No. 4.

61. CURTISS, L.E., HIRSCHOWITZ, B.I., PETERS, C.W. A long fiberscope for internal medical examinations, *J. Opt. Soc. of America*, 1957, **47**, No. 1.

62. DAWE, A.C. Vidicon camera tube with a fiber optics window, *Industrial Electronics*, 1963, **1**, No. 11.

63. DAWE, A.C. Special types of vidicon camera tubes, *Industrial Electronics*, 1963, **1**, No. 14.

64. DAY, R., KRAUSS, D.M. Fiber optics yields a new scanner concept, *Control Engineering*, 1961, **8**, No. 12.

65. DRESSLER, R. The PDF chromatron-single or multi-gun tricolor cathode ray tube, *Proc. IRE*, 1953, **41**.

66. DROUGARD, R. Optical transfer properties of fiber bundles, *J. Opt. Soc. of America*, 1964, **54**, No. 7.

67. *Electronics News*, 1963, No. 360.

68. *Electronic Design*, 1964, March 16, **12**, No. 6.

69. Eriksson, P.O. Nagot om bildförstär-kartekniken och dess tillämpning inom höghastighetsfotografian, *IVA medd.*, 1963, No. 136.

70. Ettinger, G.M. Some properties of light guides, *Rev. Sci. Instr.*, 1955, **26**, No. 8.

71. Fiber optics in new C—R tube, *Radio-Electronics*, 1961, **32**, No. 10.

72. Fiber optics promises radical new display gear, *Control Engineering*, 1961, **8**, No. 10.

73. Fiber optics for high-speed photography, *Industrial Electronics*, 1963, **1**, No. 4.

74. Flaschen, S.S., Pearson, A.D., Northover, W.R. Low melting sulfide–halogen inorganic glasses, *J. Appl. Phys.*, 1960, **31**, No. 1.

75. Forkner, J.F., Jewitt, J.W., James, W.A. A nonframing fiber optics motion-picture camera, *Photogr. Sci. and Engng.*, 1964, **8**, No. 1.

76. Fox, A.G., Li, T. Resonant modes in a maser interferometer, *Bell Syst. Techn. Journ.*, 1961, **40**, No. 2.

77. Frank, R., Stow, R.L. Securing rocksalt windows to sealedoff vacuum tubes, *Rev. Sci. Instr.*, 1954, **25**, No. 5.

78. Gilleo, M.A., Last, J.T. Optical coupling: new approach to microcircuit interconnections, *Electronics*, 1963, **36**, No. 47.

79. Goos, F., Hänchen, H. Ein neuer und fundamentaler Versuch zur Totalreflexion, *Ann. Phys.*, 1947, **1**.

80. Goos, F., Lindberg–Hänchen, H. Neumessung des Strahlversetzungseffektes bei Totalreflexion, *Ann. Phys.*, 1949, **5**.

81. Grim, E.D. Fiber optics applied to control and sensing devices, *Electro-technology*, 1962, **70**, No. 4.

82. Harris, K.D. Review of recent developments. Lasers, *Electron. Technol.*, 1962, **39**, No. 3.

83. Heet, J.H., Curtiss, L.E. Fiber optics duodenoscope and ureterscope, *J. Opt. Soc. of America*, 1961, **51**, No. 5.

84. Hicks, J.W., Kiritsy, P. Principles of fiber optics, *Space-Aeronautics*, 1961, **36**, No. 6.

85. Hicks, J.W., Kiritsy, P. Fiber optics, *Glass Industry*, 1962, **44**, No. 4.

86. Hirschowitz, B.I., Curtiss, L.E., Peters, C.W. Fiber-Scope, *Gastroenterology*, 1958, **35**, No. 1.

87. Hirschowitz, B.I. Endoscopic examination of the stomach and duodenal cap with the fiberscope, *Lancet*, 1961, No. 1.

88. Hopkins, H.H., Kapany, N.S. A flexible fiberscope, using static scanning, *Nature*, 1954, **173**, No. 4392.

89. Hyde, W.L. *Fiber Optics*, American Optical Company, Research Department, Southbridge, Mass., 1959.

90. Hyde, J.B. Transmission of ultra-violet rays in quartz fibers, *Nature*, 1963, **200**, No. 4909.

91. Image orthicons, *Electronic Industries*, 1961, **20**, No. 11.

92. Kapany, N.S. Fiber optics V. Light leakage due to frustrated total reflection, *J. Opt. Soc. of America*, 1959, **49**, No. 8.

93. KAPANY, N.S. Fiber optics I. Optical properties of certain dielectric cylinders, *J. Opt. Soc. of America*, 1957, **47**, No. 5.

94. KAPANY, N.S. Appendix "N" in *Concepts of Classical Optics*, ed. by J. STRONG, W.H. Freeman and Company, San Francisco, 1958.

95. KAPANY, N.S., BURKE, J.J., SHAW, C.C. Fiber optics IX. Waveguide effects, *J. Opt. Soc. of America*, 1961, **51**, No. 10.

96. KAPANY, N.S., BURKE, J.J., SHAW, C.C. Fiber optics X. Evanescent boundary wave propagation, *J. Opt. Soc. of America*, 1963, **53**, No. 8.

97. KAPANY, N.S. Fiber optics VI. Image quality and optical insulation, *J. Opt. Soc. of America*, 1959, **49**, No. 8.

98. KAPANY, N.S. High resolution fiber optics using sub-micron multiple fibers. *Nature*, 1959, **184**, No. 4690.

99. KAPANY, N.S. *Method and means for transmitting images through a bundle of transparent fibers*, U.S.A. Patent, class 88–1, No. 3016785.

100. KAPANY, N.S., EYER, J.A., KEIM, R.E. Fiber optics II. Image transfer on static and dynamic scanning with fiber bundles, *J. Opt. Soc. of America*, 1957, **47**, No. 5.

101. KAPANY, N.S., PIKE, J.N. Scanning integrator for producing sinusoidal test objects, *J. Opt. Soc. of America*, 1956, **46**, No. 10.

102. KAPANY, N.S., EYER, J.A., SHANNON, R.R. Production of sinusoidal test charts, *J. Opt. Soc. of America*, 1957, **47**, No. 1.

103. KAPANY, N.S. Fiber optics VIII. The focon, *J. Opt. Soc. of America*, 1961, **51**, No. 1.

104. KAPANY, N.S. Fiber optics, *Scientific American*, 1960, **203**, No. 5.

105. KAPANY, N.S., HOPKINS, R.E. Fiber optics III. Field flatteners, *J. Opt. Soc. of America*, 1957, **47**, No. 7.

106. KAPANY, N.S. A light funnel for stellar spectrograph. *Proceedings of a Symposium on Astronomical Optics and Related Subjects*, Manchester University, April 19–22, 1955, North-Holland Publishing Company, Amsterdam, 1956.

107. KAPANY, N.S., PONTARELLY, D.A. On dielectric cylinder photorefractometer, *J. Opt. Soc. of America*, 1958, **48**, No. 11.

108. KAPANY, N.S., PIKE, J.N. Fiber optics IV. A photorefractometer, *J. Opt. Soc. of America*, 1957, **47**, No. 12.

109. KAPANY, N.S., CAPELLARO, D.F. Fiber optics VII. Image transfer from Lambertian emitters, *J. Opt. Soc. of America*, 1961, **51**, No. 1.

110. KAPANY, N.S. Electro-optical systems using fiber optics, *Optica Acta*, 1960, **7**, No. 3.

111. KAPANY, N.S. Fiber optics activity expanding, *Electronic News*, 1961, **6**, No. 265.

112. KAPANY, N.S. Role of fiber optics in ultra-high-speed photography, *J. SMPTE*, 1962, **71**, No. 2.

113. KEYES, R.J., QUIST, T.M. Recombination radiation emitted from gallium arsenide, *Proc. IRE*, 1962, **50**, No. 8.

114. KOESTER, C.J., SNITZER, E. Amplification in a fiber laser, *Appl. Optics*, 1964, **3**, No. 10.

115. KRAKOWS, W.B., BAKER, A.M. Applying fiber optics, *Space-Aeronautics*, 1961, **36**, No. 6.

116. KROLAK, L.J., SIEGMUND, W.P., NEUHAUSER, R.G. Fiber optics—a new tool in electronics, *J. SMPTE*, 1960, **69**, No. 10.

117. KROLAK, L.J. Fiber optics and its use in electro-optical devices, *Light and Heat Sensing*, Pergamon Press, Oxford, London, N.Y., Paris, 1963.
118. LA FOND, C.D. Laser, fiber optics technologies join, *Missiles and Rockets*, 1961, **8**, No. 19.
119. LEURGANS, P.J. The impedance concept in thin film optics, *J. Opt. Soc. of America*, 1951, **41**, No. 10.
120. LONG, C., BRUSHENKO, A., PONTARELLY, D.A. The fiber optics hypodermic microscope, *Appl. Optics*, 1964, **3**, No. 9.
121. LUNCH, CH.J. Now in production fiber optics grows up, *Product Engng.*, 1961, **32**, No. 40.
122. MANNING, T.O. Fiber optic light transmission, *Wisconsin Engng.*, 1962, **66**, No. 4.
123. MCDERMOTT, J.R. Fiber optics: designing in a new light, *Electronic Design*, 1963, No. 11.
124. MILLER, B. Fiber optics reconnaissance explored, *Aviation Week*, 1960, **73**, No. 21.
125. MILLER, A.J., HARTSHORNE, R. Fiber optics in motion-picture printing, *J. SMPTE*, 1961, **70**, No. 9.
126. MORRISON, R.E. Reflectivity and optical constants of indium arsenide, indium antimonide and gallium arsenide, *Phys. Rev.*, 1961, **124**, No. 5.
127. New tubes use fiber optics, *Radio-Electronics*, 1962, **33**, No. 2.
128. NOVOTNY, G.V. Fiber optics for electronics engineers, *Electronics*, 1962, **35**, No. 22.
129. OLDEN, R.G. A thin window cathode ray tube for highspeed printing with "Electrofax", *RCA Review*, 1957, **18**, No. 3.
130. PAY, R. Instrument and control applications of glass fiber optics, *Instrument Practice*, 1961, **15**, No. 10.
131. POTTER, R.J. Transmission properties of optical fibers, *J. Opt. Soc. of America*, 1961, **51**, No. 10.
132. POTTER, R.J., DONATH, E., TYNAN, R. Light-collecting properties of a perfect circular optical fiber, *J. Opt. Soc. of America*, 1963, **53**, No. 2.
133. POTTER, R.J. Absorption measurements on a plastic scintillator, *Rev. Sci. Instrum.*, 1961, **32**, No. 3.
134. POTTER, R.J., HOPKINS, R.E. The optical coupling of a scintillation chamber to an image-intensifying tube, *IRE Trans. on Nuclear Science*, 1960, **NS-7**, No. 2–3.
135. RIEDFEEL, L., KAPANY, N.S. Some considerations on luminescent fiber chambers and intensifier screens, *Rev. Sci. Instrum.*, 1960, **31**, No. 10.
136. REYNOLDS, G.T., CONDON, P.E. Filament scintillation counter, *Rev. Sci. Instrum.*, 1957, **28**, No. 12.
137. ROELTING, R.C., GANLEY, W.P. Observation on the dynamic frequency response of fiber bundles, *J. Opt. Soc. of America*, 1962, **52**, No. 1.
138. SALATI, O.M. Handling light with fiber optics, *Electronic Industries*, 1961, **20**, No. 12.
139. SCHELKUNOFF, S.A. *Electromagnetic Waves*, D. Van Nostrand Comp., Inc., Princeton, New Jersey, 1943.
140. SCHLESINGER, K., KING, D.D. Dielectric image lines, *IRE Trans.*, 1958, M.T.T.-6.
141. SCHLESINGER, K. A microspot tube with very high resolution, *IRE Trans. Electron. Devices*, 1962, **9**, No. 3.

142. SHANNON, C.E. *Bell Telephone System Monograph*, 1948, B-1598.
143. SHAWLOW, A.L., TOWNES, C.H. Infrared and optical masers, *Phys. Rev.*, 1958, **112**, No. 6.
144. SHRIEVER, O.D. Elektromagnetische Wellen in dielektrischen Drähten, *Ann. Phys.*, 1920, **63**.
145. SIEGMUND, W.P. Fiber optics, *Glass Industry*, 1960, **42**, No. 9.
146. SIEGMUND, W.P. Fiber optics-principles, properties and design considerations, in *Light and Heat Sensing*, Pergamon Press, Oxford, London, N.Y., Paris, 1963.
147. SNITZER, E. Proposed fiber cavities for optical masers, *J. Appl. Phys.*, 1961, **32,** No. 1.
148. SNITZER, E. Dielectric optical waveguides, *Advances in Quantum Electronics*, ed. by J. SINGER, New-York–London, 1961.
149. SNITZER, E., OSTERBERG, H. Observed dielectric waveguide modes in the visible spectrum, *J. Opt. Soc. of America*, 1961, **51**, No. 5.
150. STOW, R.L. Fiber optics and their application to electronic tubes, *Electronic Design News*, 1961, **6**, No. 12.
151. TAYLOR, J.E. Point target autocollimator, *Rev. Sci. Instr.*, 1963, **34**, No. 2.
152. TIPPET, J.T., PUTHOFF, H.E. Optical computers approach reality, *Electronics*, 1963, **36**, No. 18.
153. TOLOPKO, L. Du Pont Div. Evolves Fiber Optical CRT's, *Electronic News*, 1961, **6**, No. 266.
154. TOLOPKO, L. Fiber optics seen moving beyond lab developments, *Electronic News*, 1961, **6**, No. 288.
155. TORADO DI FRANCIA, G. Resolving power and information, *J. Opt. Soc. of America*, 1955, **45**, No. 5.
156. TOVE, P.A. Scintillation spectrometry, using long light guides, *Rev. Sci. Instr.*, 1956, **27**, No. 3.
157. TURNER, A.F. Some current developments in multilayer optical films, *J. Phys. Rad.*, 1950, **11**, No. 7.
158. Two image orthicons, fiber-optic face, *Electronics*, 1961, **34**, No. 40.
159. VAN HEEL, A.C.S. *De Ingenieur*, 1953, **65**, No. 24.
160. VAN HEEL, A.C.S. A new method of transporting optical images without abberation, *Nature*, 1954, **173**, No. 4392.
161. WALTER, L. Fiber optics, *Text. Manufacturer*, 1963, **89**, No. 1061.
162. WEICHEL–MOORE, E.J., POTTER, R.J. Fiber optical properties of ulexite, *Nature*, 1963, **200**, No. 4912.
163. WEINSTEIN, R.M., BRADT, H.V. Attenuation lens in filament scintillators, *Rev. Sci. Instr.*, 1959, **30**, No. 3.
164. WOODLEY, W.A., ROGERS, D. High resolution and fiber-optic cathode ray tubes, *British Communications and Electronics*, 1963, **10**, No. 9.
165. WUNDERMAN, I. Principles and applications of the opto-electronical devices, *Electronics*, 1964, **37**, No. 21.
166. YOUNG, A.F.B., STEELE, J.G.C. High-speed and ultra-high-speed photography in electrical engineering, *J. Photographic Science*, 1957, **5**, No. 5.

Subject Index

LIBRARY COPY
WITHDRAWN
JUL 8 - 1976
LANGLEY RESEAR[illegible] CENTER
LIBRARY, NASA
HAMPTON, VIRGINIA